Statistical Field Theory of Ion-Molecular Fluids

Yury A. Budkov · Nikolai N. Kalikin

Statistical Field Theory of Ion-Molecular Fluids

Fundamentals and Applications in Physical Chemistry and Electrochemistry

Springer

Yury A. Budkov
Laboratory of Computational Physics
and School of Applied Mathematics
HSE University
Moscow, Russia

Laboratory of Multiscale Modeling
of Molecular Systems
G. A. Krestov Institute of Solution
Chemistry of the Russian Academy
of Science
Ivanovo, Russia

Nikolai N. Kalikin
Laboratory of Multiscale Modeling
of Molecular Systems
G. A. Krestov Institute of Solution
Chemistry of the Russian Academy
of Science
Ivanovo, Russia

ISBN 978-3-031-68363-3 ISBN 978-3-031-68364-0 (eBook)
https://doi.org/10.1007/978-3-031-68364-0

This Springer imprint is published by the registered company Springer Nature Switzerland AG
The registered company address is: Gewerbestrasse 11, 6330 Cham, Switzerland

Preface

Modern chemical engineering, physical chemistry and electrochemistry require computational methods to quickly and accurately predict the behavior of thermodynamic parameters in molecular and ion-molecular fluids. Self-consistent field theory is a useful tool in this regard, as it has been successfully applied to describe the thermodynamic properties of both simple liquids and macromolecular systems. However, for the theory of a self-consistent field to seriously compete with modern, computationally expensive simulation methods like molecular dynamics and Monte Carlo, which are widely used to model the behavior of liquids in equilibrium states both in the bulk and at phase boundaries, it is necessary to consider not only universal intermolecular interactions (dispersion and steric interactions), which are typically valid for simple liquids and synthetic polymers, but also specific interactions related to the characteristics of the chemical structure of the components. These include multipole moments, electronic and molecular polarizability, hydrophobicity/hydrophilicity (in the case of aqueous systems), asymmetry of molecule shape, and other factors. This monograph discusses examples of theoretical descriptions of ion-molecular systems where this type of consideration has been carried out. The purpose of this monograph is to demonstrate the application of mathematical tools from statistical physics based on statistical field theory for the analytical and numerical description of different ion-molecular systems. These systems include electrolyte solutions, molten salts, polyelectrolyte solutions, micellar aggregates, and metal complexes. A key feature of these systems that makes them difficult to describe theoretically is the need to take into account specific features of their molecular structure and long-range electrostatic interactions.

Until now, there have been no monographs dedicated to applying statistical field theory to these problems in physical chemistry. This monograph aims to fill this gap and provide a comprehensive overview of the subject. It should be emphasized that this book is not intended to be a textbook on statistical physics. Therefore, it is assumed that the reader has a basic understanding of this field of theoretical physics. At the moment, there are many excellent textbooks available on statistical physics and thermodynamics. We have included references to some of these textbooks in our monograph.

The book contains necessary mathematical introduction that the reader can consult to clarify any mathematical concepts or details omitted in the main text to avoid unnecessary overload. The authors have tried to focus on solving specific problems and formulating the necessary theoretical foundations as needed. To read the book, the reader should have a basic knowledge of calculus, linear algebra and probability theory, which can be obtained through standard university courses. The authors hope that this book will be a valuable resource for physical chemists, physicists, and chemical engineers who work with molecular systems. It will also be useful for undergraduate and graduate students studying specialized topics in these fields. The book provides a range of methods and approaches that can be used to solve problems in physical chemistry and condensed matter physics.

Moscow\Ivanovo, Russia Yury A. Budkov
Ivanovo, Russia Nikolai N. Kalikin

Contents

Abbreviations

DC	Differential capacitance
DFT	Density functional theory
DH	Debye-Hückel
EDL	Electric double layer
GA	Gaussian approximation
GC	Gouy-Chapman
GCDM	Gaussian core dipolar model
GCM	Gaussian core model
GER	Gaussian equivalent representation
HCOCP	Hard-core one-component plasma
HS	Hubbard-Stratonovich
MF	Mean-field
OCP	One-component plasma
OS	Onsager-Samaras
PB	Poisson-Boltzmann

Notations

T	Temperature
k_B	Boltzmann constant
μ	Chemical potential
P	Pressure
F	Helmholtz free energy
S	Entropy
Ω	Grand thermodynamic potential
V	Volume system
N	Number of particles
β	$(k_B T)^{-1}$

f	Free energy density
Q_N	Canonical partition function of fluid of N molecules
Ξ	Grand canonical partition function of fluid
$\Lambda = h/(2\pi m k_B T)$	De Broglie thermal wavelength
$z = \Lambda^{-3} e^{\beta\mu}$	Activity
$U(\mathbf{r})$	Pair interaction potential
$\hat{n}(\mathbf{r})$	Microscopic concentration
$u(\mathbf{r})$	External potential
$n(\mathbf{r})$	Local number density of fluid
$\delta(x)$	Delta function
$\delta(\mathbf{r}) = \delta(x)\delta(y)\delta(z)$	3D delta function
$-k_B T c^{(1)}(\mathbf{r})$	One-particle potential of fluid
$\bar{\mu}(\mathbf{r})$	Intrinsic chemical potential of fluid
$c_2(\mathbf{r}, \mathbf{r}')$	Direct correlation function
$h_2(\mathbf{r}, \mathbf{r}') = g_2(\mathbf{r}, \mathbf{r}') - 1$	Total correlation function
$g_2(\mathbf{r}, \mathbf{r}')$	Pair correlation function
$S(\mathbf{k})$	Structure factor of fluid
χ	Isothermal compressibility
$\tilde{c}_2(\mathbf{k})$	Fourier-image of direct correlation function of bulk fluid
$\tilde{h}_2(\mathbf{k})$	Fourier-image of total correlation function of bulk fluid
$\tilde{U}(\mathbf{k})$	Fourier-image of pair potential
$trA = \int d\mathbf{r} A(\mathbf{r}, \mathbf{r})$	Trace of integral operator with kernel $A(r, r')$
η	Volume fraction of fluid
$\kappa = r_D^{-1}$	Inverse Debye radius (screening constant)
$d\mathbf{k} = dk_x dk_y dk_z$	Elementary volume in inverse space
$\psi(z)$	Local electrostatic potential
ψ_0	Surface potential
C	Differential capacitance
$q = e$	Elementary charge
Π	Disjoining pressure
$\mathcal{E} = -\psi'(z)$	Local electric field
ω	Density of grand thermodynamic potential
$\Delta = \nabla^2$	Laplacian
σ_{ik}	Stress tensor
$\partial_i = \frac{\partial}{\partial x_i}$	Partial derivative with respect to coordinate x_i
$\psi_{.i}$	$\partial_i \psi$
$\psi_{.ik}$	$\partial_i \partial_k \psi$
$DetA$	Functional determinant of integral operator A
$(f_1 f_2)$	$\int d\mathbf{r} f_1(\mathbf{r}) f_2(\mathbf{r})$
$g(\mathbf{r})$	Probability distribution function of distance between charged centers
$g(\mathbf{k})$	Fourier-image of probability distribution function of distance between charged centers
$d\mathbf{r} = dxdydz$	Elementary volume
$\varphi(\mathbf{r})$	Fluctuating electrostatic potential

$\rho(\mathbf{r})$	Local density of charge
$\rho_{ext}(\mathbf{r})$	Density of external charges
I	Ionic strength
$\varkappa(\mathbf{k})$	Screening function
L_Q	Quadrupole length
ϕ_0	Volume fraction of polymer in the bulk solution
b	Bond length of polyelectrolyte chain
ε	Permittivity of medium
$\epsilon(\mathbf{r})$	Local permittivity of ionic fluid
p	Dipole moment
γ	Static polarizability
v	Volume of cell of the lattice gas

Chapter 1
Mathematical Introduction

In this chapter, we discuss the mathematical concepts that are essential to an understanding of the subsequent material in the book. We delve into fundamental concepts such as the functional, functional derivatives, and functional integrals. Additionally, we explore Gaussian functional integrals and integrals over a Gaussian measure. We introduce the cumulant expansion method and prove the Bogoliubov's inequality. Furthermore, we present methods for approximating functional integrals over Gaussian measures, including Gaussian approximation, variation method, and Gaussian equivalent representation.

1.1 Gaussian Integrals

Let us consider a Gaussian integral

$$I = \int_{-\infty}^{+\infty} dx\, e^{-x^2}. \tag{1.1}$$

To calculate such integral, we first square it

$$I^2 = \int_{-\infty}^{+\infty} dx\, e^{-x^2} \cdot \int_{-\infty}^{+\infty} dy\, e^{-y^2} = 4 \int_{0}^{+\infty} dx \int_{0}^{+\infty} dy\, e^{-(x^2+y^2)}. \tag{1.2}$$

Then letting $y = xt$, we obtain

Y. A. Budkov and N. N. Kalikin, *Statistical Field Theory of Ion-Molecular Fluids*,
https://doi.org/10.1007/978-3-031-68364-0_1

$$I^2 = 4 \int_0^{+\infty} dt \int_0^{+\infty} dx x\, e^{-x^2(1+t^2)} = 2 \int_0^{+\infty} \frac{dt}{1+t^2} = \pi. \tag{1.3}$$

Thus, we finally obtain that

$$I = \sqrt{\pi}. \tag{1.4}$$

Similar, utilizing the substitution $ax^2 = y^2$, we obtain

$$I = \int_{-\infty}^{+\infty} dx\, e^{-ax^2} = \sqrt{\frac{\pi}{a}}. \tag{1.5}$$

Let us now consider a more complex expression within the exponent

$$I = \int_{-\infty}^{+\infty} dx\, e^{-ax^2+bx}. \tag{1.6}$$

By completing the square and subsequently integrating the obtained expression, we arrive at

$$I = \sqrt{\frac{\pi}{a}} e^{\frac{b^2}{4a}}. \tag{1.7}$$

Let us now derive a useful expression for our forthcoming discussion

$$\int_{-\infty}^{+\infty} dx\, x^n e^{-ax^2+bx} = \frac{d^n}{db^n} \int_{-\infty}^{+\infty} dx\, e^{-ax^2+bx} = \sqrt{\frac{\pi}{a}} \frac{d^n}{db^n} \left(e^{\frac{b^2}{4a}} \right), \tag{1.8}$$

which for the case $n = 1$ yields

$$\int_{-\infty}^{+\infty} dx\, x e^{-ax^2+bx} = \frac{\sqrt{\pi} b}{2a^{3/2}} e^{\frac{b^2}{4a}}. \tag{1.9}$$

Let us now consider the case of the multivariate Gaussian integral. Let $\mathbf{x} \in \mathbb{R}^d$, where $\mathbf{x} = (x_1, x_2, \ldots, x_d)$. Then

$$I = \int_{-\infty}^{+\infty} dx_1 \int_{-\infty}^{+\infty} dx_2 \ldots \int_{-\infty}^{+\infty} dx_d\, e^{-\frac{1}{2}\sum\limits_{ij} x_i A_{ij} x_j + \sum\limits_k b_k x_k}, \tag{1.10}$$

where we assume that the matrix in quadratic form is symmetric, i.e., $A_{ij} = A_{ji}$. To calculate the value of such an integral, we eliminate the linear form in the exponent by transitioning to the region of the extremum of the function

$$\frac{\partial}{\partial x_l}\left(-\frac{1}{2}\sum_{ij} x_i A_{ij} x_j + \sum_k b_k x_k\right) = 0. \tag{1.11}$$

We proceed by differentiating this expression, recalling that

$$\frac{\partial x_i}{\partial x_l} = \delta_{il} = \begin{cases} 1, & i = l, \\ 0, & i \neq l. \end{cases} \tag{1.12}$$

This yields a system of linear equations

$$\sum_i A_{li} x_i = b_l, \tag{1.13}$$

the solution of which is

$$\bar{x}_i = \sum_k A^{-1}_{ik} b_k \tag{1.14}$$

with the inverse matrix, A^{-1}_{ij}, determined by definition as

$$\sum_k A_{ik} A^{-1}_{kj} = \delta_{ij}. \tag{1.15}$$

Let us now consider a deviation from the extremum, $x_i = y_i + \bar{x}_i = y_i + \sum_k A^{-1}_{ik} b_k$, then we can write

$$-\frac{1}{2}\sum_{ij} x_i A_{ij} x_j + \sum_k b_k x_k$$
$$= -\frac{1}{2}\sum_{ij} A_{ij}\left(y_i + \sum_k A^{-1}_{ik} b_k\right)\left(y_j + \sum_l A^{-1}_{jl} b_l\right) + \sum_k b_k \left(y_k + \sum_l A^{-1}_{kl} b_l\right). \tag{1.16}$$

Expanding the brackets, utilizing expression (1.15) and combining like terms, we can obtain

$$-\frac{1}{2}\sum_{ij} x_i A_{ij} x_j + \sum_k b_k x_k = -\frac{1}{2}(yAy) + \frac{1}{2}\left(bA^{-1}b\right), \tag{1.17}$$

where we introduce the following notations

$$(yAy) = \sum_{ij} y_i A_{ij} y_j, \\ \left(bA^{-1}b\right) = \sum_{kl} b_k A_{kl}^{-1} b_l. \tag{1.18}$$

Thus, we obtain the following:

$$I = \int_{-\infty}^{+\infty} dy_1 \ldots \int_{-\infty}^{+\infty} dy_d \exp\left\{-\frac{1}{2}\sum_{ij} y_i A_{ij} y_j + \frac{1}{2}\sum_{ij} b_i A_{ij}^{-1} b_j\right\} \\ = e^{\frac{1}{2}\sum_{ij} b_i A_{ij}^{-1} b_j} \int_{-\infty}^{+\infty} dy_1 \ldots \int_{-\infty}^{+\infty} dy_d \, e^{-\frac{1}{2}\sum_{ij} y_i A_{ij} y_j}. \tag{1.19}$$

To advance with further integration, it is first necessary to diagonalize the quadratic form, $(yAy) = \sum_k \lambda_k z_k^2$, where λ_k represent the diagonal matrix elements. Thus, we get

$$I = e^{\frac{1}{2}(bA^{-1}b)} \int_{-\infty}^{+\infty} dz_1 \ldots \int_{-\infty}^{+\infty} dz_d \, e^{-\frac{1}{2}\sum_k \lambda_k z_k^2} = e^{\frac{1}{2}(bA^{-1}b)} \prod_{k=1}^{d} \int_{-\infty}^{+\infty} dz_k e^{-\frac{1}{2}\lambda_k z_k^2} \\ = \frac{(2\pi)^{d/2}}{\sqrt{det A}} e^{\frac{1}{2}(bA^{-1}b)}, \tag{1.20}$$

where we take into account that the Jacobian of the orthogonal transformation is unity. The final expression for d-dimensional Gaussian integral is as follows:

$$\int_{-\infty}^{+\infty} dx_1 \int_{-\infty}^{+\infty} dx_2 \ldots \int_{-\infty}^{+\infty} dx_d \, e^{-\frac{1}{2}\sum_{ij} x_i A_{ij} x_j + \sum_k b_k x_k} = \frac{(2\pi)^{d/2}}{\sqrt{det A}} e^{\frac{1}{2}\sum_{ij} b_i A_{ij}^{-1} b_j}. \tag{1.21}$$

Now let us assume $b = 0$. Thus, we obtain

$$\sqrt{det A} \int_{-\infty}^{+\infty} \frac{dx_1}{\sqrt{2\pi}} \ldots \int_{-\infty}^{+\infty} \frac{dx_d}{\sqrt{2\pi}} e^{-\frac{1}{2}\sum_{ij} x_i A_{ij} x_j} = 1. \tag{1.22}$$

If $\mathbf{x} = (x_1, x_2, \ldots, x_d)$ is a random vector in $\mathbb{R}^d$, we can introduce a Gaussian measure

$$d\mu(x_1, x_2, \ldots, x_d) = P(x_1, x_2, \ldots, x_d)\, dx_1 dx_2 \ldots dx_d, \tag{1.23}$$

where

$$P(x_1, x_2, \dots, x_d) = \sqrt{det A}\,(2\pi)^{-d/2}\, e^{-\frac{1}{2}\sum_{ij} x_i A_{ij} x_j} \tag{1.24}$$

is the Gaussian distribution. The measure is normalized:

$$\int d\mu(x_1, x_2, \dots, x_d) = \int \dots \int dx_1 dx_2 \dots dx_d\, P(x_1, x_2, \dots, x_d) = 1. \tag{1.25}$$

The average of some function f can then be calculated as

$$\langle f(x_1, x_2, \dots, x_d)\rangle = \int d\mu\,(x_1, x_2, \dots, x_d)\, f\,(x_1, x_2, \dots, x_d)\,. \tag{1.26}$$

We can write then

$$\int d\mu(x_1, x_2, \dots, x_d) \exp\left[\sum_{k=1}^{d} b_k x_k\right] = \left\langle e^{\sum_k b_k x_k}\right\rangle = e^{\frac{1}{2}\sum_{ij} b_i \Delta_{ij} b_j}\,, \tag{1.27}$$

where we introduce the notation, $\Delta_{ij} = A^{-1}_{ij}$. Recalling the trick used in (1.8), we can find the average for any component of the random vector $\mathbf{x}$ as

$$\begin{aligned}
\langle x_i\rangle &= \int d\mu(x_1, \dots, x_d)\, x_i = \frac{\partial}{\partial b_i}\int d\mu \exp\left[\sum_k b_k x_k\right]\Bigg|_{\mathbf{b}=0} \\
&= \frac{\partial}{\partial b_i}\exp\left[\frac{1}{2}\sum_{kl} b_k \Delta_{kl} b_l\right]\Bigg|_{\mathbf{b}=0} = \exp\left[\frac{1}{2}\sum_{kl} b_k \Delta_{kl} b_l\right]\frac{\partial}{\partial b_i}\left[\frac{1}{2}\sum_{kl} b_k \Delta_{kl} b_l\right]\Bigg|_{\mathbf{b}=0} \\
&= \exp\left[\frac{1}{2}\sum_{kl} b_k \Delta_{kl} b_l\right]\frac{1}{2}\sum_{kl}(\delta_{ik}\Delta_{kl} b_l + b_k \Delta_{kl}\delta_{li})\Bigg|_{\mathbf{b}=0} = 0.
\end{aligned} \tag{1.28}$$

Thus, the so-called “correlator” of two coordinates of a random vector can be calculated as follows:

$$\begin{aligned}
\langle x_i x_j\rangle &= \int d\mu(x_1, \dots, x_d)\, x_i x_j = \frac{\partial^2}{\partial b_i \partial b_j}\int d\mu \exp\left[\sum_k b_k x_k\right]\Bigg|_{\mathbf{b}=0} \\
&= \frac{\partial}{\partial b_i}\frac{\partial}{\partial b_j}\exp\left[\frac{1}{2}\sum_{kl} b_k \Delta_{kl} b_l\right]\Bigg|_{\mathbf{b}=0} = \frac{\partial}{\partial b_j}\left\{e^{\frac{1}{2}(\mathbf{b}\Delta\mathbf{b})}\frac{1}{2}\sum_{kl}(\delta_{ik}\Delta_{kl} b_l + b_k\Delta_{kl}\delta_{li})\right\}\Bigg|_{\mathbf{b}=0} \\
&= \frac{1}{2}\sum_{kl}\left(\delta_{ik}\delta_{lj}\Delta_{kl} + \delta_{kj}\delta_{li}\Delta_{kl}\right) = \Delta_{ij} = A^{-1}_{ij},
\end{aligned} \tag{1.29}$$

where A^{-1}_{ij} denotes the correlation matrix.

The generating function is defined as a function of the following form:

$$Z(\mathbf{b}) = \left\langle e^{(\mathbf{bx})} \right\rangle = \int d\mu\, e^{(\mathbf{bx})}, \tag{1.30}$$

where $(\mathbf{bx}) = \sum_k b_k x_k$ represents the scalar product of two vectors. This function generates the average values of products of random variables

$$\begin{aligned} \left\langle x_{k_1} x_{k_2} \dots x_{k_l} \right\rangle &= \frac{\partial}{\partial b_{k_1}} \frac{\partial}{\partial b_{k_2}} \dots \frac{\partial}{\partial b_{k_l}} Z(\mathbf{b}) \bigg|_{\mathbf{b}=0} \\ &= \frac{\partial}{\partial b_{k_1}} \frac{\partial}{\partial b_{k_2}} \dots \frac{\partial}{\partial b_{k_l}} \exp\left[\frac{1}{2} \sum_{ij} b_i \Delta_{ij} b_j \right] \Bigg|_{\mathbf{b}=0}. \end{aligned} \tag{1.31}$$

If $F(\mathbf{x}) = F(x_1, x_2, \dots x_d)$ is an analytical function in the vicinity of $\mathbf{x} = 0$, i.e.

$$F(\mathbf{x}) = F(0) + \sum_{i=1}^{d} \frac{\partial F(0)}{\partial x_i} x_i + \frac{1}{2} \sum_{ij} \frac{\partial^2 F(0)}{\partial x_i \partial x_j} x_i x_j + \frac{1}{3!} \sum_{ijk} \frac{\partial^3 F(0)}{\partial x_i \partial x_j \partial x_k} x_i x_j x_k + \dots, \tag{1.32}$$

then it appears that

$$\langle F(\mathbf{x}) \rangle = F\left(\frac{\partial}{\partial \mathbf{b}}\right) \exp\left[\frac{1}{2} \sum_{ij} b_i \Delta_{ij} b_j \right] \Bigg|_{\mathbf{b}=0}, \tag{1.33}$$

where $\frac{\partial}{\partial \mathbf{b}} = \left(\frac{\partial}{\partial b_1}, \frac{\partial}{\partial b_2}, \dots, \frac{\partial}{\partial b_d} \right)$. Let us prove it in the following way:

$$\begin{aligned} \langle F(\mathbf{x}) \rangle &= \left\langle F(0) + \sum_{i=1}^{d} \frac{\partial F(0)}{\partial x_i} x_i + \frac{1}{2} \sum_{ij} \frac{\partial^2 F(0)}{\partial x_i \partial x_j} x_i x_j + \frac{1}{3!} \sum_{ijk} \frac{\partial^3 F(0)}{\partial x_i \partial x_j \partial x_k} x_i x_j x_k + \dots \right\rangle \\ &= \langle F(0) \rangle + \sum_i \frac{\partial F(0)}{\partial x_i} \langle x_i \rangle + \frac{1}{2} \sum_{ij} \frac{\partial^2 F(0)}{\partial x_i \partial x_j} \langle x_i x_j \rangle + \frac{1}{3!} \sum_{ijk} \frac{\partial^3 F(0)}{\partial x_i \partial x_j \partial x_k} \langle x_i x_j x_k \rangle + \dots \\ &= F(0) + \sum_i \frac{\partial F(0)}{\partial x_i} \frac{\partial}{\partial b_i} Z(\mathbf{b}) \bigg|_{\mathbf{b}=0} + \frac{1}{2} \sum_{ij} \frac{\partial^2 F(0)}{\partial x_i \partial x_j} \frac{\partial^2}{\partial b_i \partial b_j} Z(\mathbf{b}) \bigg|_{\mathbf{b}=0} + \dots \\ &= \left[F(0) + \sum_i \frac{\partial F(0)}{\partial x_i} \frac{\partial}{\partial b_i} + \frac{1}{2} \sum_{ij} \frac{\partial^2 F(0)}{\partial x_i \partial x_j} \frac{\partial^2}{\partial b_i \partial b_j} + \dots \right] Z(\mathbf{b}) \bigg|_{\mathbf{b}=0} \\ &= F\left(\frac{\partial}{\partial \mathbf{b}}\right) \exp\left[\frac{1}{2} (\mathbf{b} \Delta \mathbf{b}) \right] \bigg|_{\mathbf{b}=0}. \end{aligned} \tag{1.34}$$

The last identity leads to a significant statement known as the Wick's theorem: *the average value of the product $x_{k_1} x_{k_2} \dots x_{k_l}$ over a Gaussian measure can be calculated as follows: one needs to pair the indices $k_1 k_2 \dots k_l$ (where l is an even number) in* *all possible combinations, at the same time each pair of indices $k_p k_q$ corresponds to the matrix element* $\Delta_{k_q k_p}$. In other words:

$$\langle x_{k_1} x_{k_2} \dots x_{k_l} \rangle = \sum_{\substack{\text{all} \\ \text{pairs}}} \Delta_{k_{p_1} k_{p_2}} \Delta_{k_{p_3} k_{p_4}} \dots \Delta_{k_{p_{l-1}} k_{p_l}}$$
$$= \sum \langle x_{k_{p_1}} x_{k_{p_2}} \rangle \langle x_{k_{p_3}} x_{k_{p_4}} \rangle \dots \langle x_{k_{p_{l-1}}} x_{k_{p_l}} \rangle . \quad (1.35)$$

In particular, we have:

1.
$$\langle x_{i_1} x_{i_2} \rangle = \Delta_{i_1 i_2}, \quad (1.36)$$

2.
$$\langle x_{i_1} x_{i_2} x_{i_3} x_{i_4} \rangle = \Delta_{i_1 i_2} \Delta_{i_3 i_4} + \Delta_{i_1 i_3} \Delta_{i_2 i_4} + \Delta_{i_1 i_4} \Delta_{i_2 i_3}. \quad (1.37)$$

Let us also highlight a useful rule: when computing the Gaussian average of the product of $2p$ variables, the result should comprise the sum of $(2p-1) \times (2p-3) \times (2p-5) \cdots \times 5 \times 3 \times 1$ terms [1]. Additionally, let us present another useful identity:

$$\langle x_i F(\mathbf{x}) \rangle = \int d\mu x_i F(\mathbf{x}) = \sum_{l=1}^{d} \Delta_{il} \left\langle \frac{\partial F}{\partial x_l} \right\rangle . \quad (1.38)$$

Let us prove it. First, we recall that

$$\int d\mu (\dots) = (2\pi)^{-\frac{d}{2}} \sqrt{det A} \int dx_1 \dots \int dx_d \, e^{-\frac{1}{2} \sum_{ij} x_i A_{ij} x_j} (\dots) . \quad (1.39)$$

Now, let us consider the following expression:

$$\frac{\partial}{\partial x_l} \exp \left[-\frac{1}{2} \sum_{jk} x_j A_{jk} x_k \right] = \exp \left[-\frac{1}{2} \sum_{jk} x_j A_{jk} x_k \right] \frac{\partial}{\partial x_l} \left(-\frac{1}{2} \sum_{jk} x_j A_{jk} x_k \right)$$
$$= - \exp \left[-\frac{1}{2} \sum_{jk} x_k A_{jk} x_k \right] \sum_k A_{lk} x_k = - \exp \left[-\frac{1}{2} (\mathbf{x} A \mathbf{x}) \right] \sum_k A_{lk} x_k . \quad (1.40)$$

Then we multiply this equality by Δ_{il} (where Δ and A are the inverse matrices according to the introduced notation), and sum over the index l

$$\sum_l \Delta_{il} \frac{\partial}{\partial x_l} \exp \left[-\frac{1}{2} (\mathbf{x} A \mathbf{x}) \right] = - \exp \left[-\frac{1}{2} (\mathbf{x} A \mathbf{x}) \right] \sum_k \sum_l A_{lk} \Delta_{il} x_k . \quad (1.41)$$

After contraction over indices on the right-hand side, we arrive at the following:

$$\sum_l \Delta_{il} \frac{\partial}{\partial x_l} \left[e^{-\frac{1}{2}(\mathbf{x}A\mathbf{x})} \right] = -x_i e^{-\frac{1}{2}(\mathbf{x}A\mathbf{x})}. \tag{1.42}$$

Now let us apply this to

$$\begin{aligned}
\langle x_i F(\mathbf{x}) \rangle &= (2\pi)^{-\frac{d}{2}} \sqrt{det A} \int_{-\infty}^{+\infty} \dots \int_{-\infty}^{+\infty} dx_1 \dots dx_d \, e^{-\frac{1}{2}(\mathbf{x}A\mathbf{x})} x_i F(\mathbf{x}) \\
&= -(2\pi)^{-\frac{d}{2}} \sqrt{det A} \int_{-\infty}^{+\infty} \dots \int_{-\infty}^{+\infty} dx_1 \dots dx_d \sum_{l=1}^{d} \Delta_{il} \frac{\partial}{\partial x_l} \left[e^{-\frac{1}{2}(\mathbf{x}A\mathbf{x})} \right] F(\mathbf{x}) \\
&= -(2\pi)^{-\frac{d}{2}} \sqrt{det A} \sum_l \Delta_{il} \int_{-\infty}^{+\infty} \dots \int_{-\infty}^{+\infty} dx_1 \dots dx_d \, \frac{\partial}{\partial x_l} \left[e^{-\frac{1}{2}(\mathbf{x}A\mathbf{x})} \right] F(\mathbf{x}).
\end{aligned} \tag{1.43}$$

Then, we integrate by parts as

$$\begin{aligned}
u &= F(\mathbf{x}) & dv &= \frac{\partial}{\partial x_l} \left[e^{-\frac{1}{2}(\mathbf{x}A\mathbf{x})} \right] dx_l \\
du &= \frac{\partial F}{\partial x_l} dx_l & v &= e^{-\frac{1}{2}(\mathbf{x}A\mathbf{x})}.
\end{aligned} \tag{1.44}$$

and, thus, finally obtain:

$$\begin{aligned}
\langle x_i F(\mathbf{x}) \rangle &= -(2\pi)^{-\frac{d}{2}} \sqrt{det A} \sum_l \Delta_{il} \\
&\times \int_{-\infty}^{+\infty} \dots \int_{-\infty}^{+\infty} dx_1 \dots dx_{l-1} dx_{l+1} \dots dx_d \Bigg[F(\mathbf{x}) e^{-\frac{1}{2}(\mathbf{x}A\mathbf{x})} \Big|_{x_l=-\infty}^{x_l=+\infty} \\
&- \int_{-\infty}^{+\infty} dx_l \, e^{-\frac{1}{2}(\mathbf{x}A\mathbf{x})} \frac{\partial F}{\partial x_l} \Bigg] = \sum_l \Delta_{il} (2\pi)^{-\frac{d}{2}} \sqrt{det A} \int_{-\infty}^{+\infty} dx_1 \dots \int_{-\infty}^{+\infty} dx_d \, e^{-\frac{1}{2}(\mathbf{x}A\mathbf{x})} \frac{\partial F}{\partial x_l} \\
&= \sum_{l=1}^{d} \Delta_{il} \left\langle \frac{\partial F}{\partial x_l} \right\rangle.
\end{aligned} \tag{1.45}$$

1.2 Generalized Functions and Green's Function

Generalized functions. Let us consider a set of functions for which the following condition is satisfied:

$$\int_{-\infty}^{+\infty} \Phi_1(x) \, dx = 1, \tag{1.46}$$

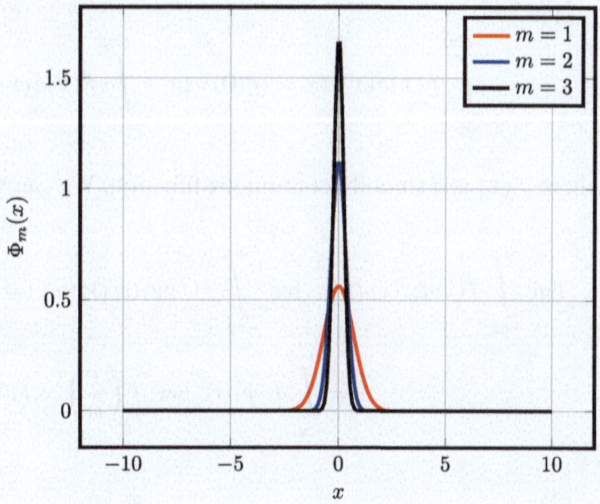

Fig. 1.1 Graphs of the function $\Phi_m(x)$ for $m = 1, 2, 3$

for instance,

$$\Phi_1(x) = \frac{1}{\pi}\frac{1}{1+x^2} \tag{1.47}$$

or

$$\Phi_1(x) = \frac{1}{\sqrt{\pi}}e^{-x^2}. \tag{1.48}$$

We can then construct the functions

$$\Phi_m(x) = m\Phi_1(mx), \tag{1.49}$$

whose graphs (for $m = 1, 2, 3$) can be seen in Fig. 1.1.

The above condition is also satisfied for them:

$$\int_{-\infty}^{+\infty} \Phi_m(x)\,dx = \int_{-\infty}^{+\infty} \Phi_1(mx)\,d(mx) = \left| mx = t \right| = \int_{-\infty}^{+\infty} \Phi_1(t)\,dt = 1. \tag{1.50}$$

If we now assume that $m \to \infty$, we obtain a function that satisfies the following conditions:

1. $\Phi(x) = 0$, for $x \neq 0$,
2. $\Phi(x) = \infty$, for $x = 0$,
3. $\int_{-\infty}^{+\infty} dx\, \Phi(x) = 1$.

Let us refer to the function $\Phi(x) = \lim_{m\to\infty} \Phi_m(x) = \delta(x)$ as the Dirac delta function. Now, let us explore the properties of the Dirac δ-function:

(1)

$$\int_{-\infty}^{+\infty} f(x)\delta(x)\,dx = f(0) \quad \text{or} \quad \int_{-\infty}^{+\infty} f(x)\delta(x-a)\,dx = f(a),$$

where $f(x)$ is a smooth, continuous function. We can verify this directly as

$$\lim_{m\to\infty}\int_{-\infty}^{+\infty} f(x)\Phi_m(x)dx = \lim_{m\to\infty}\int_{-\infty}^{+\infty} f(x)\Phi_1(mx)d(mx) = \lim_{m\to\infty}\int_{-\infty}^{+\infty} dt\, f\left(\frac{t}{m}\right)\Phi_1(t)$$
$$= \int_{-\infty}^{+\infty} dt\,\Phi_1(t)\lim_{m\to\infty} f\left(\frac{t}{m}\right) = f(0)\int_{-\infty}^{+\infty} dt\,\Phi_1(t) = f(0). \quad (1.51)$$

(2)

$$\delta(ax) = \frac{1}{|a|}\delta(x). \quad (1.52)$$

This property can be proven using the substitution

$$\int_{-\infty}^{+\infty} dx\,\delta(ax) = \left| ax = y \right| = \frac{1}{|a|}\int_{-\infty}^{+\infty} dy\,\delta(y) = \frac{1}{|a|}, \quad (1.53)$$

from which we can obtain the expression (1.52).

(3)

$$\delta(\varphi(x)) = \frac{1}{|\varphi'(a)|}\delta(x-a), \quad \text{where } \varphi(a) = 0. \quad (1.54)$$

In a more general way:

$$\delta(\varphi(x)) = \sum_{i=1}^{n}\frac{1}{|\varphi'(a_i)|}\delta(x-a_i) \quad (1.55)$$

where a_i are the roots of the function $\varphi(x)$.

We prove this by expanding the function $\varphi(x)$ into a series in the vicinity of $x = a$, and using the previous property:

$$\delta(\varphi(x)) = \delta\left(\varphi(a) + \varphi'(a)(x-a)\right) = \delta\left(\varphi'(a)(x-a)\right) = \frac{1}{|\varphi'(a)|}\delta(x-a). \quad (1.56)$$

We can also introduce a d-dimensional δ-function:

$$\delta^{(d)}(\mathbf{x}) = \prod_{i=1}^{d}\delta(x_i), \quad \text{where } \mathbf{x} = (x_1, x_2, \ldots, x_d). \quad (1.57)$$

Let us now represent the δ-function in the form of a Fourier integral

$$\Phi_1^{(d)}(\mathbf{x}) = \frac{1}{\pi^{d/2}} e^{-\mathbf{x}^2}, \quad \mathbf{x}^2 = x_1^2 + x_2^2 + \cdots + x_d^2. \tag{1.58}$$

$$\Phi_m^{(d)}(\mathbf{x}) = m^d \Phi_1^{(d)}(\mathbf{x}). \tag{1.59}$$

Let us write expressions for the direct and inverse Fourier transforms:

$$\Phi(\mathbf{x}) = \int \frac{d\mathbf{k}}{(2\pi)^d} e^{i\mathbf{kx}} \tilde{\Phi}(\mathbf{k}), \tag{1.60}$$

$$\tilde{\Phi}(\mathbf{k}) = \int d\mathbf{x}\, e^{-i\mathbf{kx}} \Phi(\mathbf{x}), \tag{1.61}$$

where we utilize the notations for the scalar product, $\mathbf{kx} = k_1 x_1 + k_2 x_2 + \cdots + k_d x_d$. Then,

$$\frac{1}{\pi^{d/2}} \int d\mathbf{x}\, e^{-x^2 - i\mathbf{kx}} = \left[\frac{1}{\pi^{1/2}} \int_{-\infty}^{+\infty} dx_1\, e^{-x_1^2 - ik_1 x_1} \right]^d, \tag{1.62}$$

$$\int_{-\infty}^{+\infty} dx\, e^{-x^2 - ikx} = e^{-\frac{k^2}{4}} \int_{-\infty}^{+\infty} dx\, e^{-(x + \frac{ik}{2})^2} = e^{-\frac{k^2}{4}} \int_{-\infty}^{+\infty} dy\, e^{-y^2} = \sqrt{\pi} e^{-\frac{k^2}{4}}. \tag{1.63}$$

$$\tilde{\Phi}(\mathbf{k}) = e^{-\frac{k^2}{4}}, \qquad \tilde{\Phi}_m(\mathbf{k}) = e^{-\frac{k^2}{4m}}. \tag{1.64}$$

$$\lim_{m\to\infty} \Phi_m^{(d)}(\mathbf{x}) = \delta^{(d)}(\mathbf{x}), \qquad \lim_{m\to\infty} \tilde{\Phi}_m^{(d)}(\mathbf{k}) = 1. \tag{1.65}$$

Finally, we obtain

$$\delta^{(d)}(\mathbf{x}) = \int \frac{d\mathbf{k}}{(2\pi)^d} e^{i\mathbf{kx}} \cdot 1, \tag{1.66}$$

and, hence, in the one-dimensional case:

$$\delta(x) = \int_{-\infty}^{+\infty} \frac{dk}{2\pi} e^{ikx}. \tag{1.67}$$

Let us also write down the definition of the Heaviside step-function used hereafter:

$$\Theta(x) = \int_{-\infty}^{x} \delta(\xi) d\xi = \begin{cases} 1, & x > 0, \\ 1/2, & x = 0, \\ 0, & x < 0. \end{cases} \tag{1.68}$$

The derivative of Θ-function is δ-function:

$$\Theta'(x) = \delta(x). \tag{1.69}$$

Indeed,

$$\int_{-\infty}^{\infty} dx\, \varphi(x)\Theta'(x) = \varphi(x)\Theta(x)\Big|_{-\infty}^{\infty} - \int_{-\infty}^{\infty} dx\, \varphi'(x)\Theta(x)$$
$$= \varphi(\infty) - \int_{0}^{\infty} dx\, \varphi'(x) = \varphi(\infty) - \varphi(x)\Big|_{0}^{\infty} = \varphi(0). \tag{1.70}$$

Green's function. Let us now also discuss the idea behind a Green's function, which can be easily defined using the delta function. Consider the following operator equation:

$$Ly = f, \tag{1.71}$$

where L denotes some linear operator and f represents the source. Then the general solution is:

$$y(x) = y_0(x) + L^{-1}f(x) = y_0(x) + \int d\xi\, L^{-1}\delta(x-\xi)f(\xi), \tag{1.72}$$

where $L^{-1}\delta(x-\xi) = G(x,\xi)$ is the Green's function, and $y_0(x)$ is the general solution of the homogeneous equation, i.e., Eq. (1.71) with $f = 0$. Thus, we have the following equation for the Green's function:

$$L_x G(x,\xi) = \delta(x-\xi), \tag{1.73}$$

where index x means that operator L acts on argument x of delta-function. We will utilize the concept of Green's function in the subsequent discussion.

1.3 Functionals. Functional Differentiation and Integration

Let us define a functional $F[y]$ as a mapping of a class of functions $y(x)$ defined on the interval $x \in (a, b)$ to the set of real or complex numbers. Put simply, a functional represents a rule that assigns a value of some scalar parameter to each function within a specified class. The simplest example of a functional is the definite integral of the function $y(x)$

$$F[y] = \int_a^b y(x)\, dx, \tag{1.74}$$

or

$$F[y] = \int_a^b f(x)y(x)\, dx = (fy), \tag{1.75}$$

where $f(x)$ is a real-valued function, and we introduce the notation (fy) for the scalar product. It's important to note that these functionals are linear, meaning they satisfy the following condition:

$$F[C_1 y_1 + C_2 y_2] = C_1 F[y_1] + C_2 F[y_2], \tag{1.76}$$

where C_1, C_2 denote constants, and y_1, y_2 represent functions. Considering a function of two variables, $K(x_1, x_2)$, we can now illustrate an example of a quadratic (bilinear) functional:

$$F[y] = \iint y(x_1)K(x_1, x_2)y(x_2)\, dx_1 dx_2 = (yKy). \tag{1.77}$$

It's worth noting that a functional can be also interpreted as a function with an infinite number of arguments. One way to visualize this is by expanding the argument of the functional into a series over a complete orthonormalized system of functions $\varphi_n(x)$ as:

$$F[y] = F\left[\sum_n u_n \varphi_n\right] = \mathcal{F}(u_1, u_2, \ldots). \tag{1.78}$$

Subsequently, for linear and bilinear functionals, respectively, we arrive at:

$$\mathcal{F}(u_1, u_2, \ldots) = \sum_n u_n \int dx\, \varphi_n(x) f(x) \tag{1.79}$$

and

$$\mathcal{F}(u_1, u_2, \ldots) = \sum_{nm} u_n u_m \iint dx_1 dx_2\, \varphi_n(x_1)K(x_1, x_2)\varphi_m(x_2). \tag{1.80}$$

Some functionals can also be represented as series, which terms themselves are also functionals. Similar to representing a function as a power series, we can write:

$$F[y] = \sum_{n=0}^{\infty} \frac{1}{n!} \int dx_1 \int dx_2 \ldots \int dx_n\, f(x_1, x_2, \ldots, x_n)y(x_1)y(x_2)\ldots y(x_n). \tag{1.81}$$

For instance, if we set $f(x_1, x_2, \ldots, x_n) = \prod_{s=1}^{n} \varphi(x_s)$, we obtain an exponential series:

$$\sum_{n=0}^{\infty} \frac{1}{n!} \left[\int dx\, \varphi(x)y(x)\right]^n = \exp\left[\int dx\, \varphi(x)y(x)\right] = e^{(\varphi y)}, \tag{1.82}$$

or, similarly, a Gaussian series:

$$\sum_{n=o}^{\infty} \frac{1}{n!} \left[\frac{1}{2} \iint dx_1 dx_2\, y(x_1)K(x_1, x_2)y(x_2)\right]^n = \exp\left[\frac{1}{2} \iint dx_1 dx_2\, y(x_1)K(x_1, x_2)y(x_2)\right]$$

$$= e^{\frac{1}{2}(yKy)}. \tag{1.83}$$

The functional (or variational) derivative is defined similarly to the partial derivative of a multivariable function. Let us give a function $y(x)$ a small variation $\delta y(x)$, then the variation of the functional with respect to $\delta y(x)$ within the first order can be expressed as follows:

$$\delta F = F[y + \delta y] - F[y] = \int dx\, f(x)\delta y(x) = \int dx\, \frac{\delta F}{\delta y(x)}\delta y(x). \tag{1.84}$$

We then designate the object $\delta F/\delta y(x)$ in the integrand expression as the functional derivative. Now, let us compare this result with the expression for the total differential of a multivariable function $u(x_1, \ldots, x_n)$:

$$du = \sum_s \frac{\partial u}{\partial x_s} dx_s. \tag{1.85}$$

We can draw the following analogies:

$$\delta F \leftrightarrow du, \qquad x \leftrightarrow u, \qquad y(x) \leftrightarrow x_s \tag{1.86}$$

$$\frac{\delta F}{\delta y(x)} \leftrightarrow \frac{\partial u}{\partial x_s}, \qquad \int dx \leftrightarrow \sum_s. \tag{1.87}$$

Consider the differentiation of the functional

$$F_n[y] = \int \cdots \int dx_1 \ldots dx_n\, f(x_1, \ldots, x_n) y(x_1) \ldots y(x_n), \tag{1.88}$$

then the functional derivative can be written as

$$\begin{aligned} \frac{\delta F_n}{\delta y(z)} &= \int \cdots \int dx_2 \ldots dx_n\, f(z, x_2, \ldots, x_n) y(x_2) \ldots y(x_n) \\ &+ \int \cdots \int dx_1 dx_3 \ldots dx_n\, f(x_1, z, \ldots, x_n) y(x_1) y(x_3) \ldots y(x_n) + \ldots \\ &+ \int \cdots \int dx_1 \ldots dx_{n-1}\, f(x_1, x_2, \ldots, z) y(x_1) \ldots y(x_{n-1}), \end{aligned} \tag{1.89}$$

which, for a function $f(x_1, \ldots, x_n)$, symmetric with respect to its arguments, can be rewritten as

$$\frac{\delta F_n}{\delta y(z)} = n \int \cdots \int dx_1 \ldots dx_{n-1}\, f(x_1, x_2, \ldots, x_{n-1}, z) y(x_1) \ldots y(x_{n-1}). \tag{1.90}$$

Similarly, we can evaluate higher-order functional derivatives. Thus, for the second-order derivative, we obtain:

$$\frac{\delta^2 F_n}{\delta y(z_1)\delta y(z_2)} = n(n-1)\int \ldots \int dx_1 \ldots dx_{n-2}\, f(x_1, \ldots, x_{n-2}, z_1, z_2) y(x_1) \ldots y(x_{n-2}), \tag{1.91}$$

and more generally, for the m-th-order derivative, we have:

$$\frac{\delta^m F_n}{\delta y(z_1) \ldots \delta y(z_m)} = \frac{n!}{(n-m)!}\int \ldots \int dx_1 \ldots dx_{n-m}\, f(x_1, \ldots, x_{n-m}, z_1, \ldots, z_m) y(x_1) \ldots y(x_{n-m}). \tag{1.92}$$

Thus, we can formulate a following useful identity

$$\frac{\delta y(x_1)}{\delta y(x_2)} = \delta(x_1 - x_2), \tag{1.93}$$

where $\delta(x_1 - x_2)$ represents the Dirac delta function. Equation (1.93) serves as a continuous analog of the equality (1.12), with delta-function acting as the continuous counterpart of the Kronecker symbol.

Let us explore several useful examples of evaluating functional derivatives.

1. Let us consider a functional $F[q]$ given by the form

$$F[q] = \int \cdots \int_\Gamma f(q(x_1, \ldots, x_n))\, dx_1 \ldots dx_n, \tag{1.94}$$

where $f(q)$ is a single-variable function, and $q(x_1, \ldots, x_n)$ is a function of n variables defined in the domain $\Gamma \subseteq \mathbb{R}^n$. Then, the variation of this functional is expressed as

$$\delta F = \int \cdots \int_\Gamma \frac{df(q(x_1, \ldots, x_n))}{dq(x_1, \ldots, x_n)} \delta q(x_1, \ldots, x_n)\, dx_1 \ldots dx_n, \tag{1.95}$$

and, thus, finding the variational derivative simplifies to calculating the standard derivative of the integrand function:

$$\frac{\delta F[q]}{\delta q(x_1, \ldots, x_n)} = \frac{df(q(x_1, \ldots, x_n))}{dq(x_1, \ldots, x_n)}. \tag{1.96}$$

2. Let us now consider the case where the functional $F[q]$ also depends on the partial derivatives of the function q:

$$F[q] = \int \cdots \int_\Gamma f\left(q(x_1, \ldots, x_n), \frac{\partial q}{\partial x_1}, \ldots, \frac{\partial q}{\partial x_n}\right) dx_1 \ldots dx_n. \tag{1.97}$$

A variation of this functional is

$$\delta F = \int \cdots \int_\Gamma dx_1 \ldots dx_n \left[\frac{\partial f}{\partial q}\delta q(x_1, \ldots, x_n) + \sum_{k=1}^{n} \frac{\partial f}{\partial \left(\frac{\partial q}{\partial x_k}\right)} \delta\left(\frac{\partial q}{\partial x_k}\right)\right]. \tag{1.98}$$

Integrating by parts for each variable x_k and assuming that the variations of the function q are zero at the boundary of the region Γ, we obtain:

$$\delta F = \int \cdots \int_{\Gamma} dx_1 \ldots dx_n \left\{ \frac{\partial f}{\partial q} - \sum_{k=1}^{n} \frac{\partial}{\partial x_k} \left[\frac{\partial f}{\partial \left(\frac{\partial q}{\partial x_k} \right)} \right] \right\} \delta q(x_1, \ldots, x_n). \tag{1.99}$$

Thus, the variational derivative is as follows:

$$\frac{\delta F[q]}{\delta q(x_1, \ldots, x_n)} = \frac{\partial f}{\partial q} - \sum_{k=1}^{n} \frac{\partial}{\partial x_k} \left[\frac{\partial f}{\partial \left(\frac{\partial q}{\partial x_k} \right)} \right]. \tag{1.100}$$

Functional integration. If we adopt the definition of a functional as a function of an infinite number of variables, then the functional integral, at a physical level of rigor, is represented by

$$\begin{aligned} \int F[y]\mathcal{D}y &= \lim_{n \to \infty} \int \cdots \int f(u_1, \ldots, u_n) du_1 \ldots du_n \\ &= \int \cdots \int f(u_1, \ldots, u_n, \ldots) du_1 \ldots du_n \ldots, \end{aligned} \tag{1.101}$$

provided that the limit in the right-hand side of the expression exists. Below, we will delve into some specific examples of functional integrals.

Functionals of scalar fields. Let us write down the functional

$$I[y] = \int_a^b F(x, y, y')dx, \tag{1.102}$$

where $y = y(x)$ is a single-variable function. Now, let us transition from the function of a single variable $y(x)$ to functions of multiple variables, i.e., scalar fields

$$\varphi = \varphi(\mathbf{x}) = \varphi(x_1, x_2, \ldots, x_n), \tag{1.103}$$

where $\varphi \in \mathbb{R}$, $\mathbf{x} = (x_1, x_2, \ldots, x_n) \in \Gamma \subseteq \mathbb{R}^n$, $n \geq 1$. Then by analogy, we can determine

$$I[\varphi] = \int_{\Gamma} d\mathbf{x}\, F(\mathbf{x}, \varphi, \nabla\varphi), \tag{1.104}$$

as a functional defined on the space of scalar fields $\varphi = \varphi(\mathbf{x})$, where $d\mathbf{x} = dx_1 dx_2 \ldots dx_n$, and $\nabla\varphi = \left(\frac{\partial \varphi}{\partial x_1}, \frac{\partial \varphi}{\partial x_2}, \ldots, \frac{\partial \varphi}{\partial x_n} \right) = (\partial_1\varphi, \partial_2\varphi, \ldots, \partial_n\varphi)$ represents the gradient of the field φ. Suppose that the field φ satisfies certain boundary conditions at the boundary $\partial\Gamma$. Let us express the change in the functional $\Delta I[\varphi]$ resulting from the variation of the scalar field $\delta\varphi(\mathbf{x})$, with the constraint $\delta\varphi|_{\partial\Gamma} = 0$:

$$\Delta I[\varphi] = I[\varphi + \delta\varphi] - I[\varphi] = \int_\Gamma d\mathbf{x}\, [F(\mathbf{x}, \varphi + \delta\varphi, \nabla(\varphi + \delta\varphi)) - F(\mathbf{x}, \varphi, \nabla\varphi)]\,. \tag{1.105}$$

Let us expand the first term of the integrand into a series:

$$F(\mathbf{x}, \varphi + \delta\varphi, \nabla(\varphi + \delta\varphi)) = F(\mathbf{x}, \varphi, \nabla\varphi) + \delta\varphi\frac{\partial F}{\partial\varphi} + \sum_{k=1}^{n}\frac{\partial F}{\partial(\partial_k\varphi)}\partial_k(\delta\varphi) + \ldots \tag{1.106}$$

Everywhere below, we'll adhere to Einstein's convention, implying summation over repeated coordinate indices. For example:

$$\sum_{k=1}^{n}\frac{\partial F}{\partial(\partial_k\varphi)}(\partial_k\delta\varphi) = \frac{\partial F}{\partial(\partial_k\varphi)}\partial_k(\delta\varphi). \tag{1.107}$$

Then,

$$\Delta I[\varphi] = \int_\Gamma d\mathbf{x}\left(\delta\varphi\frac{\partial F}{\partial\varphi} + \frac{\partial F}{\partial(\partial_k\varphi)}\partial_k(\delta\varphi) + \ldots\right) = \delta I[\varphi] + \ldots \tag{1.108}$$

Thus, let us write a final expression for the linear part of the increment of the functional—the variation of the functional

$$\delta I[\varphi] = \int_\Gamma d\mathbf{x}\left(\delta\varphi\frac{\partial F}{\partial\varphi} + \frac{\partial F}{\partial(\partial_k\varphi)}\partial_k(\delta\varphi)\right) \tag{1.109}$$

In the one-dimensional case it yields

$$\delta I[y] = \int_a^b dx\left(\frac{\partial F}{\partial y}\delta y + \frac{\partial F}{\partial y'}\delta y'\right). \tag{1.110}$$

Let us examine the second term in the integrand in detail by introducing the vector $\mathbf{A} = \left(\frac{\partial F}{\partial(\partial_1\varphi)}, \frac{\partial F}{\partial(\partial_2\varphi)}, \ldots, \frac{\partial F}{\partial(\partial_n\varphi)}\right)$, i.e.

$$\int_\Gamma d\mathbf{x}\,\frac{\partial F}{\partial(\partial_k\varphi)}\partial_k(\delta\varphi) = \int_\Gamma d\mathbf{x}\,\mathbf{A}\cdot\nabla(\delta\varphi). \tag{1.111}$$

Now we can apply the product rule:

$$\frac{\partial F}{\partial(\partial_k\varphi)}\partial_k(\delta\varphi) = \partial_k\left(\delta\varphi\frac{\partial F}{\partial(\partial_k\varphi)}\right) - \delta\varphi\partial_k\left(\frac{\partial F}{\partial(\partial_k\varphi)}\right), \tag{1.112}$$

or in alternative notations:

$$\mathbf{A}\cdot\nabla\delta\varphi = \nabla\cdot(\mathbf{A}\delta\varphi) - \delta\varphi\nabla\cdot\mathbf{A}. \tag{1.113}$$

Then,

$$\int_\Gamma d\mathbf{x}\,(\nabla\cdot(\mathbf{A}\delta\varphi)-\delta\varphi\nabla\cdot\mathbf{A})=\int_\Gamma d\mathbf{x}\,\nabla\cdot(\mathbf{A}\delta\varphi)-\int_\Gamma d\mathbf{x}\,\delta\varphi\nabla\cdot\mathbf{A}. \tag{1.114}$$

Using the divergence theorem (Gauss-Ostrogradsky's theorem)

$$\int_\Gamma d\mathbf{x}\,\nabla\cdot\mathbf{B}=\oint_{\partial\Gamma} dS\,\mathbf{B}\cdot\mathbf{n}, \tag{1.115}$$

where $\mathbf{n}$ denotes the normal vector, we notice that

$$\int_\Gamma \nabla\cdot(\mathbf{A}\delta\varphi)\,d\mathbf{x}=\oint_{\partial\Gamma} dS\,(\mathbf{n}\cdot\mathbf{A})\delta\varphi\equiv 0, \tag{1.116}$$

due to $\delta\varphi|_{\partial\Gamma}=0$. Then, we obtain:

$$\int_\Gamma d\mathbf{x}\,\frac{\partial F}{\partial(\partial_k\varphi)}\partial_k(\delta\varphi)=-\int_\Gamma d\mathbf{x}\,\delta\varphi\partial_k\left(\frac{\partial F}{\partial(\partial_k\varphi)}\right), \tag{1.117}$$

and the final expression for the variation of the functional takes the form:

$$\delta I[\varphi]=\int_\Gamma d\mathbf{x}\left(\frac{\partial F}{\partial\varphi}-\partial_k\left(\frac{\partial F}{\partial(\partial_k\varphi)}\right)\right)\delta\varphi(\mathbf{x}). \tag{1.118}$$

Thus, the variational (or functional) derivative is

$$\frac{\delta I[\varphi]}{\delta\varphi(\mathbf{x})}=\frac{\partial F}{\partial\varphi}-\partial_k\left(\frac{\partial F}{\partial(\partial_k\varphi)}\right), \tag{1.119}$$

or in the one-dimensional case:

$$\frac{\delta I[y]}{\delta y(x)}=\frac{\partial F}{\partial y}-\frac{d}{dx}\frac{\partial F}{\partial y'}. \tag{1.120}$$

Applying the necessary condition for an extremum, $\frac{\delta I[\varphi]}{\delta\varphi(\mathbf{x})}=0$, we arrive at the Euler-Lagrange equations:

$$\frac{\partial F}{\partial\varphi}-\partial_k\left(\frac{\partial F}{\partial(\partial_k\varphi)}\right)=0. \tag{1.121}$$

From there, with the boundary conditions on $\partial\Gamma$ also known, we can determine the extremal of the functional $\varphi(\mathbf{x})=\bar{\varphi}(\mathbf{x})$.

Constrained extremum of the functional. Below, we'll encounter problems related to finding the constrained extremum of a functional. Let us focus on this matter further. First, we revisit the analogy with functions of multiple variables. For simplicity, we consider a function of two variables $z=f(x,y)$. Now, let us impose a certain constraint linking the variables x and y, namely, $F(x, y) = h$, where h is a constant. We're interested in finding the extremum of the function z subject to this constraint. If the functional form $F(x, y)$ allows expressing one variable in terms of the other, the problem can be reduced to finding the extremum of a single-variable function. However, in practice, this is often not achievable. In such cases, we resort to the method of Lagrange multipliers. Therefore, at the extremum point, we have:

$$\frac{dz}{dx} = 0 = \frac{\partial f}{\partial x} + \frac{\partial f}{\partial y}\frac{dy}{dx} = f_x + f_y\frac{dy}{dx}. \tag{1.122}$$

Then, also at the extremum point, the following equality holds:

$$\frac{dy}{dx} = -\frac{f_x}{f_y}. \tag{1.123}$$

On the other hand, we can also differentiate the condition $F(x, y) = h$ with respect to x:

$$F_x + F_y\frac{dy}{dx} = 0 \quad \Rightarrow \quad \frac{dy}{dx} = -\frac{F_x}{F_y}, \tag{1.124}$$

and this equality holds for any point. Therefore, at the extremum point, we can write:

$$\frac{f_x}{f_y} = \frac{F_x}{F_y} \quad \Leftrightarrow \quad \frac{f_x}{F_x} = \frac{f_y}{F_y} = -\lambda. \tag{1.125}$$

Then, we can express the Lagrange function (or Lagrange polynomial) as:

$$\tilde{f}(x, y, \lambda) = f(x, y) + \lambda F(x, y), \tag{1.126}$$

where the coefficient λ is referred to as the Lagrange multiplier. Thus, to solve the constrained extremum problem, we need to solve the following system of equations:

$$\begin{cases} f_x + \lambda F_x = 0 \\ f_y + \lambda F_y = 0 \\ F(x, y) = h \end{cases} \quad \Leftrightarrow \quad \begin{cases} \tilde{f}_x(x, y, \lambda) = 0 \\ \tilde{f}_y(x, y, \lambda) = 0 \\ F(x, y) = h, \end{cases} \tag{1.127}$$

which leads us to the extremum point and the Lagrange multiplier, $(\bar{x}, \bar{y}, \lambda)$. It's worth separately discussing the "physical" significance of the Lagrange multiplier. We notice that, strictly speaking, $\bar{x} = \bar{x}(h), \bar{y} = \bar{y}(h), \bar{z} = \bar{z}(h) = f(\bar{x}(h), \bar{y}(h))$ are functions of the constant h. Consider then the following derivative:

$$\frac{d\bar{z}}{dh} = f_x\frac{d\bar{x}}{dh} + f_y\frac{d\bar{y}}{dh} = -\lambda F_x\frac{d\bar{x}}{dh} - \lambda F_y\frac{d\bar{y}}{dh} = -\lambda\left(F_x\frac{d\bar{x}}{dh} + F_y\frac{d\bar{y}}{dh}\right) = -\lambda\frac{dF}{dh} = -\lambda. \tag{1.128}$$

So, $\lambda = -\frac{d\bar{z}}{dh}$ characterizes the rate of change of the extreme value of the function $z = f(x, y)$ with respect to the parameter h.

Let us generalize the above result to the case of a function of $n > 2$ variables. Suppose we have a function $u = f(x_1, x_2, \ldots, x_n)$ subject to k constraints:

$$\begin{aligned}
&F_1(x_1, x_2, \ldots, x_n) = h_1 \\
&F_2(x_1, x_2, \ldots, x_n) = h_2 \\
&\ldots \\
&F_k(x_1, x_2, \ldots, x_n) = h_k, \qquad (k \leq n).
\end{aligned}$$

Let us find the extremum of this function taking into account the given constraints. In this case, the Lagrange function takes the form

$$\tilde{u} = f + \lambda_1 F_1 + \lambda_2 F_2 + \cdots + \lambda_k F_k, \tag{1.129}$$

and the system of equations used to determine the extremum point and Lagrange multipliers is

$$\begin{cases}
\tilde{u}_{x_i} = 0, \qquad i = 1, 2, \ldots, n \\
F_1 = h_1 \\
\ldots \\
F_k = h_k,
\end{cases} \tag{1.130}$$

where $\tilde{u}_{x_i} = \partial\tilde{u}/\partial x_i$. We end up with a set $(\bar{x}_1, \bar{x}_2, \ldots, \bar{x}_n; \lambda_1, \lambda_2, \ldots, \lambda_k)$, where

$$\lambda_i = -\frac{\partial \bar{u}}{\partial h_i} \tag{1.131}$$

with $\bar{u} = f(\bar{x}_1, \ldots, \bar{x}_n)$ representing the function extremum.

Finally, let us generalize these concepts by analogy to the case of functionals. This can be achieved since, as we previously noted, a functional can be considered a function of the variables, which number tends to infinity. Let us consider the following functional:

$$I[y] = \int_a^b dx\, F(x, y, y') \tag{1.132}$$

with a constrain

$$\int_a^b dx\, G(x, y, y') = H. \tag{1.133}$$

Consider the same problem of finding the extremum of the functional $I[y]$ under the given constraint $\int_a^b dx\, G = H$. Now, let us construct the Lagrange functional following a similar approach as the one described above:

$$\tilde{I}[y] = \int_a^b dx\, F(x, y, y') + \lambda \int_a^b dx\, G(x, y, y') = \int_a^b dx\, \left(F(x, y, y') + \lambda G(x, y, y')\right). \tag{1.134}$$

Utilizing the extremum condition

$$\frac{\delta \tilde{I}[y]}{\delta y(x)} = 0, \tag{1.135}$$

we can find the extremal of the functional $y = \bar{y}$. The Lagrange multiplier is then

$$\lambda = -\frac{d\bar{I}}{dH}, \tag{1.136}$$

and the extremal value of the functional is determined as $\bar{I} = I[\bar{y}]$.

1.4 Gaussian Functional Integrals

Let us one more time establish the obtained expression for the d-dimensional Gaussian integral

$$\int_{-\infty}^{+\infty} \frac{dx_1}{\sqrt{2\pi}} \cdots \int_{-\infty}^{+\infty} \frac{dx_d}{\sqrt{2\pi}} e^{-\frac{1}{2}\sum\limits_{ij} x_i A_{ij}^{-1} x_j + \sum\limits_{k} b_k x_k}$$
$$= (2\pi)^{-\frac{d}{2}} \int_{-\infty}^{+\infty} dx_1 \ldots \int_{-\infty}^{+\infty} dx_d\, e^{-\frac{1}{2}(\mathbf{x}A^{-1}\mathbf{x}) + (\mathbf{b}\mathbf{x})} = \sqrt{det A} e^{\frac{1}{2}(\mathbf{b}A\mathbf{b})}, \tag{1.137}$$

where

$$(\mathbf{b}A\mathbf{b}) = \sum_{ij} b_i A_{ij} b_j. \tag{1.138}$$

Let us then denote the following expression as the Hubbard-Stratonovich transformation (formula):

$$(2\pi)^{-\frac{d}{2}} \int_{-\infty}^{+\infty} \cdots \int_{-\infty}^{+\infty} dx_1 \ldots dx_d \frac{1}{\sqrt{det A}} e^{-\frac{1}{2}(\mathbf{x}A^{-1}\mathbf{x}) + (\mathbf{b}\mathbf{x})} = e^{\frac{1}{2}(\mathbf{b}A\mathbf{b})}. \tag{1.139}$$

Let us now extend our constructions to the case of an infinite-dimensional space by introducing the Gaussian functional integral. Consider $\mathbf{x} \in \Gamma \subseteq \mathbb{R}^d$, and let scalar fields $\varphi(\mathbf{x}) = \varphi(x_1, x_2, \ldots x_d)$ be defined within this region Γ, subject to certain boundary conditions on $\partial\Gamma$. We assume that these fields are quadratically integrable, i.e.,

$$\int_\Gamma d\mathbf{x}\, \varphi^2(\mathbf{x}) < \infty, \tag{1.140}$$

where we assume that $d\mathbf{x} = dx_1 dx_2 \dots dx_d$ represents a d-dimensional infinitesimal volume. Consider a complete system of orthonormalized functions defined in $\Gamma \subseteq \mathbb{R}^d$:

$$\{\varphi_n(\mathbf{x})\} : \begin{cases} \int\limits_\Gamma d\mathbf{x}\, \varphi_n(\mathbf{x})\varphi_m(\mathbf{x}) = \delta_{nm} & \text{– orthonormality} \\ \sum\limits_n \varphi_n(\mathbf{x})\varphi_n(\mathbf{x}') = \delta(\mathbf{x} - \mathbf{x}') & \text{– completeness.} \end{cases} \tag{1.141}$$

We can now expand our field φ using this complete orthonormalized system

$$\varphi(\mathbf{x}) = \sum_n u_n \varphi_n(\mathbf{x}) = \begin{pmatrix} u_1 \\ u_2 \\ \vdots \end{pmatrix}, \tag{1.142}$$

where $u_n = \int\limits_\Gamma d\mathbf{x}\, \varphi(\mathbf{x})\varphi_n(\mathbf{x}) = (\varphi\, \varphi_n)$ denotes the Fourier coefficients. So, the function can be interpreted as an infinite-dimensional vector in the basis $\varphi_n(\mathbf{x})$. An arbitrary function can then be expressed as:

$$b(\mathbf{x}) = \sum_n b_n \varphi_n(\mathbf{x}), \tag{1.143}$$

and

$$\begin{aligned} (b\,\varphi) &= \int\limits_\Gamma d\mathbf{x}\, b(\mathbf{x})\varphi(\mathbf{x}) = \int\limits_\Gamma d\mathbf{x} \sum_n b_n \varphi_n(\mathbf{x}) \sum_m u_m \varphi_m(\mathbf{x}) \\ &= \sum_n \sum_m b_n u_m \int\limits_\Gamma d\mathbf{x}\, \varphi_n(\mathbf{x})\varphi_m(\mathbf{x}) = \sum_n b_n u_n. \end{aligned} \tag{1.144}$$

Let us define a linear operator A acting on the function space φ as follows:

$$A\varphi(\mathbf{x}) = \int\limits_\Gamma d\mathbf{y}\, A(\mathbf{x}, \mathbf{y})\varphi(\mathbf{y}), \tag{1.145}$$

where $A(\mathbf{x}, \mathbf{y})$ represents the symmetric kernel of the linear operator, i.e., $A(\mathbf{x}, \mathbf{y}) = A(\mathbf{y}, \mathbf{x})$. Now, let us consider the following quadratic form:

$$\begin{aligned} (\varphi A \varphi) &= \int\limits_\Gamma d\mathbf{x} \int\limits_\Gamma d\mathbf{y}\, \varphi(\mathbf{x}) A(\mathbf{x}, \mathbf{y})\varphi(\mathbf{y}) = \int\limits_\Gamma d\mathbf{x} \int\limits_\Gamma d\mathbf{y}\, A(\mathbf{x}, \mathbf{y}) \sum_n u_n \varphi_n(\mathbf{x}) \sum_m u_m \varphi_m(\mathbf{y}) \\ &= \sum_n \sum_m u_n u_m \int\limits_\Gamma d\mathbf{x} \int\limits_\Gamma d\mathbf{y}\, \varphi_n(\mathbf{x}) A(\mathbf{x}, \mathbf{y})\varphi_m(\mathbf{y}) = \sum_n \sum_m u_n u_m (\varphi_n A \varphi_m) = \sum_n \sum_m A_{nm} u_n u_m, \end{aligned} \tag{1.146}$$

where $A_{nm} = A_{mn}$ represents the corresponding matrix element of the operator A. Now, let us introduce the function $\tilde{b}$ and the operator $\tilde{A}$, limiting the number of elements to N:

$$\tilde{b}(\mathbf{x}) = \sum_{n \leq N} b_n \varphi_n(\mathbf{x}), \tag{1.147}$$

$$\tilde{A}^{(N)} = \begin{pmatrix} A_{11} & A_{12} & \dots & A_{1N} \\ A_{21} & A_{22} & \dots & A_{2N} \\ \vdots & \vdots & \ddots & \vdots \\ A_{N1} & A_{N2} & \dots & A_{NN} \end{pmatrix}, \tag{1.148}$$

and write the N-th-dimensional Gaussian integral

$$(2\pi)^{-\frac{N}{2}} \int_{-\infty}^{+\infty} du_1 \dots \int_{-\infty}^{+\infty} du_N\, e^{-\frac{1}{2}\sum\limits_{n,m=1}^{N} u_n \tilde{A}^{-1}_{nm} u_m + \sum\limits_{n=1}^{N} b_n u_n} = \sqrt{det \tilde{A}^{(N)}}\, e^{\frac{1}{2}\sum\limits_{n,m=1}^{N} b_n \tilde{A}^{(N)}_{nm} b_m} = I_N. \tag{1.149}$$

If there exists a limit $\lim\limits_{N\to\infty} I_N = I$, then it is referred to as the Gaussian functional integral, which we denote as follows:

$$I = \int \mathcal{D}\varphi\, e^{-\frac{1}{2}(\varphi A^{-1}\varphi)+(b\varphi)} = \sqrt{Det A}\, e^{\frac{1}{2}(bAb)}, \tag{1.150}$$

where the functional determinant is defined as $Det A = \lim\limits_{N\to\infty} det \tilde{A}^{(N)}$. Let us define introduced short-hand notations:

$$(\varphi A^{-1}\varphi) = \int\limits_{\Gamma} d\mathbf{x} \int\limits_{\Gamma} d\mathbf{y} \varphi(\mathbf{x}) A^{-1}(\mathbf{x}, \mathbf{y}) \varphi(\mathbf{y}), \tag{1.151}$$

where the inverse operator can be determined through the following:

$$A^{-1}A = I \sim \int\limits_{\Gamma} d\mathbf{z}\, A^{-1}(\mathbf{x}, \mathbf{z}) A(\mathbf{z}, \mathbf{y}) = \delta(\mathbf{x} - \mathbf{y}) \tag{1.152}$$

with the kernel of the unit operator I defined by the δ-function. Further,

$$(b\varphi) = \int\limits_{\Gamma} d\mathbf{x}\, b(\mathbf{x}) \varphi(\mathbf{x}) \tag{1.153}$$

and, respectively,

$$(bAb) = \int\limits_{\Gamma} d\mathbf{x} \int\limits_{\Gamma} d\mathbf{y}\, b(\mathbf{x}) A(\mathbf{x}, \mathbf{y}) b(\mathbf{y}). \tag{1.154}$$

Let us now also introduce the *functional* Hubbard-Stratonovich transformation

$$e^{\frac{1}{2}(bAb)} = \int \frac{\mathcal{D}\varphi}{C_A}\, e^{-\frac{1}{2}(\varphi A^{-1}\varphi)+(b\varphi)}, \tag{1.155}$$

where $C_A \sim \sqrt{Det\, A}$ is a normalization constant. Let us introduce the corresponding normalized Gaussian measure:

$$\int d\sigma_A[\varphi]\,(\ldots) = \int \frac{\mathcal{D}\varphi}{C_A} e^{-\frac{1}{2}(\varphi A^{-1}\varphi)}(\ldots), \tag{1.156}$$

i.e.,

$$\int d\sigma_A[\varphi] = 1, \tag{1.157}$$

while we refer to the function $A(\mathbf{x}, \mathbf{y})$ as the Green's function (or kernel) of the Gaussian measure. It's important to note that the definition of $A(\mathbf{x}, \mathbf{y})$ is derived from the Green's function property of the operator, as discussed above:

$$L_{\mathbf{x}} A(\mathbf{x}, \mathbf{y}) = \delta(\mathbf{x} - \mathbf{y}), \quad L_{\mathbf{x}} = A_{\mathbf{x}}^{-1}. \tag{1.158}$$

Then, we can rewrite (1.155) as

$$\int d\sigma_A[\varphi]\, e^{(b\varphi)} = e^{\frac{1}{2}(bAb)}. \tag{1.159}$$

By analogy with the discrete case, we can now compute the average over the Gaussian measure from the product of $\varphi(\mathbf{x})$ and $\varphi(\mathbf{y})$

$$\begin{aligned} \int d\sigma_A[\varphi]\,\varphi(\mathbf{x})\varphi(\mathbf{y}) &= \langle \varphi(\mathbf{x})\varphi(\mathbf{y}) \rangle = \int d\sigma_A[\varphi]\, e^{(b\varphi)} \varphi(\mathbf{x})\varphi(\mathbf{y}) \Big|_{b=0} \\ &= \frac{\delta^2}{\delta b(\mathbf{x})\delta b(\mathbf{y})} \int d\sigma_A[\varphi]\, e^{(b\varphi)} \Big|_{b=0} = \frac{\delta^2}{\delta b(\mathbf{x})\delta b(\mathbf{y})} e^{\frac{1}{2}(bAb)} \Big|_{b=0}. \end{aligned} \tag{1.160}$$

Let us evaluate the derivative:

$$\begin{aligned} &\frac{\delta^2}{\delta b(\mathbf{x})\delta b(\mathbf{y})} \exp\left[\frac{1}{2}\int d\mathbf{x}_1 \int d\mathbf{y}_1 b(\mathbf{x}_1) A(\mathbf{x}_1, \mathbf{y}_1) b(\mathbf{y}_1)\right]\Big|_{\mathbf{b}=0} \\ &= \frac{\delta}{\delta b(\mathbf{x})} \left\{ \exp\left[\frac{1}{2}(bAb)\right] \frac{\delta}{\delta b(\mathbf{y})} \left[\frac{1}{2}\int d\mathbf{x}_1 \int d\mathbf{y}_1 b(\mathbf{x}_1) A(\mathbf{x}_1, \mathbf{y}_1) b(\mathbf{y}_1)\right]\right\}\Big|_{\mathbf{b}=0}. \end{aligned} \tag{1.161}$$

Then, recalling that $\delta b(\mathbf{x}_1)/\delta b(\mathbf{y}) = \delta(\mathbf{x}_1 - \mathbf{y})$, we obtain

$$\frac{\delta}{\delta b(\mathbf{y})}\left[\frac{1}{2}\int d\mathbf{x}_1\int d\mathbf{y}_1 b(\mathbf{x}_1)A(\mathbf{x}_1,\mathbf{y}_1)b(\mathbf{y}_1)\right]$$
$$=\frac{1}{2}\int d\mathbf{x}_1\int d\mathbf{y}_1 A(\mathbf{x}_1,\mathbf{y}_1)\left[\delta(\mathbf{x}_1-\mathbf{y})b(\mathbf{y}_1)+\delta(\mathbf{y}_1-\mathbf{y})b(\mathbf{x}_1)\right]$$
$$=\frac{1}{2}\int d\mathbf{y}_1 A(\mathbf{y},\mathbf{y}_1)b(\mathbf{y}_1)+\frac{1}{2}\int d\mathbf{x}_1 A(\mathbf{x}_1,\mathbf{y})b(\mathbf{x}_1)=\int d\mathbf{y}_1\, A(\mathbf{y},\mathbf{y}_1)b(\mathbf{y}_1). \tag{1.162}$$

Thus,

$$\frac{\delta}{\delta b(\mathbf{x})}\left\{e^{\frac{1}{2}(\mathbf{bAb})}\int d\mathbf{y}_1\, A(\mathbf{y},\mathbf{y}_1)b(\mathbf{y}_1)\right\}\Bigg|_{\mathbf{b}=0}=\int d\mathbf{y}_1\, A(\mathbf{y},\mathbf{y}_1)\delta(\mathbf{x}-\mathbf{y}_1)=A(\mathbf{x},\mathbf{y}). \tag{1.163}$$

As a result, similar to the expression (1.29) for the correlator in the case of finite-dimensional Gaussian integrals, $\langle x_i x_j\rangle=\Delta_{ij}$, we obtain

$$\langle\varphi(\mathbf{x})\varphi(\mathbf{y})\rangle=A(\mathbf{x},\mathbf{y}). \tag{1.164}$$

If $\varphi(\mathbf{x})$ is a random field with a Gaussian distribution, then the Green's function $A(\mathbf{x},\mathbf{y})$ can be interpreted as a two-point correlation function of this field. Analogously to the finite-dimensional case, one can formulate the Wick's theorem:

$$\langle\varphi(\mathbf{x}_1)\varphi(\mathbf{x}_2)\ldots\varphi(\mathbf{x}_n)\rangle=\sum_{\substack{\text{all}\\ \text{pairs}}}\langle\varphi(\mathbf{x}_{i_1})\varphi(\mathbf{x}_{i_2})\rangle\langle\varphi(\mathbf{x}_{i_3})\varphi(\mathbf{x}_{i_4})\rangle\ldots\langle\varphi(\mathbf{x}_{i_{n-1}})\varphi(\mathbf{x}_{i_n})\rangle$$
$$=\sum A(\mathbf{x}_{i_1},\mathbf{x}_{i_2})A(\mathbf{x}_{i_3},\mathbf{x}_{i_4})\ldots A(\mathbf{x}_{i_{n-1}},\mathbf{x}_{i_n}). \tag{1.165}$$

Moreover, similar to the case of finite-dimensional spaces (expression 1.38), for a functional $F[\varphi]$, the following formula holds:

$$\langle\varphi(\mathbf{x})F[\varphi]\rangle=\int_\Gamma d\mathbf{y}\,\langle\varphi(\mathbf{x})\varphi(\mathbf{y})\rangle\left\langle\frac{\delta F[\varphi]}{\delta\varphi(\mathbf{y})}\right\rangle=\int_\Gamma d\mathbf{y}\,A(\mathbf{x},\mathbf{y})\left\langle\frac{\delta F[\varphi]}{\delta\varphi(\mathbf{y})}\right\rangle. \tag{1.166}$$

It's worth noting that the Hubbard-Stratonovich transformation can be extended to cases where functional integration is performed over the functional space of n-component vector fields $\Phi_\alpha(x)$, where $\alpha=1,2,\ldots,n$, namely:

$$\int\frac{\mathcal{D}\Phi}{C_A}e^{-\frac{1}{2}(\Phi\mathbf{A}^{-1}\Phi)+(B\Phi)}=e^{\frac{1}{2}(B\mathbf{A}B)}, \tag{1.167}$$

where the notations are as follows:

$$(\Phi\mathbf{A}^{-1}\Phi)=\int_\Gamma d\mathbf{x}\int_\Gamma d\mathbf{y}\sum_{\alpha\gamma}\Phi_\alpha(\mathbf{x})A^{-1}_{\alpha\gamma}(\mathbf{x},\mathbf{y})\Phi_\gamma(\mathbf{y}), \tag{1.168}$$

$$(B\Phi) = \int_{\Gamma} d\mathbf{x} \sum_{\alpha} B_{\alpha}(\mathbf{x})\Phi_{\alpha}(\mathbf{x}), \tag{1.169}$$

$$(B\mathbf{A}B) = \int_{\Gamma} d\mathbf{x} \int_{\Gamma} d\mathbf{y} \sum_{\alpha\gamma} B_{\alpha}(\mathbf{x})A_{\alpha\gamma}(\mathbf{x}, \mathbf{y})B_{\gamma}(\mathbf{y}) \tag{1.170}$$

with $B_{\alpha}(\mathbf{x})$ representing the fixed vector field, and the kernels of the integral operators $\mathbf{A}$ and $\mathbf{A}^{-1}$ satisfying the relation

$$\int_{\Gamma} d\mathbf{z} \sum_{\gamma} A^{-1}_{\alpha\gamma}(\mathbf{x}, \mathbf{z})A_{\gamma\beta}(\mathbf{z}, \mathbf{y}) = \delta_{\alpha\beta}\delta(\mathbf{x} - \mathbf{y}), \tag{1.171}$$

where $\delta_{\alpha\beta}$ denotes the Kronecker's delta.

1.5 Cumulant Expansion and Bogoliubov's Inequality

Cumulant expansion. Let us consider a positive, unit-normalized measure $d\sigma(X)$ defined on the set of elements X, i.e., such that $\int d\sigma(X) = 1$. Suppose that we need to compute the integral

$$I(g) = \int d\sigma(X) \exp\left[gW(X)\right], \tag{1.172}$$

where $W(X)$ is a function and g is a constant. Assuming that $g \ll 1$, and expanding the integrand exponent into a series, we obtain

$$I(g) = \sum_{n=0}^{\infty} \frac{g^n}{n!} \left\langle W^n(X) \right\rangle, \tag{1.173}$$

where we introduce the symbol for averaging over the measure as $\langle(\ldots)\rangle = \int d\sigma(X)(\ldots)$. Now let us represent the integral (1.172) in the following form:

$$I(g) = \exp\left[\sum_{n=1}^{\infty} \frac{g^n}{n!} C_n\right], \tag{1.174}$$

where C_n represents unknown coefficients, commonly referred to as cumulants [2]. By expanding (1.174) into a series over the powers of g and equating coefficients at the same powers with those from (1.172), we derive the following expressions for the cumulants:

$$C_1 = \langle W(X) \rangle ,$$

$$C_2 = \langle W^2(X)\rangle - \langle W(X)\rangle^2, \tag{1.175}$$
$$C_3 = \langle W^3(X)\rangle - 3\langle W^2(X)\rangle\langle W(X)\rangle + 2\langle W(X)\rangle^3, \ldots$$

Thus, we have the following:

$$I(g) = \exp\left[\sum_{n=1}^{N} \frac{g^n}{n!} C_n + O(g^{N+1})\right] \tag{1.176}$$

Expression (1.176) defines the formula for the cumulant expansion, widely employed in various fields of physics for constructing perturbation theory based on a small parameter g. It's important to note that we have addressed the formal aspect of the cumulant expansion without delving into the convergence of the resulting series. The issue of convergence for the cumulant series must be addressed on a case-by-case basis.

Bogoliubov's inequality. Now, let us consider Bogoliubov's inequality. Suppose the function $F(s)$ is twice continuously differentiable and convex, i.e. $F''(s) > 0$ for all $s \in (-\infty, +\infty)$, and the functional $W(X)$ is real. In this case, the following inequality holds:

$$\int d\sigma(X) F(W(X)) \geq F\left(\int d\sigma(X) W(X)\right). \tag{1.177}$$

In particular, Bogoliubov's inequality must be satisfied

$$\int d\sigma(X) e^{W(X)} \geq e^{\int d\sigma(X) W(X)}. \tag{1.178}$$

To prove the inequality (1.177), we need to utilize the fact that a convex function is always at least as large as its tangent, i.e.

$$F(s) \geq F(t) + (s-t)F'(t), \quad s, t \in (-\infty, +\infty). \tag{1.179}$$

Then the following equality holds:

$$\int d\sigma(X) F(W(X)) \geq \max_u \int d\sigma(X) \left(F(u) + (W(X) - u)F'(u)\right) = F\left(\int d\sigma(X) W(X)\right). \tag{1.180}$$

The application of Bogoliubov's inequality forms the basis for the variational method, employed to compute integrals over real normalized measures of various kinds when constructing perturbation theory is not feasible (due to the absence of a small parameter). Let us consider the integral:

$$I = \int d\sigma(X) e^{W(X)}, \tag{1.181}$$

which cannot be calculated exactly. Let us represent the integrand exponent in the following form:

$$W(X) = W_0(X; a) + (W(X) - W_0(X; a)), \tag{1.182}$$

so that the trial functional $W_0(X; a)$ depends on the set of (variational) parameters a, and the integral

$$I_0(a) = \int d\sigma(X) e^{W_0(X;a)} \tag{1.183}$$

can be calculated exactly. Thus, we arrive at

$$I = I_0(a) \left\langle e^{W(X)-W_0(X;a)} \right\rangle_0 \geq \max_a I_0(a) e^{\langle W(X)-W_0(X;a)\rangle_0}, \tag{1.184}$$

where we introduce the averaging over a new measure

$$\langle(\ldots)\rangle_0 = I_0^{-1} \int d\sigma(X) e^{W_0(X;a)}(\ldots). \tag{1.185}$$

Thus, by finding the maximum of the right-hand side of expression (1.184) with respect to the parameters a, we can determine the best approximation of the integral for the trial functional $W_0(X; a)$. Below we apply the variational method to the computation of functional integrals over Gaussian measure.

1.6 Functional Integrals over Gaussian Measure and Gaussian Approximation

Functional integrals over Gaussian measure. In numerous physical applications, including the subject matter of this book, we encounter the so-called functional integrals over a Gaussian measure, which have the following form:

$$I(g) = \int d\sigma_A[\varphi]\, e^{gW[\varphi]}, \tag{1.186}$$

where

$$\int d\sigma_A[\varphi](\ldots) = \int \frac{\mathcal{D}\varphi}{C_A} e^{-\frac{1}{2}(\varphi A^{-1}\varphi)}(\ldots) \tag{1.187}$$

defines the Gaussian measure; g is the dimensionless constant, and the functional, "perturbing" the Gaussian measure, has the form

$$W[\varphi] = \int d\mu_\eta e^{(\eta\varphi)}, \tag{1.188}$$

where $\eta(\mathbf{x})$ represents a fixed generalized function, and $\int d\mu_\eta(\dots)$ denotes a measure defined within the region $\Gamma \subset \mathbb{R}^d$; $(\eta\,\varphi) = \int_\Gamma d\mathbf{x}\,\eta(\mathbf{x})\varphi(\mathbf{x})$. The field variable of integration, $\varphi(\mathbf{x})$, is subject to specific boundary conditions at the boundary of Γ.

Gaussian approximation. Generally, exact computation of functional integrals over a Gaussian measure is infeasible. Nonetheless, up to date, various methods have been devised for approximate evaluation of such integrals, typically relying on the computation of Gaussian integrals. One noteworthy approach is the Gaussian approximation.

Let us rewrite the functional integral (1.186) in the form

$$I = \int \frac{\mathcal{D}\varphi}{C_A} \exp\left[-S[\varphi]\right], \tag{1.189}$$

where we introduce the following auxiliary functional

$$S[\varphi] = \frac{1}{2}\left(\varphi A^{-1}\varphi\right) - gW[\varphi]. \tag{1.190}$$

Further, we aim to determine the extremal of the integrand functional, represented by the scalar field φ_0, commonly referred to as the mean field, which satisfies the Euler-Lagrange equation

$$\left.\frac{\delta S[\varphi]}{\delta\varphi(\mathbf{x})}\right|_{\varphi=\varphi_0} = 0 \tag{1.191}$$

or

$$\int_\Gamma d\mathbf{y} A^{-1}(\mathbf{x},\mathbf{y})\varphi_0(y) = U(\mathbf{x}|\varphi_0), \tag{1.192}$$

where

$$U(\mathbf{x}|\varphi_0) = \left.\frac{\delta W[\varphi]}{\delta\varphi(\mathbf{x})}\right|_{\varphi=\varphi_0} = g\int d\mu_\eta\,\eta(\mathbf{x})e^{(\eta\varphi_0)}. \tag{1.193}$$

Then, by expanding the integrand into a functional series around the mean field φ_0 and neglecting terms of degree higher than the second, we obtain the following approximation

$$I(g) \approx \exp\left[-S[\varphi_0]\right]\int \frac{\mathcal{D}\varphi}{C_A}\exp\left[-\frac{1}{2}\left(\varphi B^{-1}\varphi\right)\right], \tag{1.194}$$

where the Green's function of new Gaussian measure satisfies the relation

$$B^{-1}(\mathbf{x},\mathbf{y}) = A^{-1}(\mathbf{x},\mathbf{y}) - \left.\frac{\delta^2 W[\varphi]}{\delta\varphi(\mathbf{x})\delta\varphi(\mathbf{y})}\right|_{\varphi=\varphi_0} = A^{-1}(\mathbf{x},\mathbf{y}) - g\int d\mu_\eta\,\eta(\mathbf{x})\eta(\mathbf{y})e^{(\eta\varphi_0)}. \tag{1.195}$$

Thus, evaluating the integral (1.194), we obtain

$$I(g) \approx \exp[-S[\varphi_0]]\sqrt{\frac{Det\,B}{Det\,A}} = \exp\left[-S[\varphi_0] + \frac{1}{2}(tr\ln B - tr\ln A)\right], \quad (1.196)$$

where we employed the relation $\ln Det\,A = tr\ln A$, alongside the definition of the trace of the integral operator

$$tr(C) = \int d\mathbf{x}C(\mathbf{x}, \mathbf{x}), \quad (1.197)$$

where $C(\mathbf{x}, \mathbf{y})$ represents the kernel of the operator C. The first term in the integrand exponent in (1.196) defines the mean-field approximation, while the second term describes the fluctuation correction.

Evaluation of functional determinants. Let us proceed to derive a convenient expression for computing the fluctuation correction[3]. Therefore, we obtain

$$\begin{aligned}
tr\ln B - tr\ln A &= \sum_n \ln\frac{a_n}{b_n} \\
&= \int_0^\infty ds \sum_n \left(\frac{1}{s+b_n} - \frac{1}{s+a_n}\right) \\
&= \int_\Gamma d\mathbf{x} \int_0^\infty ds \sum_n \left(\frac{1}{s+b_n}V_n(\mathbf{x})V_n(\mathbf{x}) - \frac{1}{s+a_n}U_n(\mathbf{x})U_n(\mathbf{x})\right) \\
&= \int_\Gamma d\mathbf{x} \int_0^\infty ds \left(\left(s+B^{-1}\right)^{-1} - \left(s+A^{-1}\right)^{-1}\right)\delta(\mathbf{x}-\mathbf{y})\Big|_{\mathbf{y}=\mathbf{x}} \\
&= \int_\Gamma d\mathbf{x} \int_0^\infty ds\,(G_B(\mathbf{x}, \mathbf{x}|s) - G_A(\mathbf{x}, \mathbf{x}|s)), \quad (1.198)
\end{aligned}$$

where the Green's functions G_A and G_B are defined by the relations

$$(s + A^{-1})G_A(\mathbf{x}, \mathbf{y}|s) = \delta(\mathbf{x} - \mathbf{y}) \quad (1.199)$$

and

$$(s + B^{-1})G_B(\mathbf{x}, \mathbf{y}|s) = \delta(\mathbf{x} - \mathbf{y}) \quad (1.200)$$

with appropriate boundary conditions. During the evaluation process, we introduce the eigenvalues of the operators A^{-1} and B^{-1}, denoted as a_n and b_n respectively, alongside orthonormalized complete systems of eigenfunctions U_n and V_n corresponding to these eigenvalues.

Useful identity. Let us consider the functional integral

$$I = \int \mathcal{D}\varphi e^{-S[\varphi]} F[\varphi], \tag{1.201}$$

where $S[\varphi]$ and $F[\varphi]$ represent functionals. We proceed by making an infinitesimal shift of the field variable $\varphi(\mathbf{r}) \to \varphi(\mathbf{r}) + \delta\varphi(\mathbf{r})$, and express the variation of the functional integral as a consequence of this shift

$$\delta I = \int d\mathbf{r}\, \delta\varphi(\mathbf{r}) \int \mathcal{D}\varphi e^{-S[\varphi]} \left[\frac{\delta F[\varphi]}{\delta\varphi(\mathbf{r})} - F[\varphi] \frac{\delta S[\varphi]}{\delta\varphi(\mathbf{r})} \right]. \tag{1.202}$$

From the condition that the variation of the functional integral is equal to zero, we arrive at the Schwinger-Dyson equation [4–6]

$$\left\langle \frac{\delta F[\varphi]}{\delta\varphi(\mathbf{r})} \right\rangle = \left\langle F[\varphi] \frac{\delta S[\varphi]}{\delta\varphi(\mathbf{r})} \right\rangle, \tag{1.203}$$

where the average can be determined as

$$\langle (\ldots) \rangle = \frac{\int \mathcal{D}\varphi e^{-S[\varphi]} (\ldots)}{\int \mathcal{D}\varphi e^{-S[\varphi]}}. \tag{1.204}$$

In particular, Eq. (1.203) yields useful relations

$$\left\langle \frac{\delta S[\varphi]}{\delta\varphi(\mathbf{r})} \right\rangle = 0, \quad \left\langle \frac{\delta S[\varphi]}{\delta\varphi(\mathbf{r})} \varphi(\mathbf{r}') \right\rangle = \delta(\mathbf{r} - \mathbf{r}'). \tag{1.205}$$

1.7 Variational Method for Computing Functional Integrals over a Gaussian Measure

Let us consider the functional integral over a Gaussian measure

$$I(g) = \int d\sigma_A[\varphi] e^{gW[\varphi]}, \tag{1.206}$$

where as above

$$\int d\sigma_A[\varphi](\ldots) = \int \frac{\mathcal{D}\varphi}{C_A} e^{-\frac{1}{2}(\varphi A^{-1} \varphi)} (\ldots) \tag{1.207}$$

denotes the Gaussian measure, and the functional "perturbing" the Gaussian measure can be represented in the form of

$$gW[\varphi] = g \int d\mu_\eta \, e^{i(\eta\varphi)}. \tag{1.208}$$

Let us formulate a variational method for evaluating the functional integral (1.206) based on the application of Bogoliubov's inequality as formulated above.

First we replace the integration variable

$$\varphi \to \varphi + \varphi_0, \tag{1.209}$$

where φ_0 is a fixed function. Then, we introduce new Green's function of the Gaussian measure $B(\mathbf{x}, \mathbf{y})$:

$$I(g) = \frac{C_B}{C_A} \int \frac{\mathcal{D}\varphi}{C_B} e^{-\frac{1}{2}(\varphi B^{-1}\varphi)-(\varphi A^{-1}\varphi_0)-\frac{1}{2}(\varphi[A^{-1}-B^{-1}]\varphi)-\frac{1}{2}(\varphi_0 A^{-1}\varphi_0)+gW[\varphi+\varphi_0]}. \tag{1.210}$$

Now let us apply the Bogoliubov's inequality:

$$\begin{aligned} I(g) \geq \max_{\varphi_0, B} \exp\Bigg[\ln\sqrt{\frac{Det\, B}{Det\, A}} + g \int d\sigma_B W[\varphi + \varphi_0] \\ - \frac{1}{2} \int d\sigma_B \left((\varphi[A^{-1} - B^{-1}]\varphi) - \frac{1}{2}(\varphi_0 A^{-1}\varphi_0) - (\varphi A^{-1}\varphi_0) \right) \Bigg] = \exp W[\varphi_0; B], \end{aligned} \tag{1.211}$$

where we define the following variational functional

$$W[\varphi_0; B] = \ln\sqrt{\frac{Det\, B}{Det\, A}} - \frac{1}{2} tr\left(B[A^{-1} - B^{-1}]\right) - \frac{1}{2}(\varphi_0 A^{-1}\varphi_0) + g \int d\mu_\eta \, e^{i(\eta\varphi_0) - \frac{1}{2}(\eta B\eta)}, \tag{1.212}$$

which can be obtained after the integration over new Gaussian measure $d\sigma_B$.

We determine the variational functions φ_0 and $B(\mathbf{x}, \mathbf{y})$ from the following equations:

$$\frac{\delta W}{\delta\varphi_0(\mathbf{x})} = 0, \quad \frac{\delta W}{\delta B(\mathbf{x}, \mathbf{y})} = 0. \tag{1.213}$$

The first one can be rewritten in the form of

$$\int_\Gamma d\mathbf{y} A^{-1}(\mathbf{x}, \mathbf{y})\varphi_0(\mathbf{y}) = ig \int d\mu_\eta \, e^{i(\eta\varphi_0) - \frac{1}{2}(\eta B\eta)} \eta(\mathbf{x}). \tag{1.214}$$

The second one can be expressed as (see also Chap. 2)

$$B^{-1}(\mathbf{x}, \mathbf{y}) - A^{-1}(\mathbf{x}, \mathbf{y}) = g \int d\mu_\eta \, e^{i(\eta\varphi_0) - \frac{1}{2}(\eta B\eta)} \eta(\mathbf{x})\eta(\mathbf{y}). \tag{1.215}$$

Solving the self-consistency equations (1.214) and (1.215) allows us to calculate the variational functional $W[\varphi_0; B]$ and thus calculate the variational estimate of the functional integral. Solving the self-consistency equations requires a separate approach for each individual case [7]. These individual cases will be addressed in the subsequent discussions.

Note that Eqs. (1.214) and (1.215) can be obtained from the Schwinger-Dyson equations (1.205) [6]. Indeed, using the Wick's theorem and calculating the Gaussian functional integral, we have

$$\left\langle \frac{\delta S[\varphi]}{\delta \varphi(\mathbf{x})} \right\rangle \approx \frac{\int D\varphi e^{-\frac{1}{2}(\Lambda B^{-1}\Lambda)} \frac{\delta S[\varphi]}{\delta \varphi(\mathbf{x})}}{\int D\varphi e^{-\frac{1}{2}(\Lambda B^{-1}\Lambda)}} = \int_\Gamma d\mathbf{y} A^{-1}(\mathbf{x}, \mathbf{y})\varphi_0(\mathbf{y}) - ig \int d\mu_\eta \, e^{i(\eta\varphi_0)-\frac{1}{2}(\eta B\eta)} \eta(\mathbf{x}) = 0, \tag{1.216}$$

where $\Lambda(\mathbf{r}) = \varphi(\mathbf{r}) - \varphi_0(\mathbf{r})$ and $S[\varphi]$ is determined by

$$S[\varphi] = \frac{1}{2}\left(\varphi A^{-1}\varphi\right) - gW[\varphi]. \tag{1.217}$$

With using (1.166), the second equation yields

$$\left\langle \varphi(\mathbf{x}) \frac{\delta S[\varphi]}{\delta \varphi(\mathbf{y})} \right\rangle \approx \frac{\int D\varphi e^{-\frac{1}{2}(\Lambda B^{-1}\Lambda)} \varphi(\mathbf{x}) \frac{\delta S[\varphi]}{\delta \varphi(\mathbf{y})}}{\int D\varphi e^{-\frac{1}{2}(\Lambda B^{-1}\Lambda)}} = \int_\Gamma d\mathbf{z} A^{-1}(\mathbf{x}, \mathbf{z}) B(\mathbf{z}, \mathbf{y})$$
$$+ g \int d\mu_\eta \int d\mathbf{z} \, e^{i(\eta\varphi_0)-\frac{1}{2}(\eta B\eta)} \eta(\mathbf{x})\eta(\mathbf{z}) B(\mathbf{x}, \mathbf{z}) = \delta(\mathbf{x} - \mathbf{y}). \tag{1.218}$$

Thus, multiplying both parts of equation

$$\int_\Gamma d\mathbf{z} A^{-1}(\mathbf{x}, \mathbf{z}) B(\mathbf{z}, \mathbf{y}) + g \int d\mu_\eta \int d\mathbf{z} \, e^{i(\eta\varphi_0)-\frac{1}{2}(\eta B\eta)} \eta(\mathbf{x})\eta(\mathbf{z}) B(\mathbf{x}, \mathbf{z}) = \delta(\mathbf{x} - \mathbf{y}) \tag{1.219}$$

by $B^{-1}(\mathbf{x}, \mathbf{x}')$, integrating over $\mathbf{x}$, taking into account that

$$\int d\mathbf{x} B^{-1}(\mathbf{x}, \mathbf{x}') B(\mathbf{x}, \mathbf{y}) = \delta(\mathbf{x}' - \mathbf{y}), \tag{1.220}$$

and then performing replacement $\mathbf{x}' \to \mathbf{x}$, we arrive at Eq. (1.215).

It should be noted that the variational estimate provides an approximate value for the functional integral with uncontrolled accuracy within the variational method framework. Thus, there is no procedure within this framework to systematically address corrections to the variational estimate. In the subsequent section, we introduce one such method: the Gaussian equivalent representation.

1.8 Gaussian Equivalent Representation Method

In this section, we briefly outline the Gaussian equivalent representation (GER) method for evaluating integrals over the Gaussian measure, initially formulated in the works of G.V. Efimov and his coauthors [7–9]. The GER is similar to Variational Perturbation Theory developed by Kleinert [10].

Let us consider a functional integral over a Gaussian measure with the Green's function $V(\mathbf{x}, \mathbf{y})$

$$I(g) = \int \frac{\mathcal{D}\varphi}{C_V} e^{-\frac{1}{2}(\varphi V^{-1}\varphi)+gW[\varphi]}. \tag{1.221}$$

The integral (1.221) is defined in the space of functions $\varphi(\mathbf{x})$, which defined in the region $\Gamma \subseteq \mathbb{R}^d$, where $\mathbb{R}^d$ represents a d-dimensional space. The functions $\varphi(\mathbf{x})$ satisfy specific boundary conditions at the boundary $\partial\Gamma$. As previously mentioned, we employ the following notations for the Gaussian measure:

$$d\sigma_V[\varphi] = \frac{\mathcal{D}\varphi}{C_V} e^{-\frac{1}{2}(\varphi V^{-1}\varphi)}. \tag{1.222}$$

Gaussian measure always satisfies the normalization condition:

$$\int d\sigma_V[\varphi] = 1. \tag{1.223}$$

Similarly as above, let us consider that the functional $W[\varphi]$ can be represented in the following form:

$$W[\varphi] = \int d\mu_\eta e^{i(\eta\varphi)}, \tag{1.224}$$

where $d\mu_\eta$ denotes an integration measure, which depends on certain function $\eta(\mathbf{x})$; $(\eta\varphi) = \int_\Gamma d\mathbf{x}\eta(\mathbf{x})\varphi(\mathbf{x})$. To proceed further, let us introduce a few auxiliary notions.

Normal form of the functional. Let us introduce the concept of the normal form of the functional [7]. We begin by writing the identity:

$$\int d\sigma_V[\varphi] e^{i(\eta\varphi)} = e^{-\frac{1}{2}(\eta V\eta)} \tag{1.225}$$

and introduce the following notation

$$: e^{i(\eta\varphi)} :_V = e^{i(\eta\varphi)} e^{\frac{1}{2}(\eta V\eta)}, \tag{1.226}$$

so

$$\int d\sigma_V[\varphi] : e^{i(\eta\varphi)} :_V = 1. \tag{1.227}$$

Next, let us expand both the left and right parts of equation (1.226) into power series and then equate the coefficients corresponding to the same powers of η:

$$
\begin{aligned}
&: 1 + i(\eta\varphi) - \frac{1}{2}(\eta\varphi)^2 + \frac{i^3}{3!}(\eta\varphi)^3 + \cdots :_V \\
&= \left(1 + \frac{1}{2}(\eta V \eta) + \ldots\right)\left(1 + i(\eta\varphi) - \frac{1}{2}(\eta\varphi)^2 + \frac{i^3}{3!}(\eta\varphi)^3 + \ldots\right).
\end{aligned} \tag{1.228}
$$

Thus, we obtain

$$
\begin{aligned}
&: \varphi(\mathbf{x}) :_V = \varphi(\mathbf{x}), \\
&: \varphi(\mathbf{x}_1)\varphi(\mathbf{x}_2) :_V = \varphi(\mathbf{x}_1)\varphi(\mathbf{x}_2) - V(\mathbf{x}_1, \mathbf{x}_2), \\
&: \varphi(\mathbf{x}_1)\varphi(\mathbf{x}_2)\varphi(\mathbf{x}_3) :_V = \varphi(\mathbf{x}_1)\varphi(\mathbf{x}_2)\varphi(\mathbf{x}_3) - \varphi(\mathbf{x}_1)V(\mathbf{x}_2, \mathbf{x}_3) - \varphi(\mathbf{x}_2)V(\mathbf{x}_1, \mathbf{x}_3) - \varphi(\mathbf{x}_3)V(\mathbf{x}_1, \mathbf{x}_2),
\end{aligned} \tag{1.229}
$$

and so on.

The quantities under the symbol $: (\ldots) :_V$ are termed as the ones written in the normal form with respect to the Gaussian measure with the Green's function V. It's worth noting that the concept of the normal product was initially introduced within quantum theory in the context of ordering the creation and annihilation operators within the framework of second quantization [7, 11].

Let us define an analytical functional $S[\varphi]$, i.e. the functional, which can be formally expanded into the following functional power series:

$$
S[\varphi] = \sum_{n=0}^{\infty} \int_\Gamma d\mathbf{x}_1 \ldots \int_\Gamma d\mathbf{x}_n C_n(\mathbf{x}_1, \ldots, \mathbf{x}_n)\varphi(\mathbf{x}_1)\ldots\varphi(\mathbf{x}_n), \tag{1.230}
$$

where $C_n(\mathbf{x}_1, \ldots, \mathbf{x}_n)$ represent certain functions. The functional $S[\varphi]$, expressed in normal form with respect to a Gaussian measure with Green's function V, is defined as follows:

$$
: S[\varphi] :_V = \sum_{n=0}^{\infty} \int_\Gamma d\mathbf{x}_1 \ldots \int_\Gamma d\mathbf{x}_n C_n(\mathbf{x}_1, \ldots, \mathbf{x}_n) : \varphi(\mathbf{x}_1)\ldots\varphi(\mathbf{x}_n) :_V \tag{1.231}
$$

where each term of the series is a monomial $: \varphi(\mathbf{x}_1)\ldots\varphi(\mathbf{x}_n) :_V$, defined according to the rule (1.8).

Similarly, the following relation holds:

$$
\int d\sigma_V[\varphi] : \varphi(\mathbf{x}_1)\ldots\varphi(\mathbf{x}_n) :_V = 0, \tag{1.232}
$$

which directly follows from the relation

$$
\int d\sigma_V[\varphi] : e^{i(\eta\varphi)} :_V = 1. \tag{1.233}
$$

Gaussian equivalent representation. Now, let us revisit the functional integral (1.221). Exact evaluation of this integral for arbitrary functional W is infeasible, thus, there is a task of approximate calculation of the functional integrals over the Gaussian measure. The GER method essentially generalizes the Gaussian approximation, enabling systematic corrections to it. Importantly, unlike the Gaussian approximation, the GER method facilitates finding the best approximation of the integrand functional with the quadratic functional around the mean-field configuration.

Therefore, let us perform a variable shift and transition to a new Green's function, such that $\varphi \to \varphi + \varphi_0$ and $V \to D$:

$$\begin{aligned} I(g) &= \frac{C_D}{C_V} \int \frac{\mathcal{D}\varphi}{C_D} e^{-\frac{1}{2}(\varphi D^{-1}\varphi) - \frac{1}{2}(\varphi[V^{-1}-D^{-1}]\varphi) - (\varphi_0 V^{-1}\varphi) - \frac{1}{2}(\varphi_0 V^{-1}\varphi_0) + gW[\varphi+\varphi_0]} \\ &= e^{\ln\left(\frac{C_D}{C_V}\right) - \frac{1}{2}(\varphi_0 V^{-1}\varphi_0)} \int d\mu_D[\varphi] \exp\left[-\frac{1}{2}(\varphi[V^{-1} - D^{-1}]\varphi) + (\varphi_0 V^{-1}\varphi) + g\int d\mu_\eta e^{i(\eta\varphi_0)} e^{i(\eta\varphi)}\right]. \end{aligned} \tag{1.234}$$

Then, let us convert the integrand in the last expression into the normal form with respect to the Gaussian measure with the Green's function D:

$$\begin{aligned} -\frac{1}{2}(\varphi[V^{-1} - D^{-1}]\varphi) &= -\frac{1}{2} : (\varphi[V^{-1} - D^{-1}]\varphi) :_D -\frac{1}{2} tr(D[V^{-1} - D^{-1}]), \\ e^{i(\eta\varphi)} &=: e^{i(\eta\varphi)} :_D e^{-\frac{1}{2}(\eta D\eta)}, \end{aligned} \tag{1.235}$$

where

$$tr(D[V^{-1} - D^{-1}]) = \int_\Gamma d\mathbf{x} \int_\Gamma d\mathbf{y} D(\mathbf{x}, \mathbf{y})(V^{-1}(\mathbf{y}, \mathbf{x}) - D^{-1}(\mathbf{y}, \mathbf{x})). \tag{1.236}$$

Therefore, we obtain:

$$\begin{aligned} I(g) &= e^{\ln\left(\frac{C_D}{C_V}\right) - \frac{1}{2}(\varphi_0 V^{-1}\varphi_0) - \frac{1}{2} tr(D[V^{-1}-D^{-1}])} \int d\sigma_D[\varphi] \exp\left[-\frac{1}{2} : (\varphi[V^{-1} - D^{-1}]\varphi) :_D -(\varphi_0 V^{-1}\varphi)\right] \\ &\times \exp\left[g\int d\mu_\eta e^{i(\eta\varphi_0) - \frac{1}{2}(\eta D\eta)} : 1 + i(\eta\varphi) - \frac{(\eta\varphi)^2}{2} :_D + g\int d\mu_\eta e^{i(\eta\varphi_0) - \frac{1}{2}(\eta D\eta)} : e_2^{i(\eta\varphi)} :_D\right], \end{aligned} \tag{1.237}$$

where $e_2^x = e^x - 1 - x - x^2/2$.

Now, let us determine the functions φ_0 and $D(\mathbf{x}, \mathbf{y})$ by requiring the cancellation of linear and quadratic terms with respect to the field $\varphi(\mathbf{x})$ in the integrand exponent. This leads to the following equations, which hold for arbitrary functions $\varphi(\mathbf{x})$:

$$\begin{aligned} (\varphi_0 V^{-1}\varphi) + g\int d\mu_\eta e^{i(\eta\varphi_0) - \frac{1}{2}(\eta D\eta)} i(\eta\varphi) &= 0, \\ -\frac{1}{2}(\varphi[V^{-1} - D^{-1}]\varphi) - \frac{g}{2}\int d\mu_\eta e^{i(\eta\varphi_0) - \frac{1}{2}(\eta D\eta)} (\eta\varphi)^2 &= 0. \end{aligned} \tag{1.238}$$

These conditions arise because the quadratic functional, defined in the vicinity of the function φ_0, provides the best approximation of the total functional within the exponent under the functional integration. By equating the coefficients of arbitrary fluctuating fields $\varphi(\mathbf{x})$ in the last two expressions, we derive a system of equations to determine the function φ_0 and the Green's function D:

$$\int_\Gamma d\mathbf{y} V^{-1}(\mathbf{x}, \mathbf{y})\varphi_0(\mathbf{y}) = -ig \int d\mu_\eta e^{i(\eta\varphi_0)-\frac{1}{2}(\eta D\eta)}\eta(\mathbf{x}),$$
$$D^{-1}(\mathbf{x}, \mathbf{y}) - V^{-1}(\mathbf{x}, \mathbf{y}) = g \int d\mu_\eta e^{i(\eta\varphi_0)-\frac{1}{2}(\eta D\eta)}\eta(\mathbf{x})\eta(\mathbf{y}). \tag{1.239}$$

It's worth noting that the resulting equations coincide with the self-consistent equations derived using the variational method. Upon solving these equations, we can approximately evaluate the functional integral using its *Gaussian equivalent representation*:

$$I(g) = e^{W_0} \int d\mu_D[\varphi] e^{g\tilde{W}[\varphi]}. \tag{1.240}$$

The zero approximation of the GER takes the following form:

$$W_0 = \ln \frac{C_D}{C_V} - \frac{1}{2} tr(D[V^{-1} - D^{-1}]) - \frac{1}{2}(\varphi_0 V^{-1}\varphi_0) + g \int d\mu_\eta e^{i\eta\varphi_0 - (\eta D\eta)}, \tag{1.241}$$

where

$$\ln \frac{C_D}{C_V} = \ln \frac{Det D}{Det V} = tr \ln D - tr \ln V \tag{1.242}$$

can be evaluated using the method described above (see (1.198)). The functional $\tilde{W}$ contains terms starting from the third power with respect to φ:

$$\tilde{W}[\varphi] = \int d\mu_\eta e^{i(\eta\varphi_0)-\frac{1}{2}(\eta D\eta)} : e_2^{i(\eta\varphi)} :_D . \tag{1.243}$$

In the zeroth approximation of the GER:

$$I(g) \approx \exp W_0. \tag{1.244}$$

The zeroth approximation of the GER coincides with the result of the variational estimate when utilizing Bogoliubov's inequality. We can enhance this approximation by conducting a cumulant expansion with respect to the functional $\tilde{W}[\varphi]$:

$$\int d\sigma_D[\varphi] e^{g\tilde{W}[\varphi]} = e^{gW_1 + \frac{g^2}{2}W_2 + ..}, \tag{1.245}$$

where

$$W_1 = \int d\mu_D[\varphi]\,\tilde{W}[\varphi] = \langle \tilde{W}[\varphi] \rangle = 0,$$
$$W_2 = \langle \tilde{W}[\varphi]^2 \rangle - \langle \tilde{W}[\varphi] \rangle^2 = \langle \tilde{W}[\varphi]^2 \rangle, \tag{1.246}$$

and so on.

Therefore, the GER can be regarded as a method that enables the evaluation of systematic corrections to the variational estimate by employing Bogoliubov's inequality.

Chapter 2
Statistical Field Theory of Simple Fluids

In this introductory chapter, we will revisit some fundamental concepts of equilibrium statistical mechanics of simple fluids, which are fluids whose molecules interact through pair-additive central potentials. We will also demonstrate the application of statistical field theory to the description of these fluids.

2.1 Some Necessary Information from Statistical Physics and Thermodynamics of Simple Fluids

Free energy and grand thermodynamic potential. As is known from statistical mechanics (see, e.g., [12]), one can determine Helmholtz free energy of the classical l-component fluid through a statistical integral (or partition function) Z within the following expression[1]

$$F(\{N_\alpha\}, V, T) = -k_B T \ln Z(\{N_\alpha\}, V, T), \tag{2.1}$$

where N_α is a number of particles of αth component ($\alpha = 1, 2, \ldots, l$), V is the volume of a system, T is the temperature, k_B is the Boltzmann constant. The statistical ensemble characterized by the Helmholtz free energy is commonly referred to as the canonical or NVT-ensemble, as the system's macrostate is defined by the constant particle number, volume, and temperature. Partition function Z can be written as

$$Z(\{N_\alpha\}, V, T) = \int \ldots \int \prod_{\alpha=1}^{l} d\Gamma_\alpha \exp\left[-\beta H_N\right], \tag{2.2}$$

[1] We consider fluid as a system of interacting particles without a long-range order. The concept of a fluid, therefore, unifies such systems as gases, liquids, supercritical fluids, liquid-phase solutions, etc.

Y. A. Budkov and N. N. Kalikin, *Statistical Field Theory of Ion-Molecular Fluids*,
https://doi.org/10.1007/978-3-031-68364-0_2

where $\beta = (k_B T)^{-1}$, and H_N is the Hamiltonian of the system, which is determined as

$$H_N = K_N + U_N, \tag{2.3}$$

where

$$K_N = \sum_{\alpha=1}^{l} \sum_{i_\alpha=1}^{N_\alpha} \frac{\mathbf{p}_{i_\alpha}^2}{2m_\alpha} \tag{2.4}$$

represents the total kinetic energy associated with the translation motion of the fluid molecules[2]; m_α is the mass of the αth molecule species; U_N denotes the total potential energy of the fluid molecules, which depends solely on the coordinates of the molecules' mass centers. The phase space element $d\Gamma_\alpha$ for the αth component is expressed as follows:

$$d\Gamma_\alpha = \frac{1}{N_\alpha! h^{3N_\alpha}} \prod_{i_\alpha=1}^{N_\alpha} d\mathbf{p}_{i_\alpha} d\mathbf{r}_{i_\alpha}, \tag{2.5}$$

where h represents the Planck constant, while $\mathbf{p}_{i_\alpha}$ and $\mathbf{r}_{i_\alpha}$ denote the momentum and the radius-vector of the mass center of the molecule belonging to the αth component with an index i_α. Further, taking a Gaussian integral over the momenta of the molecules, we can rewrite the partition function in the following manner

$$Z = Z_{id} Q_N, \tag{2.6}$$

where

$$Z_{id} = \frac{V^N}{\prod_\alpha N_\alpha! \Lambda_\alpha^{3N_\alpha}} \tag{2.7}$$

is the partition function of the ideal gas mixture; $N = \sum_\alpha N_\alpha$ is the total number of the particles in the system, $\Lambda_\alpha = h(2\pi m_\alpha k_B T)^{-1/2}$ is the de Broglie thermal wavelength of the molecule of the αth component. Configuration integral Q_N, which defines the excess properties of the fluid, can be expressed as follows

$$Q_N = \int \ldots \int \prod_{\alpha=1}^{l} \prod_{i_\alpha=1}^{N_\alpha} \frac{d\mathbf{r}_{i_\alpha}}{V} \exp\left[-\beta U_N(\{\mathbf{r}_{i\alpha}\})\right], \tag{2.8}$$

where the potential energy of the fluid molecules is given by the following relation

$$U_N(\{\mathbf{r}_{i_\alpha}\}) = \sum_\alpha \sum_{i_\alpha} U_\alpha(\mathbf{r}_{i_\alpha}) + \sum_{\alpha\gamma} \sum_{i_\alpha j_\gamma} U_{\alpha\gamma}(\mathbf{r}_{i_\alpha}, \mathbf{r}_{j_\gamma}). \tag{2.9}$$

[2] Internal degrees of freedom of molecules are not considered here.

The first term in (2.9) represents the potential energy of the particles in external potential fields, while the second one accounts for the potential energy arising from pairwise interactions among the fluid particles.[3] Therefore, the Helmholtz free energy of the system is determined by the following expression:

$$F = F_{id} + F_{ex}, \tag{2.10}$$

where the first term defines the free energy of the ideal gas mixture:

$$F_{id} = -k_B T \ln Z_{id} = \sum_{\alpha} N_\alpha k_B T \left(\ln \left(\frac{N_\alpha \Lambda_\alpha^3}{V} \right) - 1 \right), \tag{2.11}$$

while the second term determines the excess free energy of the fluid relative to the ideal gas mixture and is governed by the intermolecular interactions

$$F_{ex} = -k_B T \ln Q_N. \tag{2.12}$$

Therefore, knowing the pair potentials of interaction between the fluid particles, one can compute the partition function and subsequently derive all thermodynamic parameters of the system using well-known thermodynamic relations. Indeed, utilizing calculated Helmholtz free energy, we can ascertain the values of pressure, entropy and chemical potentials of the components via the following expressions [12]

$$P = -\left(\frac{\partial F}{\partial V} \right)_{T,\{N_\alpha\}}, \quad S = -\left(\frac{\partial F}{\partial T} \right)_{V,\{N_\alpha\}}, \quad \mu_\alpha = \left(\frac{\partial F}{\partial N_\alpha} \right)_{V,T,\{N_{\gamma \neq \alpha}\}}. \tag{2.13}$$

The computation of the configuration integral of the system Q_N relies on the understanding of the intermolecular interaction potential form. Typically, for the systems of *simple fluids*, central interaction potentials (or spherically symmetric interaction potentials) are commonly employed, i.e. when $U_{\alpha\gamma}(\mathbf{r}_{i_\alpha}, \mathbf{r}_{j_\gamma}) = U_{\alpha\gamma}(|\mathbf{r}_{i_\alpha} - \mathbf{r}_{j_\gamma}|)$. However, even with central interaction potentials, the computation of the configuration integral cannot be achieved exactly. In practice, various approximate methods are employed to address this task, some of which are discussed below.

Another useful in practice thermodynamic potential is the grand thermodynamic potential

$$\Omega(\{\mu_\alpha\}, V, T) = -k_B T \ln \Xi(\{\mu_\alpha\}, V, T), \tag{2.14}$$

which essentially serves as a characteristic function of the grand canonical ensemble, also known as the μVT-ensemble, which macrostates are defined by constant chemical potentials of the components, as well as the volume and the temperature

[3] It is assumed that interactions beyond pairwise interactions, such as three-body, four-body, and so on, are negligibly small.

of the system. The grand thermodynamic potential is determined via the following grand partition function:

$$\Xi(\{\mu_\alpha\}, V, T) = \sum_{\{N_\alpha\}} \exp\left[\beta \sum_\alpha \mu_\alpha N_\alpha\right] Z(\{N_\alpha\}, V, T), \tag{2.15}$$

where the summation symbol $\sum_{N_\alpha}$ represents the summation over all numbers of particles of each component. After integrating over momenta, we obtain the following expression for the grand partition function:

$$\Xi = \sum_{\{N_\alpha\}} Q_N \prod_\alpha \frac{(z_\alpha V)^{N_\alpha}}{N_\alpha!}, \tag{2.16}$$

where $z_\alpha = \Lambda_\alpha^{-3} \exp[\beta\mu_\alpha]$ are the activities of the system's components. By computing the grand partition function and recalling the relationship between the grand thermodynamic potential and the pressure [12], $\Omega = -PV$, we can derive the pressure as a function of the chemical potentials and temperature

$$P(\{\mu_\alpha\}, T) = -\frac{k_B T}{V} \ln \Xi(\{\mu_\alpha\}, V, T). \tag{2.17}$$

Then, we can compute the average densities of the components using the following thermodynamic relation:

$$n_\gamma(\{\mu_\alpha\}, T) = \frac{\partial P}{\partial \mu_\gamma} = z_\gamma \frac{\partial(\beta P)}{\partial z_\gamma}, \quad \gamma = 1, 2, \ldots l. \tag{2.18}$$

Equations (2.17) and (2.18) define the equation of state of the multicomponent fluid in parametric form.

Finally, let us discuss the relation between the Helmholtz free energy and grand thermodynamic potential. As is known [12], an infinitesimal change in the Helmholtz free energy is a total differential. Thus, for the NVT-ensemble, we have

$$dF = -SdT - PdV + \sum_\alpha \mu_\alpha dN_\alpha. \tag{2.19}$$

If we add and subtract $d\left(\sum_\alpha \mu_\alpha N_\alpha\right)$ in the right-hand side of Eq. (2.19) and introduce the grand thermodynamic potential as $\Omega = F - \sum_\alpha \mu_\alpha N_\alpha$, we can obtain

$$d\Omega = -SdT - PdV - \sum_\alpha N_\alpha d\mu_\alpha. \tag{2.20}$$

Thus, Helmholtz free energy is related to the grand thermodynamic potential via the Legendre transformation, converting the independent variables from T, V and

N_α to T, V and μ_α. Considering $\Omega = -PV$, it can be convenient to express the relation connecting pressure, chemical potentials, and the Helmholtz free energy density $f = F/V$ as follows:

$$P = \sum_\alpha \mu_\alpha n_\alpha - f, \tag{2.21}$$

where $n_\alpha = N_\alpha / V$, and the chemical potentials are defined as

$$\mu_\alpha = \frac{\partial f}{\partial n_\alpha}. \tag{2.22}$$

Virial equation of state of a fluid. The pressure of a weakly nonideal gas, consisting of a macroscopically large number of particles N confined within a volume V, can be expressed as a series expansion in powers of density, known in thermodynamics as the virial equation of state

$$P = \frac{Nk_BT}{V}\left(1 + \frac{N}{V}B(T) + \frac{N^2}{V^2}C(T) + \cdots\right), \tag{2.23}$$

where $B(T)$, $C(T)$, etc. denote the virial coefficients, which are temperature-dependent functions. The theory outlined above can be employed to derive a microscopic interpretation of these coefficients and establish their relationship with the intermolecular interaction potential.

Now, consider a one-component fluid with a pairwise central intermolecular interaction potential $U(r)$. To characterize such a system, we turn to the grand partition function, which can be expressed as follows:

$$\Xi = \sum_{N=0}^{\infty} \frac{z^N}{N!} Q_N, \tag{2.24}$$

where $z = \Lambda^{-3} e^{\beta\mu}\theta$ represents the activity of the gas molecules, where μ denotes the chemical potential, and θ is the internal partition function of the molecule. The configuration integral of the system is given by[4]:

$$Q_N = \int \cdots \int d\mathbf{r}_1 \ldots d\mathbf{r}_N \exp\left[-\beta U_N(\mathbf{r}_1, \ldots, \mathbf{r}_N)\right], \tag{2.25}$$

where the interaction potential energy is

$$U_N(\mathbf{r}_1, \ldots, \mathbf{r}_N) = U_{1\ldots N}(\mathbf{r}_1, \ldots, \mathbf{r}_N) = \sum_{i<j} U(|\mathbf{r}_i - \mathbf{r}_j|). \tag{2.26}$$

[4] It is worth noting that within the μVT-ensemble, it is more convenient to introduce the configuration integrals for a cluster of N particles without the $1/V^N$ prefactor.

Thus, we can write the following expression for the grand thermodynamic potential

$$\Omega = -k_B T \ln \left[1 + zV + \frac{z^2}{2!} \iint d\mathbf{r}_1 d\mathbf{r}_2 e^{-\beta U_{12}} + \frac{z^3}{3!} \iiint d\mathbf{r}_1 d\mathbf{r}_2 d\mathbf{r}_3 e^{-\beta U_{123}} + \dots \right]$$
$$= -k_B T \ln \left[1 + zV + \frac{z^2 V}{2!} \int d\mathbf{r}_2 e^{-\beta U_{12}} + \frac{z^3 V}{3!} \iint d\mathbf{r}_2 d\mathbf{r}_3 e^{-\beta U_{123}} + \dots \right], \quad (2.27)$$

where we take into account that potential energies U_{12}, U_{123}, etc. depend solely on the distance between the particles, so each term involves integration, resulting in the total volume of the system as a prefactor. Thus, expanding the logarithm into a series, we arrive at the expression:

$$P = k_B T \sum_{s=1}^{\infty} b_s z^s, \quad (2.28)$$

where

$$b_1 = 1, \; b_2 = \frac{1}{2} \int d\mathbf{r} \left(e^{-\beta U(r)} - 1 \right),$$

$$b_3 = \frac{1}{6} \int \int d\mathbf{r}_2 d\mathbf{r}_3 \left(e^{-\beta U_{123}} - e^{-\beta U_{12}} - e^{-\beta U_{13}} - e^{-\beta U_{23}} + 2 \right) \quad (2.29)$$

etc., represent group integrals [13]. Note that group integrals b_l follow a clear rule: they are nonzero if l particles are sufficiently close to each other. Therefore, the group expansion (2.28) considers correlations between the molecules in groups of l particles of a nonideal gas, arising from their pairwise interactions. To relate the group expansion to the virial expansion, we need to compute the density using Eq. (2.18), resulting in:

$$n = z \frac{\partial (\beta P)}{\partial z} = \sum_{s=1}^{\infty} b_s s z^s. \quad (2.30)$$

Together, Eqs. (2.28) and (2.30) establish the equation of state for a nonideal gas in a parametric form. By eliminating the activity z from these equations, we can derive the virial equation of state. As demonstrated in [13], within the diagrammatic technique, the virial expansion takes the following form:

$$\frac{P}{k_B T} = n + \sum_{s=1}^{\infty} B_s n^{s+1}, \quad (2.31)$$

where the virial coefficients B_s are determined by the relation

$$B_s = -\frac{s}{s+1} \beta_s, \quad (2.32)$$

and coefficients β_s are related to the group integrals through the following formulas

$$\beta_1 = 2b_2, \quad \beta_2 = 3b_3 - 6b_2^2, \quad \beta_3 = 4b_4 - 24b_2b_3 + \frac{80}{3}b_2^3, \ \ldots \tag{2.33}$$

The second virial coefficient, crucial for experimental investigations of intermolecular interactions in gases, is determined as

$$B = B_1 = \frac{1}{2}\int d\mathbf{r}\left(1 - e^{-\beta U(r)}\right). \tag{2.34}$$

For realistic interaction potentials like the Lennard-Jones potential, which provides a good description of the van der Waals interaction between the atoms in noble gases [14]

$$U(r) = 4\epsilon\left(\left(\frac{\sigma}{r}\right)^{12} - \left(\frac{\sigma}{r}\right)^{6}\right), \tag{2.35}$$

the second virial coefficient is a monotonically increasing function of temperature. At sufficiently low temperatures, it is negative; at a certain Boyle temperature, it becomes zero, and then further increases in the region of positive temperature values.

It is also instructive to investigate the behavior of the second virial coefficient for the case of *repulsive* potential, which leads to mutual repulsion of the particles at any distance. Let us consider a hypothetical system of particles with a purely repulsive potential of the form $U(r) = Ar^{-\alpha}$, where $A > 0$ and $\alpha > 0$, and calculate the second virial coefficient B for this system. The result takes the following form

$$B(T) = \frac{2\pi}{3}\left(\frac{A}{k_BT}\right)^{3/\alpha}\Gamma\left(1 - \frac{3}{\alpha}\right), \tag{2.36}$$

where $\Gamma(x)$ denotes an Euler gamma-function. As one can conclude, the second virial coefficient is determined only for the case of $\alpha > 3$, i.e., for sufficiently short-range potentials. For long-range potentials with $\alpha \leq 3$, the second virial coefficient diverges. It can be shown that for $\alpha \leq 3$, the virial coefficients of higher order also diverge. The divergence of the virial coefficients for systems with long-range repulsive potentials is explained by the absence of a *thermodynamic limit* ($N \to \infty$, $V \to \infty$, $N/V \to n$) of the configuration integral in such systems [15]. In other words, systems with long-range purely repulsive potentials cannot be thermodynamically stable. However, there are ionic systems, such as high-temperature plasma or electrolyte solutions, where the particles interact through the Coulomb potential, for which $\alpha = 1$. In such systems, due to both attraction and repulsion between the particles arising from their different charges, as well as the global electric neutrality of the system, the configuration integral still has a thermodynamic limit [16], although all virial coefficients diverge. This is because the thermodynamic behavior of ionic systems is determined by many-body *electrostatic correlations* [17], thus, meaning that the deviation from ideality cannot be captured through several terms of the virial series, even for low densities. In Chap. 3, we illustrate how to describe the thermodynamic properties of ionic fluids within the field theory framework. In this chapter, we focus on the description of non-ionic fluids.

Density fluctuations correlator in simple fluids and the Ornstein–Zernike relation. Consider a partition function of a simple fluid within the NVT-ensemble

$$Z = \int d\Gamma_N \exp\left[-\beta H_N\right], \tag{2.37}$$

where $H_N = U_N + V_N$ represents the total Hamiltonian of the interactions within the system, where U_N signifies the previously introduced total potential energy arising from pairwise interactions between the fluid particles, and V_N denotes the potential energy associated with the interaction between the particles and an external field. The latter is defined as follows:

$$V_N = \sum_{i=1}^{N} u(\mathbf{r}_i), \tag{2.38}$$

where $u(\mathbf{r})$ is the potential of the external field with the subscript N indicating the presence of N particles within the system. The integration measure over the fluid's phase space is expressed as follows:

$$\int d\Gamma_N(\ldots) = \frac{\Lambda^{-3N}}{N!}\int \ldots \int d\mathbf{r}_1 \ldots d\mathbf{r}_N\,(\ldots), \tag{2.39}$$

where the integration over momenta is performed.

The average density of the fluid particles can be defined as the mean value of the microscopic density $\hat{n}(\mathbf{r})$, i.e.

$$n(\mathbf{r}) = \left\langle \hat{n}(\mathbf{r}) \right\rangle = \frac{1}{Z_N}\int d\Gamma_N\, e^{-\beta H_N}\hat{n}(\mathbf{r}), \tag{2.40}$$

where

$$\hat{n}(\mathbf{r}) = \sum_{j=1}^{N} \delta(\mathbf{r} - \mathbf{r}_j). \tag{2.41}$$

The expression (2.40) can be also rewritten as

$$n(\mathbf{r}) = -k_B T \frac{\delta}{\delta u(\mathbf{r})} \ln Z_N = \frac{\delta F}{\delta u(\mathbf{r})}. \tag{2.42}$$

Let's analyze the change in the average particle density resulting from the variation of the external potential. We obtain:

$$\delta n(\mathbf{r}) = -\frac{\delta Z_N}{Z_N^2}\int d\Gamma_N\, e^{-\beta H_N}\hat{n}(\mathbf{r}) + \frac{1}{Z_N}\int d\Gamma_N\, e^{-\beta H_N}(-\beta\delta V_N)\hat{n}(\mathbf{r})$$
$$= -\frac{\delta Z_N}{Z_N}n(\mathbf{r}) + \frac{1}{Z_N}\int d\Gamma_N\, e^{-\beta H_N}\hat{n}(\mathbf{r})\sum_{k=1}^{N}(-\beta\delta u(\mathbf{r}_k))$$
$$= -\beta\int d\mathbf{r}'\,\delta u(\mathbf{r}')\left\langle\sum_{ik}\delta(\mathbf{r}-\mathbf{r}_i)\delta(\mathbf{r}'-\mathbf{r}_k)\right\rangle - \frac{\delta Z_N}{Z_N}n(\mathbf{r}). \quad (2.43)$$

Then, considering that

$$\frac{\delta Z_N}{Z_N} = -\beta\int d\mathbf{r}'\, n(\mathbf{r}')\delta u(\mathbf{r}'), \quad (2.44)$$

we obtain

$$\delta n(\mathbf{r}) = -\beta\int d\mathbf{r}'\, S(\mathbf{r},\mathbf{r}')\delta u(\mathbf{r}'), \quad (2.45)$$

where we introduce the correlation function of local density fluctuations of the fluid particles $\delta\hat{n}(\mathbf{r}) = \hat{n}(\mathbf{r}) - n(\mathbf{r})$ as

$$S(\mathbf{r},\mathbf{r}') = \langle\hat{n}(\mathbf{r})\hat{n}(\mathbf{r}')\rangle - n(\mathbf{r})n(\mathbf{r}') = \langle\delta\hat{n}(\mathbf{r})\delta\hat{n}(\mathbf{r}')\rangle. \quad (2.46)$$

Thus, we can now express the following:

$$\frac{\delta n(\mathbf{r})}{\delta u(\mathbf{r}')} = -\beta S(\mathbf{r},\mathbf{r}'). \quad (2.47)$$

Then, we have

$$\langle\hat{n}(\mathbf{r})\hat{n}(\mathbf{r}')\rangle = \sum_i\langle\delta(\mathbf{r}-\mathbf{r}_i)\delta(\mathbf{r}'-\mathbf{r}_i)\rangle + \sum_{i\neq k}\langle\delta(\mathbf{r}-\mathbf{r}_i)\delta(\mathbf{r}'-\mathbf{r}_k)\rangle$$
$$= \int d\mathbf{r}''\,\langle\hat{n}(\mathbf{r}'')\rangle\,\delta(\mathbf{r}-\mathbf{r}'')\delta(\mathbf{r}'-\mathbf{r}'') + n_2(\mathbf{r},\mathbf{r}') = n(\mathbf{r})\delta(\mathbf{r}-\mathbf{r}') + n_2(\mathbf{r},\mathbf{r}'), \quad (2.48)$$

where

$$n_2(\mathbf{r},\mathbf{r}') = n(\mathbf{r})n(\mathbf{r}')g_2(\mathbf{r},\mathbf{r}'). \quad (2.49)$$

Function $g_2(\mathbf{r},\mathbf{r}')$ is commonly referred to as the pair correlation function. We will elaborate on its physical meaning in more detail below. Thus, we arrive at the following relation:

$$-k_BT\frac{\delta n(\mathbf{r})}{\delta u(\mathbf{r}')} = n(\mathbf{r})\delta(\mathbf{r}-\mathbf{r}') + n(\mathbf{r})n(\mathbf{r}')h_2(\mathbf{r},\mathbf{r}'), \quad (2.50)$$

where $h_2(\mathbf{r}, \mathbf{r}') = g_2(\mathbf{r}, \mathbf{r}') - 1$ is commonly referred to as the total correlation function, and the relation (2.50) is known as the first Ivon equation. Using the introduced notations, we can rewrite the correlator of the microscopic density fluctuations in the following form

$$S(\mathbf{r}, \mathbf{r}') = n(\mathbf{r})\delta(\mathbf{r} - \mathbf{r}') + n(\mathbf{r})n(\mathbf{r}')h_2(\mathbf{r}, \mathbf{r}'). \tag{2.51}$$

In the case of an ideal gas in an external field, the partition function is given by the following expression:

$$Z = \frac{\Lambda^{-3N}}{N!}\left[\int d\mathbf{r} e^{-\beta u(\mathbf{r})}\right]^N, \tag{2.52}$$

from which, using Eq. (2.42), we can obtain the Boltzmann distribution

$$n(\mathbf{r}) = a e^{-\beta u(\mathbf{r})}, \tag{2.53}$$

where a denotes normalization constant. For the system with intermolecular interactions, we can express an analogue of the Boltzmann distribution as follows:

$$n(\mathbf{r}) = a e^{-\beta u(\mathbf{r}) + c^{(1)}(\mathbf{r})}, \tag{2.54}$$

introducing the so-called one-particle potential $-k_B T c^{(1)}(\mathbf{r})$, which is a functional of the intermolecular interaction potential. Thus, we can write

$$c^{(1)}(\mathbf{r}) = \frac{u(\mathbf{r})}{k_B T} + \ln(n(\mathbf{r})/a). \tag{2.55}$$

If we introduce the direct correlation function with the expression

$$c_2(\mathbf{r}, \mathbf{r}') = \frac{\delta c^{(1)}(\mathbf{r})}{\delta n(\mathbf{r}')}, \tag{2.56}$$

we can then define the second Ivon equation as

$$c_2(\mathbf{r}, \mathbf{r}') = \frac{1}{k_B T}\frac{\delta u(\mathbf{r})}{\delta n(\mathbf{r}')} + \frac{\delta(\mathbf{r} - \mathbf{r}')}{n(\mathbf{r})}. \tag{2.57}$$

Using the identity

$$\int d\mathbf{r}'' \frac{\delta u(\mathbf{r})}{\delta n(\mathbf{r}'')}\frac{\delta n(\mathbf{r}'')}{\delta u(\mathbf{r}')} = \delta(\mathbf{r} - \mathbf{r}') \tag{2.58}$$

and Ivon equations, we derive the Ornstein–Zernike relation

$$h_2(\mathbf{r}, \mathbf{r}') = c_2(\mathbf{r}, \mathbf{r}') + \int d\mathbf{r}'' c_2(\mathbf{r}, \mathbf{r}'') n(\mathbf{r}'') h_2(\mathbf{r}'', \mathbf{r}'). \tag{2.59}$$

The direct and total correlation functions are central to the theory of simple fluids. It's important to note that the Ornstein–Zernike relation itself is merely an identity that relates these correlation functions. To be practically useful, it must be coupled with a specific relation (closure) for the correlation functions, which is derived through some form of approximation. This allows the Ornstein–Zernike relation to be utilized as a nonlinear integral equation for determining the correlation functions. For example, common closures used in the modeling of simple fluids include the Percus–Yevick one

$$c_2(\mathbf{r}_1, \mathbf{r}_2) = g_2(\mathbf{r}_1, \mathbf{r}_2)\,(1 - \exp[\beta U(\mathbf{r}_1, \mathbf{r}_2)])\,, \tag{2.60}$$

hypernetted-chain one

$$\ln g_2(\mathbf{r}_1, \mathbf{r}_2) + \beta U(\mathbf{r}_1, \mathbf{r}_2) = \int d\mathbf{r}_3 n(\mathbf{r}_3)(h_2(\mathbf{r}_1, \mathbf{r}_3) - \ln g_2(\mathbf{r}_1, \mathbf{r}_3) - \beta U(\mathbf{r}_1, \mathbf{r}_3)) h_2(\mathbf{r}_3, \mathbf{r}_2) \tag{2.61}$$

or mean spherical approximation one

$$c(\mathbf{r}_1, \mathbf{r}_2) = -\beta U(\mathbf{r}_1, \mathbf{r}_2). \tag{2.62}$$

Since these topics are beyond the scope of this book, we recommend referring to specialized monographs on the statistical theory of simple fluids for further exploration [18, 19].

Classical density functional theory. Let us consider in more details the expression (2.55), which can be rewritten as

$$\bar{\mu}(\mathbf{r}) + u(\mathbf{r}) = \mu, \tag{2.63}$$

where we introduce the intrinsic chemical potential

$$\bar{\mu}(\mathbf{r}) = k_B T \ln(n(\mathbf{r})\Lambda^3) - k_B T c^{(1)}(\mathbf{r}). \tag{2.64}$$

The first term on the right-hand side of Eq. (2.64) represents the ideal-gas contribution to the intrinsic chemical potential. The second term defines the contribution of intermolecular interactions, so-called excess chemical potential $\mu_{ex}(\mathbf{r}) = -k_B T c^{(1)}(\mathbf{r})$. In the case of a spatially-homogeneous fluid ($u = 0$), where $n(\mathbf{r}) = n = N/V$, this equation should yield the expression for the chemical potential (see (2.22))

$$\mu = \frac{\partial f}{\partial n} = k_B T \ln(n\Lambda^3) + \mu_{ex}(n, T). \tag{2.65}$$

Analogously to the formula for the chemical potential of a homogeneous fluid (2.65), which is determined as the partial derivative of the free energy density with respect to the density, we can introduce a relation for the intrinsic chemical potential of an inhomogeneous fluid as the functional derivative

$$\bar{\mu}(\mathbf{r}) = \frac{\delta F[n]}{\delta n(\mathbf{r})}, \tag{2.66}$$

where $F[n]$ represents the Helmholtz free energy, a functional of the average fluid density $n(\mathbf{r})$. Hence, Eq. (2.63) can be restated as an extremum condition of the functional [19, 20]

$$\Omega[n] = F[n] + \int d\mathbf{r}\, n(\mathbf{r})u(\mathbf{r}) - \mu \int d\mathbf{r}\, n(\mathbf{r}), \tag{2.67}$$

where μ acts as the Lagrange multiplier, corresponding to the condition of the fixed number of particles in the system

$$\int d\mathbf{r}\, n(\mathbf{r}) = N. \tag{2.68}$$

The formulation of the theory of simple fluids described above is known as classical density functional theory (cDFT). Unfortunately, the functional $F[n]$ cannot be determined in the general case and is only known within certain approximations for a limited number of model systems [21]. The functional $\Omega[n]$ can be identified with the grand thermodynamic potential of the system. It's worth noting that an analogous theory, formulated within the quantum chemistry framework and known as quantum density functional theory (qDFT), is used to determine the structure of electron density in atoms and molecules.

Let us derive a number of useful relations for the free energy functional. For the ideal gas case, when $c^{(1)}(\mathbf{r}) = 0$, we have

$$\frac{\delta F[n]}{\delta n(\mathbf{r})} = k_B T \ln(\Lambda^3 n(\mathbf{r})). \tag{2.69}$$

Let's introduce the density $n(\mathbf{r}; \tau)$, which linearly changes from zero to $n(\mathbf{r})$ when parameter τ varies from zero to one, i.e.,

$$n(\mathbf{r}; \tau) = \tau n(\mathbf{r}). \tag{2.70}$$

Integration in (2.66) over this path leads to

$$F[n] = k_B T \int_0^1 d\tau \int d\mathbf{r}\, n(\mathbf{r}) \left(\ln(\Lambda^3 n(\mathbf{r})) - 1 - c^{(1)}(\mathbf{r}; \tau n)\right). \tag{2.71}$$

Given that $c^{(1)}(\mathbf{r})$ is itself a functional integral over $n(\mathbf{r})$ of the direct correlation function $c_2(\mathbf{r}, \mathbf{r}')$, we can write

$$
\begin{aligned}
F[n] &= k_B T \int d\mathbf{r}\, n(\mathbf{r}) \left(\ln(\Lambda^3 n(\mathbf{r})) - 1\right) - k_B T \\
&\times \int_0^1 d\tau_1 \int_0^{\tau_1} d\tau_2 \int d\mathbf{r}_1 \int d\mathbf{r}_2\, n(\mathbf{r}_1) n(\mathbf{r}_2) c_2(\mathbf{r}_1, \mathbf{r}_2; \tau_2 n) \\
&= k_B T \int d\mathbf{r}\, n(\mathbf{r}) \left(\ln(\Lambda^3 n(\mathbf{r})) - 1\right) - k_B T \\
&\times \int_0^1 d\tau (1-\tau) \int d\mathbf{r}_1 \int d\mathbf{r}_2\, n(\mathbf{r}_1) n(\mathbf{r}_2) c_2(\mathbf{r}_1, \mathbf{r}_2; \tau n),
\end{aligned}
\tag{2.72}
$$

where we changed the order of integration. Thus, while it is theoretically possible to calculate the free energy of the inhomogeneous fluid, knowing the direct correlation function, in practice, this approach is of limited utility. Computation of the direct correlation function is often as complex as direct calculation of the partition function of the fluid.

Finally, let's utilize the following identity

$$
\frac{\delta F}{\delta U(\mathbf{r} - \mathbf{r}')} = \frac{1}{2} n_2(\mathbf{r}, \mathbf{r}') = \frac{1}{2} n(\mathbf{r}) n(\mathbf{r}') g_2(\mathbf{r}, \mathbf{r}'),
\tag{2.73}
$$

integrating it over the path τ from zero to unity during the transition from a certain reference intermolecular interaction potential $U_0(\mathbf{r})$ to the potential $U(\mathbf{r})$ of the fluid under consideration:

$$
U(\mathbf{r}; \tau) = \tau U(\mathbf{r}) + (1 - \tau) U_0(\mathbf{r}),
\tag{2.74}
$$

we get

$$
F[n] = F_0[n] + \frac{1}{2} \int_0^1 d\tau \int d\mathbf{r} \int d\mathbf{r}'\, n(\mathbf{r}) n(\mathbf{r}') g_2(\mathbf{r}, \mathbf{r}'; \tau) (U(\mathbf{r} - \mathbf{r}') - U_0(\mathbf{r} - \mathbf{r}')),
\tag{2.75}
$$

where $F_0[n]$ denotes the free energy of the reference system with the pairwise potential $U_0(\mathbf{r})$; the pair correlation function $g_2(\mathbf{r}, \mathbf{r}'; \tau)$ is calculated for a fluid with a pairwise potential $U(\mathbf{r}; \tau)$ assuming that its density is equal to $n(\mathbf{r})$. Equation (2.75) can serve as the foundation for developing thermodynamic perturbation theory. In practice, one often employs the mean field approximation, neglecting particle correlations, i.e., assuming $g_2(\mathbf{r}, \mathbf{r}'; \tau) = 1$. This leads to the following approximation for the free energy of the fluid:

$$F[n] = F_0[n] + \frac{1}{2}\int d\mathbf{r} \int d\mathbf{r}'\, n(\mathbf{r})(U(\mathbf{r}-\mathbf{r}') - U_0(\mathbf{r}-\mathbf{r}'))n(\mathbf{r}'). \tag{2.76}$$

If the total pairwise potential is the sum of a purely repulsive potential $U_R(\mathbf{r})$ and a purely attractive potential $-U_A(\mathbf{r})$, we obtain the van der Waals type of theory with the functional:

$$F[n] = F_R[n] - \frac{1}{2}\int d\mathbf{r} \int d\mathbf{r}'\, n(\mathbf{r})U_A(\mathbf{r}-\mathbf{r}')n(\mathbf{r}'), \tag{2.77}$$

where F_R is the free energy of the fluid with purely repulsive interaction potential, so the expression for the free energy density of the homogeneous fluid takes the form

$$f = f_R - \frac{1}{2}An^2, \tag{2.78}$$

where $A = \int d\mathbf{r}\, U_A(\mathbf{r})$. The equation of state for a fluid of van der Waals type is:

$$P = P_R - \frac{1}{2}An^2, \tag{2.79}$$

where $P_R = n\partial f_R/\partial n - f_R$ denotes the pressure of the reference system with the repulsive pairwise potential. We will further revisit this approximation within the framework of the field theoretical approach. Note that in practical calculations, the reference free energy functional of a fluid, $F_R[n]$, can be obtained from the fundamental measure theory for hard-sphere systems proposed by Rosenfeld [22] and further developed by Roth and colleagues [23, 24].

Let's note that in the general case, as it follows from the relation (2.75), A in (2.78) is a function of density and temperature, through the pair distribution function, i.e.,

$$A(n, T) = \int d\mathbf{r}\, U_A(r) \int_0^1 d\tau\, g_2(r; \tau). \tag{2.80}$$

Below we use the method of the functional integration, outlined here, to compute the functional determinants.

Structure factor of the fluid. For the case of homogeneous fluid in the thermodynamic limit, when $n(\mathbf{r}) = n = N/V$, the correlator of the density fluctuations becomes translation invariant, $S(\mathbf{r} - \mathbf{r}')$, i.e. depends only on the distance between two points. The same change happens to the direct and total correlation functions. As is known [19, 21], the Fourier transform of the correlator $S(\mathbf{r})$ defines the structure factor of the fluid

$$S(\mathbf{k}) = \int d\mathbf{r}\, e^{-i\mathbf{k}\mathbf{r}} S(\mathbf{r}), \tag{2.81}$$

which can be determined experimentally via the X-ray or neutron scattering [19]. In its turn, as it follows from (2.51), the structure factor expresses through the Fourier transform $\tilde{h}_2(\mathbf{k})$ of the total correlation function in the following way

$$S(\mathbf{k}) = n\left(1 + n\tilde{h}_2(\mathbf{k})\right). \tag{2.82}$$

Rewriting the Ornstein–Zernike relation for the case of homogeneous liquid,

$$h_2(\mathbf{r}) = c_2(\mathbf{r}) + n\int d\mathbf{r}'\, h_2(\mathbf{r} - \mathbf{r}')c_2(\mathbf{r}'), \tag{2.83}$$

in the Fourier representation

$$\tilde{h}_2(\mathbf{k}) = \frac{\tilde{c}_2(\mathbf{k})}{1 - n\tilde{c}_2(\mathbf{k})}, \tag{2.84}$$

we obtain a simple relation between the structure factor and the direct correlation function

$$S(\mathbf{k}) = \frac{n}{1 - n\tilde{c}_2(\mathbf{k})}. \tag{2.85}$$

Thus, another valuable property of the direct correlation function is that it unambiguously defines the structure factor of the simple fluid.

Let us note another important property of the structure factor: its value at zero is determined as

$$S(0) = \frac{\langle \delta N^2 \rangle}{V_s}, \tag{2.86}$$

where δN defines the fluctuation of the number of particles within some sufficiently large region of the system with a volume V_s, relative to its mean value. One can see this by integrating the Eq. (2.46) over $\mathbf{r}$ and $\mathbf{r}'$ across a sufficiently large volume V_s, considering that for a homogeneous fluid, $S(\mathbf{r}, \mathbf{r}') = S(\mathbf{r} - \mathbf{r}')$. According to the thermodynamic theory of fluctuations [12], we have

$$\langle \delta N^2 \rangle = N_s k_B T \left(\frac{\partial n}{\partial P}\right)_T, \tag{2.87}$$

where N_s represents the average number of particles within the volume V_s, leading to the formula:

$$S(0) = n^2 k_B T \chi, \tag{2.88}$$

where $\chi = n^{-1}(\partial n/\partial P)_T$ defines the isothermal compressibility of the system, considering that $n = N_s/V_s$.

Cahn–Hilliard theory. From the equation

$$\frac{\delta F}{\delta n(\mathbf{r})} = \mu - u(\mathbf{r}) \tag{2.89}$$

by employing functional derivation with respect to the density, we can obtain

$$\frac{\delta^2 F}{\delta n(\mathbf{r})\delta n(\mathbf{r}')} = -\frac{\delta u(\mathbf{r})}{\delta n(\mathbf{r}')} = k_B T S^{-1}(\mathbf{r}, \mathbf{r}'), \tag{2.90}$$

where $S^{-1}(\mathbf{r}, \mathbf{r}')$ is determined by the relation

$$\int d\mathbf{r}''\, S(\mathbf{r}, \mathbf{r}'')S^{-1}(\mathbf{r}'', \mathbf{r}') = \delta(\mathbf{r} - \mathbf{r}'). \tag{2.91}$$

For the spatially homogeneous fluid we have the following

$$\left.\frac{\delta^2 F}{\delta n(\mathbf{r})\delta n(\mathbf{r}')}\right|_{n(\mathbf{r})=n_b} = k_B T S^{-1}(\mathbf{r}, \mathbf{r}') = k_B T S^{-1}(-i\nabla_{\mathbf{r}})\delta(\mathbf{r} - \mathbf{r}'), \tag{2.92}$$

where $S^{-1}(-i\nabla_{\mathbf{r}})$ is the differential operator acting on argument $\mathbf{r}$ of delta-function $\delta(\mathbf{r} - \mathbf{r}')$[5]. Indeed,

$$\begin{aligned}\int d\mathbf{r}''\, S(\mathbf{r}, \mathbf{r}'')S^{-1}(\mathbf{r}'', \mathbf{r}') &= S^{-1}(-i\nabla_{\mathbf{r}})S(\mathbf{r} - \mathbf{r}') = \int \frac{d\mathbf{k}}{(2\pi)^3}\, S(\mathbf{k})S^{-1}(-i\nabla_{\mathbf{r}})e^{i\mathbf{k}(\mathbf{r}-\mathbf{r}')} \\ &= \int \frac{d\mathbf{k}}{(2\pi)^3}\, S(\mathbf{k})S^{-1}(\mathbf{k})e^{i\mathbf{k}(\mathbf{r}-\mathbf{r}')} = \int \frac{d\mathbf{k}}{(2\pi)^3}\, e^{i\mathbf{k}(\mathbf{r}-\mathbf{r}')} = \delta(\mathbf{r} - \mathbf{r}'),\end{aligned} \tag{2.93}$$

where we use the following identity

$$S^{-1}(-i\nabla_{\mathbf{r}})e^{i\mathbf{k}(\mathbf{r}-\mathbf{r}')} = S^{-1}(\mathbf{k})e^{i\mathbf{k}(\mathbf{r}-\mathbf{r}')}. \tag{2.94}$$

Note that $S^{-1}(-i\nabla)$ depends on the $-i\nabla$ in the same way that the inverse structure factor depends on the wave vector $\mathbf{k}$, i.e.

$$S^{-1}(\mathbf{k}) = \frac{1}{S(\mathbf{k})} = n_b^{-1}(1 - n_b\tilde{c}_2(\mathbf{k})), \tag{2.95}$$

where n_b defines the density of the homogeneous fluid.

---

[5] We understand the function of operator ∇ as analytic function $F(\mathbf{x}) = F(0) + \sum_{i=1}^{d} \frac{\partial F(0)}{\partial x_i} x_i + \frac{1}{2}\sum_{ij} \frac{\partial^2 F(0)}{\partial x_i \partial x_j} x_i x_j + \frac{1}{3!}\sum_{ijk} \frac{\partial^3 F(0)}{\partial x_i \partial x_j \partial x_k} x_i x_j x_k + \cdots$ with substitution $\mathbf{x} = \nabla$.

Restricting the expansion to the quadratic term on **k** yields

$$S^{-1}(\mathbf{k}) = \frac{1}{n_b k_B T}\frac{\partial P}{\partial n_b} + \tilde{c}_2''(0)k^2 + ... = \beta\frac{\partial \mu}{\partial n_b} + \tilde{c}_2''(0)k^2, \tag{2.96}$$

where we introduce the following notation

$$\tilde{c}_2''(0) = \frac{1}{6}\int d\mathbf{r}\, r^2 c_2(r) \tag{2.97}$$

and employ the thermodynamic identity

$$\frac{\partial P}{\partial n_b} = n_b \frac{\partial \mu}{\partial n_b}. \tag{2.98}$$

Hence, for fluids whose structure deviates weakly from that of a homogeneous fluid, we can employ the following approximation for the free energy functional

$$F[n] = \int d\mathbf{r}\left(\frac{1}{2}\varkappa(\nabla n)^2 + f(n)\right), \tag{2.99}$$

consistent with the expression (2.92) up to the second term of the expansion over the operator ∇; $f(n)$ represents the free energy density with the same dependence on the local density as in the case of the homogeneous fluid, and

$$\varkappa = \frac{k_B T}{6}\int d\mathbf{r}\, r^2 c_2(r). \tag{2.100}$$

The expression (2.99) determines the free energy of the fluid within the quadratic gradient approximation, a theory commonly referred to as the Cahn–Hilliard theory. Within the mean-spherical approximation, where $c_2(r) = -\beta U(r)$, we obtain the van der Waals approximation [25]

$$\varkappa = -\frac{1}{6}\int d\mathbf{r}\, r^2 U(r). \tag{2.101}$$

For fluids with a complex molecular structure, one can modify the functional (2.99) by introducing a density-dependent $\varkappa(n)$. Particularly, in describing polymeric fluids composed of sufficiently long polymer chains, the parameter $\varkappa$ is defined by the relation $\varkappa(n) = k_B T b^2/12n$ [26, 27], thereby

$$F[n] = \int d\mathbf{r}\left(\frac{k_B T b^2}{24n}(\nabla n)^2 + f(n)\right), \tag{2.102}$$

where $f(n)$ represents the excess free energy density of the non-connected monomeric units, disregarding their translational entropy. The connectivity of monomeric units is accounted for by the first term of the integrand in (2.102), known as the Lifshitz conformational free energy density [28]; b denotes the length of the monomer unit. It is important to note that the contribution to the free energy density of the translational motion of polymer chains as a whole is omitted in the integrand, which is justified by the presence of a sufficiently large number of monomer units in the chain.

Distribution functions. l-particle distribution functions play an important role in the theory of simple fluids, as they provide complete information about the thermodynamic state of the fluid. Distribution functions were introduced in the works of Bogolyubov, Born, Green, Kirkwood, and Yvon. Let us briefly review their theory, which is necessary for the following discussion[6].

For simplicity, let's consider the case of a one-component simple homogeneous fluid. The starting point for introducing the distribution functions is the N-particle Gibbs distribution function

$$g_N = \frac{1}{Q_N} \exp\left[-\beta U_N\right], \tag{2.103}$$

where, as mentioned earlier, Q_N denotes the configuration integral of the fluid, and U_N represents the total potential energy of pairwise interactions between the fluid particles. Integrating Eq. (2.103) over the coordinates of $N - l$ particles provides the definition of the l-particle distribution function

$$\frac{g_l}{V^l} = \frac{1}{V^N Q_N} \int \ldots \int d\mathbf{r}_{l+1} \ldots d\mathbf{r}_N \, \exp\left[-\beta U_N\right], \tag{2.104}$$

describing the probability density of the arrangement of l particles at fixed positions, given an arbitrary arrangement of the remaining $N - l$ particles. Thus, by definition

$$\frac{1}{V} \int d\mathbf{r}_{l+1} \, g_{l+1} = g_l, \quad \frac{1}{V} \int d\mathbf{r}_1 \, g_1 = 1, \quad g_1(\mathbf{r}) = 1. \tag{2.105}$$

The sequence of lth distribution functions is related to a system of coupled integro-differential equations, known as the Bogolyubov chain:

$$\nabla_1 g_l + g_l \beta \nabla_1 U_l + \beta n \int d\mathbf{r}_{n+1} \, \nabla_1 U(\mathbf{r}_1 - \mathbf{r}_{l+1}) g_{l+1} = 0, \tag{2.106}$$

where $U_l = \sum\limits_{1 \le i < j \le l} U(\mathbf{r}_i - \mathbf{r}_j)$ represents the interaction potential energy within the cluster of l particles, and ∇_1 is the gradient operator acting on the coordinates $\mathbf{r}_1$ of the particle of index 1.

[6] The following definitions of the correlation functions belong to Bogolyubov [29].

It's worth noting that the pair distribution function $g_2(\mathbf{r}_1, \mathbf{r}_2)$, previously discussed in the context of inhomogeneous fluids, plays a central role in determining the behavior of simple fluids. This function describes the probability density of two particles being located at specific positions. For homogeneous and isotropic fluids, we have

$$g_2(\mathbf{r}_1, \mathbf{r}_2) = g_2(|\mathbf{r}_1 - \mathbf{r}_2|) = e^{-\beta W(|\mathbf{r}_1 - \mathbf{r}_2|)}. \tag{2.107}$$

The deviation of g_2 from unity is caused by interparticle correlations arising from particle interactions. The function $W(r)$ is commonly referred to as the potential of mean force or the effective potential of pair interaction [19, 30]. In dense fluids, it typically differs significantly from the initial pairwise potential $U(r)$. As mentioned earlier, the pair distribution function can be determined experimentally as it appears in the expression for the fluid's structure factor, which in turn can be obtained from X-ray diffraction experiments.

The pair distribution function is important for determining the thermodynamic functions of the system. For example, the total energy of the simple fluid can be calculated as follows [29]

$$E = N\left(\frac{3k_B T}{2} + \frac{1}{2}n \int d\mathbf{r}\, U(r) g_2(r)\right). \tag{2.108}$$

By differentiating the configuration integral over the volume, we can also obtain the expression for the pressure (the equation of state) [29]

$$P = nk_B T\left(1 - \frac{1}{6}\beta n \int d\mathbf{r} g_2(r) r U'(r)\right). \tag{2.109}$$

Let's prove this relation. For this, consider that

$$P = nk_B T + k_B T \frac{\partial}{\partial V} \ln Q_N. \tag{2.110}$$

To calculate the derivative with respect to the volume, let's implement a scale transformation $\mathbf{r}_i \to \lambda \mathbf{r}_i$, $V \to \lambda^3 V$. Thus, we get

$$Q_N(\lambda) = \frac{1}{V^N} \int \ldots \int d\mathbf{r}_1 \ldots d\mathbf{r}_N \exp\left[-\beta \sum_{i<j} U(\lambda|\mathbf{r}_i - \mathbf{r}_j|)\right], \tag{2.111}$$

so we have

$$\frac{\partial}{\partial V} \ln Q_N = \frac{1}{3V Q_N} \left.\frac{\partial Q_N(\lambda)}{\partial \lambda}\right|_{\lambda=1}. \tag{2.112}$$

Considering that

$$\left.\frac{\partial Q_N(\lambda)}{\partial \lambda}\right|_{\lambda=1} = -\beta \int \ldots \int d\mathbf{r}_1 \ldots d\mathbf{r}_N\, e^{-\beta U_N} \sum_{i<j} |\mathbf{r}_i - \mathbf{r}_j| \frac{\partial U(|\mathbf{r}_i - \mathbf{r}_j|)}{\partial |\mathbf{r}_i - \mathbf{r}_j|} \tag{2.113}$$

and the definition of the pair correlation function, we obtain the expression

$$P = nk_BT - \frac{N(N-1)}{6V^3} \int d\mathbf{r}_1 \int d\mathbf{r}_2\, |\mathbf{r}_1 - \mathbf{r}_2| g_2(\mathbf{r}_1, \mathbf{r}_2) U'(|\mathbf{r}_1 - \mathbf{r}_2|), \tag{2.114}$$

which, in turn, in the thermodynamic limit transfers to the expression (2.109).

As it follows from the identity

$$\frac{g_2(r)}{V} = -\frac{1}{2\beta} \frac{\delta \ln Q_N}{\delta U(r)}, \tag{2.115}$$

for the homogeneous fluid, we have the following useful relation

$$Q_N = \exp\left[-\frac{1}{2}\beta V n^2 \int_0^1 d\tau \int d\mathbf{r}\, U(r) g_2(r;\tau)\right], \tag{2.116}$$

where τ is the parameter, which characterizes the turning on of the interaction, introduced in the same way as in the derivation of the relation (2.75).

Finally, let us write down one more helpful relation for the distribution functions [18]

$$\chi n^2 \frac{\partial}{\partial n} n^l g_l = \beta n^l \left[l g_l + n \int d\mathbf{r}_{l+1}\, (g_{l+1} - g_l) \right], \tag{2.117}$$

where χ is the previously introduced isothermal compressibility of the fluid.

2.2 Free Energy and Structure Factor of Simple Fluid with a "Soft" Repulsive Pairwise Potential. Method of Generating Functional and Gaussian Approximation

Let us now consider model fluids with repulsive central "soft" potentials $U(r) > 0$, which have a positive Fourier transform, i.e.

$$\tilde{U}(\mathbf{k}) = \int d\mathbf{r} e^{-i\mathbf{k}\mathbf{r}} U(r) > 0. \tag{2.118}$$

As shown in the work by Fisher [16], the condition of the positive definiteness of the Fourier transform of the potential guarantees the existence of the thermodynamic limit ($N \to \infty$, $V \to \infty$, $N/V \to n$) of the configuration integral of the system. Models of fluids with repulsive soft potentials are often used to describe the thermodynamic properties of macromolecular systems. For example, a system of particles with a Gaussian interaction potential, $U(r) = U_0 \exp\left[-r^2/R^2\right]$, is frequently employed to describe diluted and semi-diluted polymeric coil solutions [31]. Similarly, the system with a Yukawa potential, $U(r) = A \exp\left[-\kappa r\right]/r$, is utilized to describe the phase behavior of colloid suspensions and emulsions [32].

Let us consider a fluid consisting of N molecules at equilibrium in a volume V at temperature T. Let's introduce the following auxiliary notations. The local density of particles is determined as

$$\hat{n}(\mathbf{r}) = \sum_{i=1}^{N} \delta(\mathbf{r} - \mathbf{r}_i), \qquad (2.119)$$

where $\delta(\mathbf{r})$ represents the three-dimensional Dirac delta function[7]. The fluctuations of the density are determined as

$$\delta\hat{n}(\mathbf{r}) = \hat{n}(\mathbf{r}) - n, \qquad (2.120)$$

where $n = N/V$ is the average number density of the particles. Hamiltonian of the interactions in the system can be written in terms of the local density in the following way

$$U_N = \frac{1}{2} \sum_{i,j \neq i} U\left(|\mathbf{r}_i - \mathbf{r}_j|\right) = \frac{1}{2}(\hat{n}U\hat{n}) - U_{self}, \qquad (2.121)$$

where we introduce the short-hand notation

$$(\hat{n}U\hat{n}) = \int d\mathbf{r} \int d\mathbf{r}'\, \hat{n}(\mathbf{r})U(\mathbf{r} - \mathbf{r}')\hat{n}(\mathbf{r}') \qquad (2.122)$$

and the self-energy of the particles is given by $U_{self} = NU(0)/2$. Now, let us introduce the configuration integral

$$Q_N[\phi] = \frac{1}{V^N} \int \ldots \int d\mathbf{r}_1 \ldots d\mathbf{r}_N \exp\left[-\beta(\delta\hat{n}\phi) - \frac{\beta}{2}(\hat{n}U\hat{n})\right], \qquad (2.123)$$

where $\phi(\mathbf{r})$ is an auxiliary scalar field, satisfying the condition $\int d\mathbf{r}\phi(\mathbf{r}) = 0$ (see below), and where we introduce the short-hand notation

---

[7] Unless stated otherwise, we work in three-dimensional Euclidean space.

$$(\delta\hat{n}\phi) = \int d\mathbf{r}\,\delta\hat{n}(\mathbf{r})\phi(\mathbf{r}). \tag{2.124}$$

Using the auxiliary configuration integral introduced earlier, one can construct the *generating functional*

$$G[\phi] = \frac{Q_N[\phi]}{Q_N[0]}, \tag{2.125}$$

which allows us to calculate the averages of the products of density fluctuations at different points in space using the functional differentiation with respect to the auxiliary potential $\phi(\mathbf{r})$. Specifically, one can calculate the pair correlator of the density fluctuations with the following formula

$$S(\mathbf{r}-\mathbf{r}') = \langle\delta\hat{n}(\mathbf{r})\delta\hat{n}(\mathbf{r}')\rangle = \frac{1}{\beta^2}\frac{\delta^2 G[\phi]}{\delta\phi(\mathbf{r})\delta\phi(\mathbf{r}')}\bigg|_{\phi=0}, \tag{2.126}$$

where the averaging is performed over the microstates of the NVT-ensemble, i.e

$$\langle(\ldots)\rangle = \frac{1}{Q_N}\int\frac{d\mathbf{r}_1}{V}\ldots\int\frac{d\mathbf{r}_N}{V}\exp\left[-\beta U_N(\mathbf{r}_1,\ldots,\mathbf{r}_N)\right](\ldots). \tag{2.127}$$

Let us return to the configuration integral (2.123). Utilizing the Hubbard–Stratonovich transformation

$$e^{-\frac{\beta}{2}(\hat{n}U\hat{n})} = \int\frac{\mathcal{D}\Phi}{\mathcal{N}_U}e^{-\frac{\beta}{2}(\Phi U^{-1}\Phi)+i\beta(\hat{n}\Phi)}, \tag{2.128}$$

where, as before, the following notations are introduced

$$(\Phi U^{-1}\Phi) = \int d\mathbf{r}\int d\mathbf{r}'\,\Phi(\mathbf{r})U^{-1}(\mathbf{r}-\mathbf{r}')\Phi(\mathbf{r}'), \tag{2.129}$$

$$(\hat{n}\Phi) = \int d\mathbf{r}\,\hat{n}(\mathbf{r})\Phi(\mathbf{r}), \tag{2.130}$$

we obtain the representation of the configuration integral (2.123) in the form of the functional integral over the Gaussian measure

$$\begin{aligned}Q_N[\phi] &= \frac{1}{V^N}e^{\beta U_{self}}\int\frac{\mathcal{D}\Phi}{\mathcal{N}_U}e^{-\frac{\beta}{2}(\Phi U^{-1}\Phi)}\int d\mathbf{r}_1\cdots\int d\mathbf{r}_N\prod_{j=1}^{N}e^{i\beta(\Phi(\mathbf{r}_j)-i\phi(\mathbf{r}_j))}\\ &= e^{\beta U_{self}}\int\frac{\mathcal{D}\Phi}{\mathcal{N}_U}e^{-\frac{\beta}{2}(\Phi U^{-1}\Phi)}\left[\int\frac{d\mathbf{r}}{V}e^{i\beta(\Phi(\mathbf{r})-i\phi(\mathbf{r}))}\right]^N,\end{aligned} \tag{2.131}$$

where we introduce the normalization constant of the Gaussian measure

$$\mathcal{N}_U = \int \mathcal{D}\Phi \, e^{-\frac{\beta}{2}(\Phi U^{-1} \Phi)} \tag{2.132}$$

and the kernel of the inverse integral operator U^{-1}, satisfying the integral relation

$$\int d\mathbf{r}'' \, U(\mathbf{r} - \mathbf{r}'')U^{-1}(\mathbf{r}'' - \mathbf{r}') = \delta(\mathbf{r} - \mathbf{r}') \tag{2.133}$$

or in the Fourier representation

$$\tilde{U}(\mathbf{k})\tilde{U}^{-1}(\mathbf{k}) = 1. \tag{2.134}$$

Thus, we have $\tilde{U}^{-1}(\mathbf{k}) = 1/\tilde{U}(\mathbf{k})$, i.e., the Fourier transform of the kernel of the inverse operator is equal to the inverse Fourier transform of the operator. It is worth noting that, generally speaking, the kernel of the inverse operator is not defined in terms of regular functions, therefore, $U^{-1}(\mathbf{r} - \mathbf{r}')$ is to be considered as a generalized function of the form

$$U^{-1}(\mathbf{r} - \mathbf{r}') = \tilde{U}^{-1}(-i\nabla_{\mathbf{r}})\delta(\mathbf{r} - \mathbf{r}'). \tag{2.135}$$

Then, representing the field variable in the form of $\Phi = \Phi_0 + \Lambda$, where $\Phi_0 = V^{-1} \int d\mathbf{r} \, \Phi(\mathbf{r})$ and $\int d\mathbf{r} \, \Lambda(\mathbf{r}) = 0$, and calculating the Gaussian integral over Φ_0, we obtain

$$Q_N[\phi] = e^{\beta U_{self}} Q_N^{(MF)} \int \frac{\mathcal{D}\Lambda}{\mathcal{N}_U} e^{-\frac{\beta}{2}(\Lambda U^{-1} \Lambda)} \left[\int \frac{d\mathbf{r}}{V} e^{i\beta(\Lambda(\mathbf{r}) - i\phi(\mathbf{r}))} \right]^N, \tag{2.136}$$

where

$$Q_N^{(MF)} = \exp\left[-\frac{\beta}{2}(nUn) \right] = \exp\left[-\frac{\beta N^2 \tilde{U}(0)}{2V} \right] \tag{2.137}$$

denotes the configuration integral of the system within the mean field approximation; $\tilde{U}(0)$—Fourier-image of the pair potential at $\mathbf{k} = 0$. This name is due to the fact that the contribution to the free energy can be obtained by considering the interaction of a single isolated particle not with other individual discrete particles, but with a continuous medium "smeared" with constant density, $n = N/V$. Thus, such energy is equal to

$$n \int d\mathbf{r} \, U(\mathbf{r}) = n\tilde{U}(0). \tag{2.138}$$

To obtain the total interaction energy between the particles, one must multiply (2.138) by the total number of particles N and then divide the result by 2 to exclude double account.

Now, shifting the field variable $\Lambda \to \Lambda + i\phi$ to exclude the dependence on ϕ in the integrand functional, we arrive at the expression

$$Q_N[\phi] = e^{\beta U_{self}} Q_N^{(MF)} \exp\left[\frac{\beta}{2}(\phi U^{-1}\phi)\right] \\ \times \int \frac{\mathcal{D}\Lambda}{\mathcal{N}_U} \exp\left[-\frac{\beta}{2}(\Lambda U^{-1}\Lambda) - i\beta(\Lambda U^{-1}\phi)\right] \left[\int \frac{d\mathbf{r}}{V} \exp\left[i\beta\Lambda(\mathbf{r})\right]\right]^N . \tag{2.139}$$

Let's note that the property of the auxiliary scalar field $\int d\mathbf{r}\,\phi = 0$ allows us to make a shift $\Lambda \to \Lambda + i\phi$ without changing the properties of the fluctuating scalar field Λ.

Now, taking into account that in thermodynamic limit, using the asymbtotic

$$\left(1 + \frac{a}{x}\right)^x \simeq e^a, \; x \gg 1, \tag{2.140}$$

we obtain

$$\left[\int \frac{d\mathbf{r}}{V} \exp\left[i\beta\Lambda(\mathbf{r})\right]\right]^N = \left[1 + \int \frac{d\mathbf{r}}{V} (\exp\left[i\beta\Lambda(\mathbf{r})\right] - 1)\right]^N \\ \simeq \exp\left[n \int d\mathbf{r}\, (\exp\left[i\beta\Lambda(\mathbf{r})\right] - 1)\right], \tag{2.141}$$

we arrive at the final expression

$$Q_N[\phi] = e^{\beta U_{self}} Q_N^{(MF)} \exp\left[\frac{\beta}{2}(\phi U^{-1}\phi)\right] \int \frac{\mathcal{D}\Lambda}{\mathcal{N}_U} \exp\left[-\frac{\beta}{2}(\Lambda U^{-1}\Lambda) - i\beta(\Lambda U^{-1}\phi) + W[\Lambda]\right], \tag{2.142}$$

where we introduce the functional

$$W[\Lambda] = n \int d\mathbf{r}\, (\exp\left[i\beta\Lambda(\mathbf{r})\right] - 1). \tag{2.143}$$

Expanding the integrand in (2.143) in series to quadratic terms and considering that $\int d\mathbf{r}\,\Lambda(\mathbf{r}) = 0$, we arrive at the expression for the configuration integral within the Gaussian approximation (GA)

$$Q_N[\phi] = e^{\beta U_{self}} Q_N^{(MF)} \exp\left[\frac{\beta}{2}(\phi U^{-1}\phi)\right] \int \frac{\mathcal{D}\Lambda}{\mathcal{N}_U} \exp\left[-\frac{\beta}{2}(\Lambda D^{-1}\Lambda) - i\beta(\Lambda U^{-1}\phi)\right], \tag{2.144}$$

where the inverse Green's function is introduced:

$$D^{-1}(\mathbf{r} - \mathbf{r}') = U^{-1}(\mathbf{r} - \mathbf{r}') + \beta n \delta(\mathbf{r} - \mathbf{r}'), \tag{2.145}$$

which in Fourier transform gives the following

$$\tilde{D}(\mathbf{k}) = \frac{\tilde{U}(\mathbf{k})}{1 + \beta n \tilde{U}(\mathbf{k})}. \tag{2.146}$$

By calculating the Gaussian functional integral, we arrive at the final expression

$$Q_N[\phi] = Q_N^{(GA)} \exp\left[\frac{\beta}{2}(\phi U^{-1}\phi) - \frac{\beta}{2}(\phi U^{-1} D U^{-1}\phi)\right], \tag{2.147}$$

where

$$Q_N^{(GA)} = e^{\beta U_{self}} Q_N^{(MF)} \exp\left[\frac{1}{2}(tr \ln D - tr \ln U)\right] \tag{2.148}$$

defines the configuration integral of the system within the GA. The trace of the integral operator A, given by the kernel $A(\mathbf{r}, \mathbf{r}')$, is determined as follows

$$tr A = \int d\mathbf{r}\, A(\mathbf{r}, \mathbf{r}). \tag{2.149}$$

In the case of a translationally invariant kernel $A(\mathbf{r}, \mathbf{r}') = A(\mathbf{r} - \mathbf{r}')$, we have

$$tr A = V \int \frac{d\mathbf{k}}{(2\pi)^3} \tilde{A}(\mathbf{k}), \tag{2.150}$$

where $\tilde{A}(\mathbf{k})$ is the Fourier transform of the function $A(\mathbf{r})$. Thus, using relation (1.198), we obtain the following expression for the excess Helmholtz free energy of a system with a soft repulsive potential, within the Gaussian approximation

$$F_{ex} = -k_B T \ln Q_N[0] = \frac{V n^2 \tilde{U}(0)}{2} + \frac{V k_B T}{2} \int \frac{d\mathbf{k}}{(2\pi)^3} \left(\ln(1 + \beta n \tilde{U}(\mathbf{k})) - \beta n \tilde{U}(\mathbf{k})\right), \tag{2.151}$$

where we take into account that

$$U_{self} = \frac{N U(0)}{2} = \frac{N}{2} \int \frac{d\mathbf{k}}{(2\pi)^3} \tilde{U}(\mathbf{k}). \tag{2.152}$$

This expression was first obtained by Zubarev [33], using a collective variables approach [34], and independently by R. Storer, within a field-theoretical framework [35].

It should be mentioned that the first term in expression (2.151) determines the contribution of intermolecular interactions to the Helmholtz free energy within the mean-field approximation, while the second term determines the fluctuation correction to the mean-field approximation within the GA. By utilizing expressions (2.151) and (2.21), we can derive the expression for pressure as a function of temperature and density, i.e., the equation of state which was obtained by Storer [35]

$$P = nk_BT + \frac{n^2\tilde{U}(0)}{2} + \frac{k_BT}{2}\int \frac{d\mathbf{k}}{(2\pi)^3}\left(\frac{\beta n\tilde{U}(\mathbf{k})}{1+\beta n\tilde{U}(\mathbf{k})} - \ln(1+\beta n\tilde{U}(\mathbf{k}))\right). \tag{2.153}$$

The expression (2.147) and definition (2.126) enable us to calculate the structure factor. Indeed, we have

$$\begin{aligned} S(\mathbf{r}-\mathbf{r}') &= \frac{1}{\beta^2}\frac{\delta^2}{\delta\phi(\mathbf{r})\delta\phi(\mathbf{r}')}\exp\left[\frac{\beta}{2}(\phi U^{-1}\phi) - \frac{\beta}{2}(\phi U^{-1}DU^{-1}\phi)\right]\Bigg|_{\phi=0} \\ &= \frac{1}{\beta}\left(U^{-1}(\mathbf{r}-\mathbf{r}') - \int d\mathbf{r}_1\int d\mathbf{r}_2\, U^{-1}(\mathbf{r}-\mathbf{r}_1)D(\mathbf{r}_1-\mathbf{r}_2)U^{-1}(\mathbf{r}_2-\mathbf{r}')\right), \end{aligned} \tag{2.154}$$

or in the Fourier representation

$$\beta S(\mathbf{k}) = \frac{1}{\tilde{U}(\mathbf{k})}\left(1 - \frac{\tilde{D}(\mathbf{k})}{\tilde{U}(\mathbf{k})}\right). \tag{2.155}$$

Then, taking into account relation (2.146), we arrive at the final expression for the structure factor within the GA

$$S(\mathbf{k}) = \frac{n}{1+\beta n\tilde{U}(\mathbf{k})}. \tag{2.156}$$

Comparing expressions (2.156) and (2.85), we conclude that within the GA, the direct correlation function is determined by the expression $\tilde{c}_2(\mathbf{k}) = -\beta\tilde{U}(\mathbf{k})$, i.e., it coincides with the mean spherical closure for the Ornstein–Zernike relation. Now, we utilize expression (2.88), which leads us to

$$\frac{\partial P}{\partial n} = k_BT\left(1 + \beta n\tilde{U}(0)\right). \tag{2.157}$$

After integration, we arrive at the expression

$$P = nk_BT + \frac{n^2\tilde{U}(0)}{2}, \tag{2.158}$$

which determines the equation of state of the fluid within the mean-field approximation. The inconsistency between relations (2.153) and (2.158) arises from the fact that both the expression for the excess free energy (2.151) and the expression for the structure factor (2.156) are approximations, each with a different level of accuracy. However, in contrast to (2.158), relation ((2.153)) accounts for fluctuation correction to the mean field approximation at the GA level.

We will use the generating functional method outlined in this section in Chap. 5 to describe the solutions of molecules with complex internal electrical structure. Below, we consider the method of systematically accounting for corrections to the GA.

2.3 Free Energy of the Model Fluids with the Pairwise Gauss and Yukawa Potentials Within the Gaussian Approximation

In this section, we apply the previously derived expression (2.151) to evaluate the free energy of fluids with Gauss and Yukawa potentials, as mentioned at the beginning of the previous section. For the fluid with the Gauss potential (the Gaussian core model [31], GCM), which was first investigated in the work by Stillinger [36], the Fourier transform yields the following expression

$$\tilde{U}(\mathbf{k}) = \pi^{3/2} R^3 U_0 \exp\left[-\frac{k^2 R^2}{4}\right]. \tag{2.159}$$

Substituting the expression (2.159) into (2.151) and evaluating the integral, we arrive at the following expression for the excess free energy of GCM

$$\frac{F_{ex}}{N k_B T} = \frac{\xi}{2}\left(1 - \beta U_0 \mathcal{F}(\xi)\right), \tag{2.160}$$

where $\xi = \pi^{3/2} \beta n R^3 U_0$, and we also introduce following auxiliary function

$$\mathcal{F}(\xi) = \frac{4}{3\sqrt{\pi}} \int\limits_0^1 \frac{dt\,(-\ln t)^{3/2} t}{1 + \xi t}. \tag{2.161}$$

As shown in [37], GA provides a good approximation for the equation of state of GCM at sufficiently small values of the parameter βU_0 (rather soft cores). The Fourier transform of the Yukawa potential has the following form:

$$\tilde{U}(\mathbf{k}) = \frac{4\pi A}{k^2 + \kappa^2}, \tag{2.162}$$

so the excess free energy is determined as

$$\frac{F_{ex}}{N k_B T} = \frac{\theta}{2} - \frac{\kappa^3}{24\pi n}\left(2(1+\theta)^{3/2} - 2 - 3\theta\right), \tag{2.163}$$

where we introduce the dimensionless parameter $\theta = 4\pi \beta n A \kappa^{-2}$. It is worth noting that unlike the fluid with the Gaussian interaction potential, the fluid whose particles interact interact via the Yukawa potential can undergo phase separation into liquid and gaseous phases at sufficiently low temperatures [8, 32, 38]. Additionally, the fluid with the Yukawa potential exemplifies a system with a repulsive potential that can exhibit a liquid-gas phase separation. Moreover, the corresponding phase transition in such a system is fluctuationally-driven and cannot be captured within the mean field approximation (see expression (2.158)).

2.4 Variational Field Theory

In this section, we examine another method to describe fluids with repulsive interaction potential: the variational field theory, based on the utilization of Bogolyubov's inequality (see Chap. 1). Earlier, we focused solely on spatially-homogeneous fluids. However, in this section, we also consider the case of inhomogeneous fluids and formulate equations allowing us to calculate local fluid concentration.

Let us work in the grand canonical ensemble. The grand partition function of the fluid takes the form

$$\Xi = \sum_{N=0}^{\infty} \frac{z^N}{N!} Q_N(V,T), \tag{2.164}$$

where the configuration integral can be written in a standard way as

$$\begin{aligned} Q_N &= \int d\mathbf{r}_1 \ldots \int d\mathbf{r}_N \exp\left[-\frac{\beta}{2}\sum_{i,j\neq i} U(|\mathbf{r}_i - \mathbf{r}_j|) - \beta\sum_{i=1}^{N} u(\mathbf{r}_i) + \beta U_{self}\right] \\ &= \int d\mathbf{r}_1 \ldots \int d\mathbf{r}_N \exp\left[-\frac{\beta}{2}(\hat{n}U\hat{n}) - \beta(\hat{n}u) + \beta U_{self}\right] \\ &= \int \frac{\mathcal{D}\Phi}{\mathcal{N}_U} \exp\left[-\frac{\beta}{2}(\Phi U^{-1}\Phi)\right]\left[\int d\mathbf{r}\, e^{i\beta\Phi(\mathbf{r}) + \frac{\beta U(0)}{2} - \beta u(\mathbf{r})}\right]^N, \end{aligned} \tag{2.165}$$

where we utilized the Hubbard–Stratonovich transformation. Substituting expression (2.165) into (2.164), we arrive at the representation of the grand partition function in the form of a functional integral

$$\Xi = \int \frac{\mathcal{D}\Phi}{\mathcal{N}_U} e^{-\frac{\beta}{2}(\Phi U^{-1}\Phi) + W[\Phi]}, \tag{2.166}$$

where we introduce the short-hand notation

$$W[\Phi] = z e^{\frac{\beta U(0)}{2}} \int d\mathbf{r}\, e^{i\beta\Phi(\mathbf{r}) - \beta u(\mathbf{r})}. \tag{2.167}$$

Now, let us apply a variational method to evaluate this functional integral. The first step is to shift the field variable $\Phi \to \Phi + \Phi_0$ by some function Φ_0, i.e.,

$$\Xi = \int \frac{\mathcal{D}\Phi}{\mathcal{N}_U} e^{-\frac{\beta}{2}((\Phi+\Phi_0)U^{-1}(\Phi+\Phi_0))+W[\Phi+\Phi_0]}$$
$$= \int \frac{\mathcal{D}\Phi}{\mathcal{N}_U} e^{-\frac{\beta}{2}(\Phi U^{-1}\Phi)-\beta(\Phi_0 U^{-1}\Phi)-\frac{\beta}{2}(\Phi_0 U^{-1}\Phi_0)+W[\Phi+\Phi_0]}. \tag{2.168}$$

The second step consists of transitioning from the Green's function $U(\mathbf{r}-\mathbf{r}')$ to some new Green function $D(\mathbf{r},\mathbf{r}')$ of a Gaussian measure, i.e.,

$$\Xi = \frac{\mathcal{N}_D}{\mathcal{N}_U}\int \frac{\mathcal{D}\Phi}{\mathcal{N}_D} e^{-\frac{\beta}{2}(\Phi D^{-1}\Phi)-\frac{\beta}{2}(\Phi[U^{-1}-D^{-1}]\Phi)-\beta(\Phi_0 U^{-1}\Phi)-\frac{\beta}{2}(\Phi_0 U^{-1}\Phi_0)+W[\Phi+\Phi_0]}$$
$$= \frac{\mathcal{N}_D}{\mathcal{N}_U}\int \frac{\mathcal{D}\Phi}{\mathcal{N}_D} e^{-\frac{\beta}{2}(\Phi D^{-1}\Phi)+\tilde{W}[\Phi]}, \tag{2.169}$$

where

$$\tilde{W}[\Phi] = -\frac{\beta}{2}(\Phi[U^{-1}-D^{-1}]\Phi) - \beta(\Phi_0 U^{-1}\Phi) - \frac{\beta}{2}(\Phi_0 U^{-1}\Phi_0) + W[\Phi+\Phi_0]. \tag{2.170}$$

Now, we apply Bogolyubov's inequality

$$\Xi \geq \max_{\Phi_0, D} \exp\left[\ln\frac{\mathcal{N}_D}{\mathcal{N}_U} + \left\langle \tilde{W}[\Phi]\right\rangle_D\right], \tag{2.171}$$

where we introduce the notation for the averaging over new Gaussian measure with the Green's function D

$$\langle(\ldots)\rangle_D = \int \frac{\mathcal{D}\Phi}{\mathcal{N}_D} e^{-\frac{\beta}{2}(\Phi D^{-1}\Phi)}(\ldots). \tag{2.172}$$

Thus, the main idea of the method is to find a function Φ_0 in the vicinity of which the functional has an extremum, and to approximate the functional in the vicinity of this function with a quadratic functional with the Green's function $D(\mathbf{r},\mathbf{r}')$. Now, let's evaluate $\left\langle \tilde{W}[\Phi]\right\rangle_D$. Taking into account that

$$\left\langle \Phi(\mathbf{r})\Phi(\mathbf{r}')\right\rangle_D = \frac{1}{\beta}D(\mathbf{r},\mathbf{r}') \tag{2.173}$$

we obtain

$$
\langle(\Phi[U^{-1}-D^{-1}]\Phi)\rangle_D = \left\langle \int d\mathbf{r}\int d\mathbf{r}'\,\Phi(\mathbf{r})\Phi(\mathbf{r}')\left[U^{-1}(\mathbf{r}-\mathbf{r}')-D^{-1}(\mathbf{r},\mathbf{r}')\right]\right\rangle_D
$$
$$
=\frac{1}{\beta}\int d\mathbf{r}\int d\mathbf{r}'\,D(\mathbf{r},\mathbf{r}')\left[U^{-1}(\mathbf{r}-\mathbf{r}')-D^{-1}(\mathbf{r},\mathbf{r}')\right]=\frac{1}{\beta}tr\left(D[U^{-1}-D^{-1}]\right). \tag{2.174}
$$

Then, we have

$$
\langle(\Phi U^{-1}\Phi_0)\rangle_D = 0, \tag{2.175}
$$

due to $\langle\Phi(\mathbf{r})\rangle_D = 0$ and

$$
\langle W[\Phi+\Phi_0]\rangle_D = ze^{\frac{\beta U(0)}{2}}\int d\mathbf{r}\,e^{i\beta\Phi_0(\mathbf{r})-\beta u(\mathbf{r})}\left\langle e^{i\beta\Phi(\mathbf{r})}\right\rangle_D = z\int d\mathbf{r}\,A(\mathbf{r})e^{i\beta\Phi_0(\mathbf{r})-\beta u(\mathbf{r})}, \tag{2.176}
$$

where we introduce the following auxiliary function

$$
A(\mathbf{r}) = \exp\left[\frac{\beta}{2}(U(0)-D(\mathbf{r},\mathbf{r}))\right]. \tag{2.177}
$$

Thus, we arrive at the following variational estimation

$$
\Xi \geq \max_{\Phi_0,D}\exp\left[\ln\frac{\mathcal{N}_D}{\mathcal{N}_U}-\frac{1}{2}tr\left(D[U^{-1}-D^{-1}]\right)\right.
$$
$$
\left.-\frac{\beta}{2}(\Phi_0U^{-1}\Phi_0)+z\int d\mathbf{r}\,A(\mathbf{r})e^{i\beta\Phi_0(\mathbf{r})-\beta u(\mathbf{r})}\right] = \exp S[\Phi_0, D]. \tag{2.178}
$$

The best estimation is achieved under the following conditions

$$
\frac{\delta S}{\delta\Phi_0(\mathbf{r})} = 0, \quad \frac{\delta S}{\delta D(\mathbf{r},\mathbf{r}')} = 0. \tag{2.179}
$$

The first equation gives

$$
\frac{\delta S}{\delta\Phi_0(\mathbf{r})} = -\beta\int d\mathbf{r}'\,U^{-1}(\mathbf{r}-\mathbf{r}')\Phi_0(\mathbf{r}') + i\beta zA(\mathbf{r})e^{i\beta\Phi_0(\mathbf{r})-\beta u(\mathbf{r})} = 0. \tag{2.180}
$$

By substituting $\Phi_0 = i\psi_0$, which allows us to eliminate the imaginary unit, and inverting the integral operator, we arrive at the equation

$$
\psi_0(\mathbf{r}) = z\int d\mathbf{r}'\,U(\mathbf{r}-\mathbf{r}')A(\mathbf{r}')e^{-\beta\psi_0(\mathbf{r}')-\beta u(\mathbf{r}')}. \tag{2.181}
$$

Before evaluating the variational derivative with respect to the Green's function $D(\mathbf{r}, \mathbf{r}')$, we take into account that

$$\ln \frac{\mathcal{N}_D}{\mathcal{N}_U} = \frac{1}{2} \ln \frac{Det\, D}{Det\, U} = \frac{1}{2} (tr \ln D - tr \ln U) \,. \tag{2.182}$$

Let us remind that the trace of the operator can be evaluated as

$$tr\, A = \sum_n a_n, \tag{2.183}$$

where a_n denotes the eigenvalues of the operator in the basis of the eigenfunctions ψ_n, such that

$$a_n = \int d\mathbf{r} \int d\mathbf{r'}\, \psi_n(\mathbf{r}) A(\mathbf{r}, \mathbf{r}') \psi_n(\mathbf{r}'), \tag{2.184}$$

and the kernel of the operator can be written as

$$A(\mathbf{r}, \mathbf{r}') = \sum_n a_n \psi_n(\mathbf{r}) \psi_n(\mathbf{r}'). \tag{2.185}$$

The kernel of the inverse operator is

$$A^{-1}(\mathbf{r}, \mathbf{r}') = \sum_n \frac{1}{a_n} \psi_n(\mathbf{r}) \psi_n(\mathbf{r}'). \tag{2.186}$$

Thus, we have

$$\frac{\delta}{\delta D(\mathbf{r}, \mathbf{r}')} tr \ln D = \frac{\delta}{\delta D(\mathbf{r}, \mathbf{r}')} \sum_n \ln d_n = \sum_n \frac{1}{d_n} \frac{\delta d_n}{\delta D(\mathbf{r}, \mathbf{r}')}, \tag{2.187}$$

where d_n are the eigenvalues of the operator D.

Further, we have

$$\frac{\delta d_n}{\delta D(\mathbf{r}, \mathbf{r}')} = \frac{\delta}{\delta D(\mathbf{r}, \mathbf{r}')} \int d\mathbf{r}_1 \int d\mathbf{r}_2\, \psi_n(\mathbf{r}_1) D(\mathbf{r}_1, \mathbf{r}_2) \psi_n(\mathbf{r}_2) = \psi_n(\mathbf{r}) \psi_n(\mathbf{r}'), \tag{2.188}$$

where we take into account that

$$\frac{\delta D(\mathbf{r}_1, \mathbf{r}_2)}{\delta D(\mathbf{r}, \mathbf{r}')} = \delta(\mathbf{r}_1 - \mathbf{r}') \delta(\mathbf{r}_2 - \mathbf{r}'). \tag{2.189}$$

From which we obtain

$$\frac{\delta}{\delta D(\mathbf{r}, \mathbf{r}')} tr \ln D = \sum_n \frac{1}{d_n} \psi_n(\mathbf{r}) \psi_n(\mathbf{r}') = D^{-1}(\mathbf{r}, \mathbf{r}'). \tag{2.190}$$

Further,

$$\frac{\delta}{\delta D(\mathbf{r},\mathbf{r}')}tr\left(D[U^{-1}-D^{-1}]\right)=\frac{\delta}{\delta D(\mathbf{r},\mathbf{r}')}tr\left(DU^{-1}-I\right)=\frac{\delta}{\delta D(\mathbf{r},\mathbf{r}')}tr(DU^{-1})$$
$$=\frac{\delta}{\delta D(\mathbf{r},\mathbf{r}')}\int d\mathbf{r}_1\int d\mathbf{r}_2\,D(\mathbf{r}_1,\mathbf{r}_2)U^{-1}(\mathbf{r}_1-\mathbf{r}_2)=U^{-1}(\mathbf{r}-\mathbf{r}') \quad (2.191)$$

and

$$\frac{\delta}{\delta D(\mathbf{r},\mathbf{r}')}\left(z\int d\mathbf{x}\,A(\mathbf{x})e^{-\beta\psi_0(\mathbf{x})-\beta u(\mathbf{x})}\right)=-\frac{\beta z}{2}\int d\mathbf{x}\,A(\mathbf{x})$$
$$\times\,e^{-\beta\psi_0(\mathbf{x})-\beta u(\mathbf{x})}\delta(\mathbf{r}-\mathbf{x})\delta(\mathbf{r}'-\mathbf{x})=-\frac{\beta z}{2}A(\mathbf{r})e^{-\beta\psi_0(\mathbf{r})-\beta u(\mathbf{r})}\delta(\mathbf{r}-\mathbf{r}'), \quad (2.192)$$

where we introduce the relation

$$\frac{\delta A(\mathbf{x})}{\delta D(\mathbf{r},\mathbf{r}')}=-\frac{\beta}{2}A(\mathbf{x})\delta(\mathbf{r}-\mathbf{x})\delta(\mathbf{r}'-\mathbf{x}). \quad (2.193)$$

Thus, the second equation, providing the best variational estimation for the grand partition function, takes the form

$$D^{-1}(\mathbf{r},\mathbf{r}')-U^{-1}(\mathbf{r}-\mathbf{r}')=\Sigma(\mathbf{r},\mathbf{r}'), \quad (2.194)$$

where

$$\Sigma(\mathbf{r},\mathbf{r}')=\beta zA(\mathbf{r})e^{-\beta\psi_0(\mathbf{r})-\beta u(\mathbf{r})}\delta(\mathbf{r}-\mathbf{r}'). \quad (2.195)$$

Equation (2.194) can be rewritten in a more convenient form for numerical computation

$$D(\mathbf{r},\mathbf{r}')=U(\mathbf{r}-\mathbf{r}')-\beta z\int d\mathbf{r}''\,e^{-\beta\psi_0(\mathbf{r}'')-\beta u(\mathbf{r}'')}A(\mathbf{r}'')U(\mathbf{r}-\mathbf{r}'')D(\mathbf{r}'',\mathbf{r}'). \quad (2.196)$$

Thus, the variational estimation for the grand canonical potential takes the form

$$\Omega\le\Omega_{var}=-\frac{k_BT}{2}\left(\ln\frac{Det\,D}{Det\,U}-tr\left(D[U^{-1}-D^{-1}]\right)\right)$$
$$+\frac{1}{2}(\psi_0U^{-1}\psi_0)+zk_BT\int d\mathbf{r}\,A(\mathbf{r})e^{-\beta\psi_0(\mathbf{r})-\beta u(\mathbf{r})}. \quad (2.197)$$

Let us represent the expression (2.197) in a more computationally convenient form. To do this, we note that

$$\frac{\delta}{\delta D^{-1}(\mathbf{r},\mathbf{r}')}\ln Det\,D=-D(\mathbf{r},\mathbf{r}'), \quad (2.198)$$

which can be verified in the same way as in the derivation of the relation (2.190). Hence, we can write [39]

$$\ln \frac{Det D}{Det U} = \ln Det D - \ln Det U = -\int d\mathbf{r} \int d\mathbf{r}' \int_{U^{-1}}^{D^{-1}} D(\mathbf{r}, \mathbf{r}')\delta D^{-1}(\mathbf{r}, \mathbf{r}'), \tag{2.199}$$

where the integration is along the "path", connecting the "points" $U^{-1}(\mathbf{r}, \mathbf{r}')$ and $D^{-1}(\mathbf{r}, \mathbf{r}')$ in the space of generalized functions $D^{-1}(\mathbf{r}, \mathbf{r}')$. Let us choose the integration path by introducing the parameter $\tau \in [0; 1]$, which characterizes the turning on of the interaction, so

$$\int_{U^{-1}}^{D^{-1}} D(\mathbf{r}, \mathbf{r}')\, \delta D^{-1}(\mathbf{r}, \mathbf{r}') = \int_{0}^{1} d\tau\, D(\mathbf{r}, \mathbf{r}'; \tau)\Sigma(\mathbf{r}, \mathbf{r}'), \tag{2.200}$$

where the "intermediate" Green's function $D(\mathbf{r}, \mathbf{r}'; \tau)$ is defined by the equation

$$D^{-1}(\mathbf{r}, \mathbf{r}'; \tau) - U^{-1}(\mathbf{r} - \mathbf{r}') = \tau\Sigma(\mathbf{r}, \mathbf{r}'). \tag{2.201}$$

Applying Eq. (2.194), we can simplify the second term in the parentheses in (2.197), which gives

$$\frac{\Omega_{var}}{k_B T} = \frac{1}{2}\int d\mathbf{r} \int d\mathbf{r}'\, \Sigma(\mathbf{r}, \mathbf{r}') \int_{0}^{1} d\tau \left(D(\mathbf{r}, \mathbf{r}'; \tau) - D(\mathbf{r}, \mathbf{r}')\right)$$
$$+ \frac{1}{2}(\psi_0 U^{-1} \psi_0) + z \int d\mathbf{r}\, A(\mathbf{r}) e^{-\beta\psi_0(\mathbf{r}) - \beta u(\mathbf{r})}. \tag{2.202}$$

The last expression can be further simplified if we take into account the average density of the particles

$$n(\mathbf{r}) = \frac{\delta\Omega_{var}}{\delta u(\mathbf{r})} = z A(\mathbf{r}) e^{-\beta\psi_0(\mathbf{r}) - \beta u(\mathbf{r})}. \tag{2.203}$$

We should note that this expression is obtained under the assumption that Ω_{var} reaches an extremum at ψ_0 and $D(\mathbf{r}, \mathbf{r}')$, satisfying the system of Eqs. (2.181) and (2.194). Thus, when functionally differentiating with respect to $u(\mathbf{r})$, they can be considered constant. Hence, by considering that $z = \Lambda^{-3} e^{\beta\mu}$, one can obtain

$$\mu = k_B T \ln(\Lambda^3 n(\mathbf{r})) + \psi_0(\mathbf{r}) + \frac{1}{2}\left(D(\mathbf{r}, \mathbf{r}) - U(0)\right) + u(\mathbf{r}) = \bar{\mu}(\mathbf{r}) + u(\mathbf{r}), \tag{2.204}$$

where $\bar{\mu}(\mathbf{r})$ can be interpreted as the intrinsic chemical potential of the fluid's molecules, and the expression (2.204) itself as an equilibrium condition. The function $\psi_0(\mathbf{r})$ can be interpreted as the average interaction potential energy between the molecule and its surroundings, while $(D(\mathbf{r}, \mathbf{r}) - U(0))/2$ can be defined as the "correlation" contribution to the chemical potential arising from the fluctuations of this field due to the molecules' thermal motion. The sum of these values determines nothing more than a one-particle potential within the variational field theory approximation, i.e

$$-k_B T c^{(1)}(\mathbf{r}) = \psi_0(\mathbf{r}) + \frac{1}{2}\left(D(\mathbf{r}, \mathbf{r}) - U(0)\right). \tag{2.205}$$

Considering the obtained results (2.203) and (2.181), we have

$$\Sigma(\mathbf{r}, \mathbf{r}') = \beta n(\mathbf{r})\delta(\mathbf{r} - \mathbf{r}'), \tag{2.206}$$

$$\psi_0(\mathbf{r}) = \int d\mathbf{r}'\, U(\mathbf{r} - \mathbf{r}')n(\mathbf{r}'), \tag{2.207}$$

which, finally, leads to

$$\frac{\Omega_{var}}{k_B T} = \int d\mathbf{r}\, n(\mathbf{r}) + \frac{1}{2}(nUn) + \frac{\beta}{2}\int d\mathbf{r}\, n(\mathbf{r}) \int_0^1 d\tau \left(D(\mathbf{r}, \mathbf{r}; \tau) - D(\mathbf{r}, \mathbf{r})\right). \tag{2.208}$$

Finally, let us examine the case of the spatially-homogeneous fluid in the thermodynamic limit ($V \to \infty$). Supposing $u(\mathbf{r}) = 0$, we obtain

$$\tilde{D}(\mathbf{k}; \tau) = \frac{\tilde{U}(\mathbf{k})}{1 + \beta n \tau \tilde{U}(\mathbf{k})}, \tag{2.209}$$

$$\frac{1}{2}(nUn) = \frac{V}{2} n^2 \tilde{U}(0). \tag{2.210}$$

Also, considering that for the case of a spatially-homogeneous fluid we have $\Omega = -PV$, we arrive at the expression (2.153) for pressure, which was previously derived using the generating functional method. For convenience, we repeat it here

$$P = nk_B T + \frac{n^2 \tilde{U}(0)}{2} + \frac{k_B T}{2} \int \frac{d\mathbf{k}}{(2\pi)^3} \left(\frac{\beta n \tilde{U}(\mathbf{k})}{1 + \beta n \tilde{U}(\mathbf{k})} - \ln(1 + \beta n \tilde{U}(\mathbf{k})) \right). \tag{2.211}$$

Considering that the one-particle potential for the homogeneous fluid equals the excess chemical potential, we can use the expression (2.71) to derive the expression for the Helmholtz free energy density of the fluid

$$f = k_B T n(\ln(n\Lambda^3) - 1) + \frac{n^2 \tilde{U}(0)}{2} + \frac{k_B T}{2} \int \frac{d\mathbf{k}}{(2\pi)^3} \left(\ln(1 + \beta n \tilde{U}(\mathbf{k})) - \beta n \tilde{U}(\mathbf{k}) \right), \tag{2.212}$$

which is consistent with the expression previously obtained within the GA.

2.5 Gaussian Equivalent Representation Method. Corrections to the Gaussian Approximation

The approaches outlined above do not allow us to systematically evaluate corrections to the Gaussian Approximation. However, the method of the Gaussian Equivalent Representation (GER), introduced in the works of G. Efimov and coauthors (see, for example [8]), enables us to address such a task. In this section, we examine the GER relative to the evaluation of the thermodynamic properties of simple fluids with pairwise repulsive potentials. For simplicity, we discuss only the case of the spatially-homogeneous fluid in the thermodynamic limit. For pedagogical reasons, we work in the canonical ensemble, although similar calculations can be realized within the grand canonical ensemble, which are left as an exercise for the reader.

Configuration integral of the fluid with a pairwise intermolecular potential $U(\mathbf{r})$, having a positive Fourier transform, after the Hubbard–Stratonovich transformation has the form

$$Q_N = \frac{1}{V^N} \int \frac{\mathcal{D}\Phi}{\mathcal{N}_U} e^{-\frac{\beta}{2}(\Phi U^{-1} \Phi)} \left[\int d\mathbf{r}\, e^{\frac{\beta}{2}U(0)} e^{i\beta\Phi(\mathbf{r})} \right]^N . \tag{2.213}$$

Applying an identity

$$X^N = \frac{N!}{2\pi i} \oint \frac{dz}{z^{N+1}} e^{Xz}, \tag{2.214}$$

where closed contour integration is performed in the complex plane of the variables z around the point $z = 0$, we get

$$Q_N = \frac{N!}{2\pi i V^N} \oint \frac{dz}{z^{N+1}} I[\Phi; z], \tag{2.215}$$

where

$$I[\Phi; z] = \int \frac{\mathcal{D}\Phi}{\mathcal{N}_U} e^{-\frac{\beta}{2}(\Phi U^{-1} \Phi) + W[\Phi; z]} \tag{2.216}$$

and the auxiliary functional

$$W[\Phi; z] = z \int d\mathbf{r} : e^{i\beta\Phi(\mathbf{r})} :_U . \tag{2.217}$$

We also introduce the so-called normal form of the functional (see Chap. 1)

$$: e^{i\beta\Phi(\mathbf{r})} :_U = e^{\frac{\beta}{2}U(0)} e^{i\beta\Phi(\mathbf{r})}, \tag{2.218}$$

so

$$\int \frac{\mathcal{D}\Phi}{\mathcal{N}_U} e^{-\frac{\beta}{2}(\Phi U^{-1}\Phi)} : e^{i\beta\Phi(\mathbf{r})} :_U = 1. \tag{2.219}$$

Specifically, the following equalities are justified (see Chap. 1)

$$: \Phi(\mathbf{r}) :_U = \Phi(\mathbf{r}), \quad : \Phi(\mathbf{r})\Phi(\mathbf{r}') :_U = \Phi(\mathbf{r})\Phi(\mathbf{r}') - \frac{1}{\beta}U(\mathbf{r}-\mathbf{r}'). \tag{2.220}$$

The Gaussian equivalent form of the functional $I[\Phi; z]$ has the form

$$\begin{aligned} I[\Phi; z] = \frac{\mathcal{N}_D}{\mathcal{N}_U} \int \frac{\mathcal{D}\Phi}{\mathcal{N}_D} e^{-\frac{\beta}{2}(\Phi D^{-1}\Phi)} e^{-\frac{\beta}{2}(\Phi[U^{-1}-D^{-1}]\Phi) - \beta(\Phi_0 U^{-1}\Phi) - \frac{\beta}{2}(\Phi_0 U^{-1}\Phi_0) + z\int d\mathbf{r} e^{i\beta(\Phi_0+\Phi)} e^{\frac{\beta U(0)}{2}}} \\ = e^{W_0(z)} \int \frac{\mathcal{D}\Phi}{\mathcal{N}_D} e^{-\frac{\beta}{2}(\Phi D^{-1}\Phi) + W_{pert}[\Phi; z]}, \end{aligned} \tag{2.221}$$

where

$$W_0(z) = \frac{1}{2}\ln\frac{\mathcal{N}_D}{\mathcal{N}_U} - \frac{1}{2}tr\left(D[U^{-1} - D^{-1}]\right) - \frac{\beta}{2}(\Phi_0 U^{-1}\Phi_0) + V z e^{i\beta\Phi_0 + \frac{\beta}{2}(U(0)-D(0))}, \tag{2.222}$$

and the perturbation functional has the form

$$W_{pert}[\Phi; z] = -\frac{\beta}{2} : (\Phi[U^{-1} - D^{-1}]\Phi) :_D - \beta(\Phi U^{-1}\Phi_0) + z \int d\mathbf{r} e^{i\beta\Phi_0 + \frac{\beta}{2}[U(0)-D(0)]}(: e^{i\beta\Phi} :_D - 1), \tag{2.223}$$

where we take into account that in the thermodynamic limit, $V \to \infty$,

$$\Phi_0(\mathbf{r}) = \Phi_0 = const, \; D(\mathbf{r}, \mathbf{r}') = D(\mathbf{r} - \mathbf{r}'). \tag{2.224}$$

In accordance with the Gaussian equivalent representation method, Φ_0 and D are determined based on the conditions that there are no linear and quadratic terms of Φ in the perturbation functional, i.e.

$$-(\Phi U^{-1}\Phi_0) + i z e^{i\beta\Phi_0 + \frac{\beta}{2}(U(0)-D(0))} \int d\mathbf{r}\, \Phi(\mathbf{r}) = 0, \tag{2.225}$$

$$-\frac{1}{2} : (\Phi[U^{-1} - D^{-1}]\Phi) :_D - \frac{z\beta}{2} e^{i\beta\Phi_0 + \frac{\beta}{2}(U(0)-D(0))} \int d\mathbf{r} : \Phi^2(\mathbf{r}) :_D = 0, \tag{2.226}$$

whence, due to the arbitrariness of the fluctuating field $\Phi(\mathbf{r})$, we get

$$\Phi_0 = \frac{ic}{\beta}, \quad c = z\beta\tilde{U}(0)e^{-c+\frac{\beta}{2}(U(0)-D(0))}, \tag{2.227}$$

$$\frac{1}{\tilde{D}(\mathbf{k})} - \frac{1}{\tilde{U}(\mathbf{k})} = \frac{c}{\tilde{U}(0)}. \tag{2.228}$$

The last equation gives us

$$\tilde{D}(\mathbf{k}) = \frac{\tilde{U}(\mathbf{k})}{1 + cu(\mathbf{k})}, \quad u(\mathbf{k}) = \frac{\tilde{U}(\mathbf{k})}{\tilde{U}(0)}. \tag{2.229}$$

Equation (2.227), which relates z and c can be rewritten in the following form

$$\ln z = \ln\left(\frac{c}{\beta\tilde{U}(0)}\right) + c - \frac{\beta}{2}[U(0) - D(0)]. \tag{2.230}$$

We will use this equation further to determine an implicit complex function $c = c(z)$. Thus, the perturbation functional can be rewritten as

$$W_{pert}[\Phi; z] = \frac{c}{\beta\tilde{U}(0)} \int d\mathbf{r} : e_2^{i\beta\Phi} :_D, \tag{2.231}$$

where we introduce the following notation

$$e_2^x = e^x - 1 - x - x^2/2. \tag{2.232}$$

Now, taking into account that

$$(\Phi_0 U^{-1} \Phi_0) = \frac{Vc^2}{\beta\tilde{U}(0)}, \tag{2.233}$$

$$\frac{1}{2}\ln\frac{\mathcal{N}_D}{\mathcal{N}_U} = \frac{1}{2}(tr\ln D - tr\ln U) = \frac{V}{2}\int\frac{d\mathbf{k}}{(2\pi)^3}\ln\frac{\tilde{D}(\mathbf{k})}{\tilde{U}(\mathbf{k})}, \tag{2.234}$$

and

$$\frac{1}{2}tr\left(D[U^{-1} - D^{-1}]\right) = \frac{V}{2}\int\frac{d\mathbf{k}}{(2\pi)^3}\left(\frac{\tilde{D}(\mathbf{k})}{\tilde{U}(\mathbf{k})} - 1\right), \tag{2.235}$$

and considering written above formulas, we get

$$W_0 = \frac{V}{2}\int\frac{d\mathbf{k}}{(2\pi)^3}\left(\frac{cu(\mathbf{k})}{1 + cu(\mathbf{k})} - \ln(1 + cu(\mathbf{k}))\right) + \frac{V(c^2 + 2c)}{2\beta\tilde{U}(0)}. \tag{2.236}$$

Then, knowing that

$$U(0) - D(0) = \int \frac{d\mathbf{k}}{(2\pi)^3} \left(\tilde{U}(\mathbf{k}) - \tilde{D}(\mathbf{k}) \right), \tag{2.237}$$

and remembering expression (2.229), we arrive at

$$\ln z = \ln \frac{c}{\beta \tilde{U}(0)} + c - \frac{\beta \tilde{U}(0)}{2} \int \frac{d\mathbf{k}}{(2\pi)^3} \frac{cu^2(\mathbf{k})}{1 + cu(\mathbf{k})}. \tag{2.238}$$

Thus, configuration integral has the form

$$Q_N = \frac{1}{2\pi i} \oint \frac{dz}{z} e^{V\mathcal{F}(c(z))} \int \frac{\mathcal{D}\Phi}{\mathcal{N}_D} e^{-\frac{\beta}{2}(\Phi D^{-1}\Phi) + W_{pert}}, \tag{2.239}$$

where we introduce the auxiliary function

$$\mathcal{F}(c) = \frac{c^2 + 2c}{2\beta \tilde{U}(0)} + n \left(\ln \frac{\beta n \tilde{U}(0)}{c} - c - 1 \right)$$
$$+ \frac{1}{2} \int \frac{d\mathbf{k}}{(2\pi)^3} \left(\frac{cu(\mathbf{k})}{1 + cu(\mathbf{k})} - \ln(1 + cu(\mathbf{k})) + \frac{n\beta \tilde{U}(0) cu^2(\mathbf{k})}{1 + cu(\mathbf{k})} \right) \tag{2.240}$$

and apply Stirling's formula

$$\frac{N!}{V^N} \simeq e^{V(n \ln n - n)}, \tag{2.241}$$

where $n = N/V$ is an average number density. At zeroth approximation of the Gaussian equivalent representation the configuration integral is written in the form

$$Q_N^{(0)} = \frac{1}{2\pi i} \oint \frac{dz}{z} e^{V\mathcal{F}(c(z))}. \tag{2.242}$$

Considering that

$$\frac{dz}{z} = \frac{dc}{c} \left(c + 1 - \frac{\beta \tilde{U}(0)}{2} \int \frac{d\mathbf{k}}{(2\pi)^3} \frac{cu^2(\mathbf{k})}{(1 + cu(\mathbf{k}))^2} \right), \tag{2.243}$$

in the thermodynamic limit we arrive at

$$Q_N^{(0)} = \frac{1}{2\pi i} \oint \frac{dz}{z} e^{V\mathcal{F}(c(z))} \simeq \frac{1}{2\pi i} \oint \frac{dc}{c} e^{V\mathcal{F}(c)} \simeq e^{V\mathcal{F}(c_s)}, \tag{2.244}$$

where the saddle point c_s can be determined with the help of equation

$$\frac{d\mathcal{F}}{dc} = 0, \tag{2.245}$$

whence we obtain $c_s = \beta n \tilde{U}(0)$. Thus, we arrive at an already known expression (2.151) for the excess free energy within the zeroth approximation of the Gaussian equivalent representation

$$F_{ex}^{(0)} = \frac{Vn^2\tilde{U}(0)}{2} + \frac{Vk_BT}{2}\int \frac{d\mathbf{k}}{(2\pi)^3}\left(\ln(1+\beta n\tilde{U}(\mathbf{k})) - \beta n\tilde{U}(\mathbf{k})\right). \tag{2.246}$$

Let us evaluate next non-vanishing correction to the zeroth approximation within the cumulant expansion. Considering that, $\langle W_{pert}\rangle_D = 0$, where

$$W_{pert} = n\int d\mathbf{r} : e_2^{i\beta\Phi} :_D, \tag{2.247}$$

after the calculation of the Gaussian integrals with the help of the Wick's theorem, we get

$$F_{ex}^{(2)} = -\frac{k_BT}{2}\langle W_{pert}^2\rangle_D = -\frac{Vn^2k_BT}{2}\int d\mathbf{r} e_2^{-\beta D(\mathbf{r})}, \tag{2.248}$$

where

$$D(\mathbf{r}) = \int \frac{d\mathbf{k}}{(2\pi)^3} e^{i\mathbf{k}\mathbf{r}} \frac{\tilde{U}(\mathbf{k})}{1+\beta n\tilde{U}(\mathbf{k})} \tag{2.249}$$

and as in the previous section, we introduce the following notations for averaging

$$\langle(\ldots)\rangle_D = \int \frac{\mathcal{D}\Phi}{\mathcal{N}_D} e^{-\frac{\beta}{2}(\Phi D^{-1}\Phi)}(\ldots). \tag{2.250}$$

The condition for the applicability of the Gaussian approximation can thus be formulated as

$$|F_{ex}^{(2)}/F_{ex}^{(0)}| \ll 1. \tag{2.251}$$

In conclusion, let's examine the question regarding the pair correlation function of the fluid, which can also be represented in the form of a functional integral as

$$\frac{g_2(\mathbf{r}_1,\mathbf{r}_2)}{V^2} = \frac{1}{Q_N}\int \frac{\mathcal{D}\Phi}{\mathcal{N}_U}\exp\left[-\frac{\beta}{2}(\Phi U^{-1}\Phi)\right] : e^{i\beta\Phi(\mathbf{r}_1)} :_U : e^{i\beta\Phi(\mathbf{r}_2)} :_U \left[\int d\mathbf{r} : e^{i\beta\Phi(\mathbf{r})} :_U\right]^{N-2}. \tag{2.252}$$

Carrying out similar evaluations, we obtain within the zeroth approximation of the GER [8]

$$g_2(\mathbf{r}_1,\mathbf{r}_2) \simeq \exp[-\beta D(\mathbf{r}_1 - \mathbf{r}_2)]. \tag{2.253}$$

Thus, the function $D(\mathbf{r})$ can be interpreted as a potential of the mean force within the zeroth approximation of GER.

2.6 Self-consistent Field Theories of the Fluids with the Intermolecular Attraction

2.6.1 Ivanchenko–Lisyansky Theory

Previously, we explored model fluids with repulsive potentials. Now, let's investigate fluids whose molecules interact via a potential that comprises both repulsive $U_R(\mathbf{r})$ and attractive $-U_A(\mathbf{r})$ $(U_A(\mathbf{r}) > 0)$ contributions, i.e.

$$U(\mathbf{r}) = U_R(\mathbf{r}) - U_A(\mathbf{r}). \tag{2.254}$$

Let us assume that the potential $U_R(\mathbf{r})$ is a short-range one, while the potential $U_A(\mathbf{r})$ can have a much larger range of action.

Method, discussed below, was proposed by Ivanchenko and Lisyansky [40].

For further constructions, it will be more convenient to work in the framework of the μVT-ensemble, so that the grand partition function of the system has the form

$$\Xi = \sum_{N=0}^{\infty} \frac{z^N}{N!} Q_N, \tag{2.255}$$

where the configuration integral of the system of N particles has the following form

$$Q_N = \int \ldots \int \prod_{k=1}^{N} d\mathbf{r}_k \, e^{-\frac{\beta}{2}\sum\limits_{i\neq j} U_R(\mathbf{r}_i-\mathbf{r}_j)+\frac{\beta}{2}\sum\limits_{i\neq j} U_A(\mathbf{r}_i-\mathbf{r}_j)-\beta\sum\limits_{s=1}^{N} u(\mathbf{r}_s)}, \tag{2.256}$$

where $\varphi(\mathbf{r})$ denotes the potential energy of the external field.

Utilizing the HS transformation for the exponents in the intergrand of (2.256)

$$e^{-\frac{\beta}{2}(\hat{n}U_R\hat{n})} = \int \frac{\mathcal{D}\Phi}{\mathcal{N}_R} e^{-\frac{\beta}{2}(\Phi U_R^{-1}\Phi)+i\beta(\hat{n}\Phi)}, \tag{2.257}$$

$$e^{\frac{\beta}{2}(\hat{n}U_A\hat{n})} = \int \frac{\mathcal{D}\Psi}{\mathcal{N}_A} e^{-\frac{\beta}{2}(\Psi U_A^{-1}\Psi)+\beta(\hat{n}\Psi)} \tag{2.258}$$

and substituting them in the grand partition function, we can obtain the representation of the grand partition function in the form of functional integral over two fluctuating scalar fields

$$\Xi = \int \frac{\mathcal{D}\Phi}{\mathcal{N}_R} \int \frac{\mathcal{D}\Psi}{\mathcal{N}_A} e^{-\frac{\beta}{2}(\Phi U_R^{-1}\Phi) - \frac{\beta}{2}(\Psi U_A^{-1}\Psi)} \exp\left[\bar{z} \int d\mathbf{r}\, e^{i\beta\Phi(\mathbf{r}) + \beta\Psi(\mathbf{r}) - \beta u(\mathbf{r})}\right], \tag{2.259}$$

where $\bar{z} = ze^{\frac{\beta U(0)}{2}}$ and $\mathcal{N}_{A,R}$ are the normalization constants of the Gaussian measure. The last expression can be rewritten in the following form

$$\Xi = \int \frac{\mathcal{D}\Psi}{\mathcal{N}_A} \exp\left[-\frac{\beta}{2}(\Psi U_A^{-1}\Psi)\right] \Xi_R[\Psi - u], \tag{2.260}$$

where

$$\Xi_R[\Psi - u] = \int \frac{\mathcal{D}\Phi}{\mathcal{N}_R} \exp\left[-\frac{\beta}{2}(\Phi U_R^{-1}\Phi) + \bar{z} \int d\mathbf{r}\, e^{i\beta(\Phi(\mathbf{r}) + \Psi(\mathbf{r}) - u(\mathbf{r}))}\right] \tag{2.261}$$

is a grand partition function of the (reference) system with the repulsive potential $U_R(\mathbf{r})$, which is a functional of difference between the fluctuating scalar field $\Psi(\mathbf{r})$ and the potential energy of the external field $u(\mathbf{r})$. Applying a cumulant expansion, we get

$$\Xi_R[\Psi - u] = \exp\left[\sum_{k=1}^{\infty} \frac{\bar{z}^k}{k!} C_k[\Psi - u]\right], \tag{2.262}$$

where $C_k[\Psi - u]$ are the cumulants, evaluated from the averaging over the Gaussian measure corresponding to the repulsive part of the potential, which can be determined by the relation

$$C_k[\Psi - u] = e^{-\frac{\beta k U_R(0)}{2}} \int \ldots \int d\mathbf{r}_1 \ldots d\mathbf{r}_k B_k(\mathbf{r}_1, \ldots, \mathbf{r}_k) e^{\beta \sum_{s=1}^{k} (\Psi(\mathbf{r}_s) - u(\mathbf{r}_s))}. \tag{2.263}$$

The functions $B_k(\mathbf{r}_1, \ldots, \mathbf{r}_k)$ are determined as

$$B_1(\mathbf{r}_1) = 1, \;\; B_2(\mathbf{r}_1, \mathbf{r}_2) = e^{-\beta U_R(\mathbf{r}_1 - \mathbf{r}_2)} - 1,$$

$$B_3(\mathbf{r}_1, \ldots, \mathbf{r}_3) = e^{-\beta(U_R(\mathbf{r}_1 - \mathbf{r}_2) + U_R(\mathbf{r}_1 - \mathbf{r}_3) + U_R(\mathbf{r}_2 - \mathbf{r}_3))} - e^{-\beta U_R(\mathbf{r}_1 - \mathbf{r}_2)}$$

$$- e^{-\beta U_R(\mathbf{r}_1 - \mathbf{r}_3)} - e^{-\beta U_R(\mathbf{r}_2 - \mathbf{r}_3)} + 2, \ldots \tag{2.264}$$

It is evident that these functions are organized in such a way that they are nonzero when the values of the arguments are sufficiently close[8]. This allows us to make the following approximation for the cumulants

$$C_k[\Psi - u] \approx e^{-\frac{\beta k U_R(0)}{2}} \int \ldots \int d\mathbf{r}_1 \ldots d\mathbf{r}_k \, B_k(\mathbf{r}_1, \ldots, \mathbf{r}_k) e^{\beta k(\Psi(\mathbf{r}_1) - u(\mathbf{r}_1))}, \tag{2.265}$$

[8] Due to sufficiently small range of action of the potential $U_R(\mathbf{r})$.

which leads to the following approximation for Ξ_R

$$\Xi_R[\Psi - u] = \exp\left[\sum_{k=1}^{\infty}\left(ze^{-\frac{\beta U_A(0)}{2}}\right)^k b_k \int d\mathbf{r}\, e^{\beta k(\Psi(\mathbf{r})-u(\mathbf{r}))}\right], \tag{2.266}$$

where

$$b_k = \frac{1}{k!}\int \ldots \int d\mathbf{r}_2 \ldots d\mathbf{r}_k\, B_k(\mathbf{r}_1, \ldots, \mathbf{r}_k) \tag{2.267}$$

are the group integrals of the reference system. Thus, applying (2.28), for the exponent in (2.266) we have

$$\sum_{k=1}^{\infty}\left(ze^{-\frac{\beta U_A(0)}{2}}\right)^k b_k \int d\mathbf{r}\, e^{\beta k(\Psi(\mathbf{r})-u(\mathbf{r}))}$$
$$= \int d\mathbf{r} \sum_{k=1}^{\infty} b_k \left(\Lambda^{-3} e^{-\frac{\beta U_A(0)}{2}+\beta(\mu+\Psi-u)}\right)^k = \int d\mathbf{r}\, \beta P_R(\bar{\mu}(\mathbf{r})), \tag{2.268}$$

i.e. within this approximation, we simply obtain a system volume integral of the particles pressure, which depends on the chemical potential, shifted by the value $\Psi - u$

$$\bar{\mu}(\mathbf{r}) = \mu + \Psi(\mathbf{r}) - u(\mathbf{r}). \tag{2.269}$$

The index R denotes that the pressure dependence on the chemical potential corresponds to the reference system with the repulsive interaction potential. Thus, we arrive at the following approximation for the grand partition function.

$$\Xi = \int \frac{\mathcal{D}\Psi}{\mathcal{N}_A} \exp\left[-\beta\Omega[\Psi]\right], \tag{2.270}$$

where

$$\Omega[\Psi] = \frac{1}{2}\left(\Psi U_A^{-1} \Psi\right) - \int d\mathbf{r}\, P_R(\bar{\mu}) \tag{2.271}$$

is an effective thermodynamic potential as a functional of the scalar field $\Psi(\mathbf{r})$. Let's evaluate the grand partition function within the mean field approximation. The Euler-Lagrange equation takes the form

$$\frac{\delta\Omega[\Psi]}{\delta\Psi(\mathbf{r})} = \int d\mathbf{r}'\, U_A^{-1}(\mathbf{r}-\mathbf{r}')\Psi(\mathbf{r}') - n_R(\bar{\mu}) = 0, \tag{2.272}$$

where $n_R = \partial P_R/\partial\mu$ is the concentration of particles of the reference system. The last equation $n_R = \partial P_R/\partial\bar{\mu}$—concentration of the reference system. Utilizing the inverse operator U_A^{-1}, we can transform the last equation to

$$\Psi(\mathbf{r}) = \int d\mathbf{r}'\, U_A(\mathbf{r}-\mathbf{r}')n_R(\mu + \Psi(\mathbf{r}) - u(\mathbf{r})). \tag{2.273}$$

This equation provides a clear physical interpretation of the scalar field Ψ as the potential energy of the (molecular) mean field generated by the particles of the fluid at the point $\mathbf{r}$. Equation (2.273) represents a nonlinear integral equation for the potential energy of the self-consistent molecular field. In the case of short-range potentials, where we can assume that $U_A(\mathbf{r}-\mathbf{r}') = A\delta(\mathbf{r}-\mathbf{r}')$, with A being a positive constant, we obtain the following equation

$$\Psi(\mathbf{r}) = An_R(\mu + \Psi(\mathbf{r}) - u(\mathbf{r})). \tag{2.274}$$

In this case, instead of solving the integral Eq. (2.273), we need to solve the transcendental Eq. (2.274) for $\Psi(\mathbf{r})$ at each point $\mathbf{r}$. Then, utilizing Eq. (2.273), we derive the following expression for the grand thermodynamic potential within the mean field approximation:

$$\Omega = \frac{1}{2}(n_R\Psi) - \int d\mathbf{r}\, P_R(\mu + \Psi(\mathbf{r}) - u(\mathbf{r})), \tag{2.275}$$

where $\Psi(\mathbf{r})$ satisfies the Eq. (2.273).[9]

Bulk fluid. In the absence of external fields ($u(\mathbf{r}) = 0$), which cause the system's inhomogeneity, Eq. (2.273) takes the following form.[10]

$$\Psi(\mathbf{r}) = \int d\mathbf{r}'\, U_A(\mathbf{r}-\mathbf{r}')n_R(\mu + \Psi(\mathbf{r}')). \tag{2.276}$$

Let's seek the solution in the form of a constant $\Psi(\mathbf{r}) = \Psi_0$, which corresponds to the spatially-homogeneous phase. In this case, we have a transcendental equation for Ψ_0:

$$\Psi_0 = \tilde{U}_A(0)n_R(\mu + \Psi_0). \tag{2.277}$$

The grand thermodynamic potential of the bulk phase in this case is

$$\Omega = V\left(\frac{1}{2\tilde{U}_A(0)}\Psi_0^2 - P_R(\mu + \Psi_0)\right). \tag{2.278}$$

[9] This approximation is also known as the "local" approximation.

[10] It should be noted that the system can exhibit spatial inhomogeneity even without the external fields, particularly in the region of thermodynamic instability near phase interfaces. The theory outlined here can also describe such cases. Mean field description of such interface and calculation of surface tension can be found in [41].

So the pressure $P = -\Omega/V$ is determined as

$$P = P_R(\mu + \Psi_0) - \frac{\Psi_0^2}{2\tilde{U}_A(0)} \tag{2.279}$$

The particle number density is defined by

$$n(\mu) = \frac{\partial P}{\partial \mu} = n_R(\mu + \Psi_0). \tag{2.280}$$

Equations (2.279), (2.280), and (2.273) provide the equation of state of a spatially-homogeneous fluid in a parametric form within the mean field approximation. It is worth noting that within this approximation, the particle number density is determined by the number density of the reference system as a function of the chemical potential, shifted by Ψ_0. Thus, using the Eqs. (2.279), (2.280), and (2.273), we can express the equation of state for a fluid with intermolecular attraction in a form consistent with what we obtained earlier within the classical density functional theory

$$P(T,n) = P_0(T,n) - \frac{1}{2}An^2, \;\; A = \tilde{U}_A(0), \tag{2.281}$$

which is readily interpretable. Indeed, the first term $P_0(T,n) = P_R(\mu(T,n) + \Psi_0(T,n))$ defines the pressure of the reference system with a repulsive potential, while the second one determines the contribution of the intermolecular attraction within the mean field approximation. Note that the mean field approximation takes into account the effect of the intermolecular attraction at the level of interactions of isolated pairs. In other words, accounting for the attraction gives the contribution only to the second virial coefficient of the system.

If we use a potential of the hard spheres of diameter d as a repulsive potential:

$$U_R(r) = \begin{cases} \infty, & r \leq d \\ 0, & r > d, \end{cases} \tag{2.282}$$

then the pressure of the reference system can be calculated within the Percus–Yevick approximation [19]

$$P_0 = nk_BT\frac{1+\eta+\eta^2}{(1-\eta)^3} \tag{2.283}$$

or within the Carnahan–Starling approximation [19]

$$P_0 = nk_BT\frac{1+\eta+\eta^2-\eta^3}{(1-\eta)^3}, \tag{2.284}$$

where $\eta = \pi d^3 n/6$ is the packing fraction of the hard spheres. The best-known, albeit rather crude, approximation for the pressure of the hard sphere system is the van der Waals approximation

$$P_0 = \frac{nk_BT}{1 - nv}, \tag{2.285}$$

so the Van der Waals equation of state takes the form

$$P(T, n) = \frac{nk_BT}{1 - nv} - An^2/2, \tag{2.286}$$

where $v = 2\pi d^3/3$ is the "van der Waals" volume of the molecule. Note that for fluids with soft repulsive Gaussian-like potentials, we can utilize the expression (2.153) to describe the contribution of intermolecular repulsion.

Symmetric lattice gas. Another commonly used reference system is an ideal lattice gas [13]. First, let us consider a model of the one-component lattice gas without attractive interactions of particles (ideal lattice gas [13]). We examine a system of non-interacting particles located at the nodes of a cubic lattice with a total number of nodes M. Let us suppose that the particles are in equilibrium with a reservoir of infinite size, so the values of temperature T and chemical potential μ remain constant. In this case, the grand partition function of the system is given by

$$\Xi = \sum_{N=0}^{M} e^{\beta\mu N} Z_N, \tag{2.287}$$

where

$$Z_N = \frac{M!}{N!(M - N)!} z^N \tag{2.288}$$

is a canonical partition function of N lattice gas particles with a one-particle partition function z, related to the internal degrees of freedom of the lattice gas particles; $\beta = (k_BT)^{-1}$. Thus, substituting Eqs. 2.288 in 2.287 and summing over all nodes, we get

$$\Xi = \left(1 + ze^{\beta\mu}\right)^M. \tag{2.289}$$

Utilizing the relationship between the grand canonical partition function and the pressure, $PV = k_BT \ln \Xi$, where V represents the total volume of the system, we derive the following expression

$$P = \frac{k_BT}{v} \ln\left(1 + ze^{\beta\mu}\right), \tag{2.290}$$

where $v = V/M$ denotes the volume per lattice gas particle. Without loss of generality, we can set $z = 1$, incorporating the effects of internal degrees of freedom into the chemical potential definition.

Now, let's extend our calculation to determine the pressure of the ideal multi-component lattice gas. Suppose there are l different species present on the same lattice at temperature T and chemical potentials μ_i, where $i = 1, 2, \ldots, l$. As a result, the grand canonical partition function of the system is given by

$$\Xi = \sum_{N=0}^{M} \sum_{N_1=0}^{M} \cdots \sum_{N_l=0}^{M} e^{\beta \sum_{i=1}^{l} \mu_i N_i} \frac{M!}{(M-N)! \prod_{i=1}^{l} N_i!} \Delta\left(\sum_{i=1}^{l} N_i - N \right), \tag{2.291}$$

where N_i represents the number of particles of the ith component of the lattice gas, and the function

$$\Delta(k-m) = \delta_{km}, \tag{2.292}$$

with δ_{km} denoting the Kronecker delta, ensures the condition $N = \sum_{i=1}^{l} N_i$. Then, applying the identity

$$\Delta(N) = \frac{1}{2\pi i} \oint_C \frac{d\xi}{\xi^{N+1}}, \tag{2.293}$$

where the closed contour integration is along the path C in complex plane around the point $\xi = 0$, we can rewrite the grand canonical partition function as

$$\Xi = \frac{1}{2\pi i} \sum_{N=0}^{M} \frac{M!}{(M-N)!} \oint_C d\xi\, \xi^{N-1} \prod_{i=1}^{l} \sum_{N_i=0}^{M} \frac{1}{N_i!} \left(\frac{e^{\beta\mu_i}}{\xi} \right)^{N_i}. \tag{2.294}$$

In thermodynamic limit, i.e. at $M \gg 1$, we get

$$\Xi \simeq \frac{1}{2\pi i} \sum_{N=0}^{M} \frac{M!}{(M-N)!} \oint_C d\xi\, \xi^{N-1} e^{\frac{\theta}{\xi}}, \tag{2.295}$$

where $\theta = \sum_{i=1}^{l} e^{\beta\mu_i}$. Then,evaluating the contour integral in (2.295) utilizing the theory of residues, we arrive at

$$\Xi = (1+\theta)^M, \tag{2.296}$$

which, in turn, leads to the following expression for the pressure

$$P = \frac{k_B T}{v} \ln\left(1 + \sum_{i=1}^{l} e^{\beta\mu_i} \right). \tag{2.297}$$

Thus, the dependence of the ideal lattice gas pressure on the chemical potential has the from

$$P_R(\mu) = \frac{k_B T}{v} \ln\left(1 + e^{\beta\mu}\right), \tag{2.298}$$

where v denotes the volume of the lattice gas cell. The number density is thus determined by the expression

$$n_R(\mu) = \frac{1}{v} \frac{e^{\beta\mu}}{1 + e^{\beta\mu}}. \tag{2.299}$$

The equation of state of the ideal lattice gas takes the form

$$P_0(T, n) = -\frac{k_B T}{v} \ln(1 - nv). \tag{2.300}$$

The self-consistent field equation for the lattice gas with the intermolecular attraction is expressed as

$$\Psi(\mathbf{r}) = \int \frac{d\mathbf{r}'}{v} U_A(\mathbf{r} - \mathbf{r}') \frac{e^{\beta(\mu+\Psi(\mathbf{r}')-u(\mathbf{r}'))}}{1 + e^{\beta(\mu+\Psi(\mathbf{r}')-u(\mathbf{r}'))}}. \tag{2.301}$$

Note that presented theory can be used to describe the properties of homogeneous fluids as well as fluids in confinement and under external potential fields. In Chap. 3, we will extend the discussed self-consistent field theory to describe the ionic fluids with short-range specific interactions between the ions.

2.6.2 Hubbard–Schofield–Brilliantov Theory

Earlier, when constructing field-theoretical models of fluids with pairwise potentials containing both attraction and repulsion contributions, we employed the HS transformation. However, this method has a fundamental weakness: it requires the interaction potential of the system to have an inverse operator. Without this, uncertainty arises during such transformation. Consequently, the HS transformation, strictly speaking, cannot be applied to the systems with interaction potentials exhibiting singular behavior (e.g., hard core potential), i.e. those lacking an inverse operator. We addressed this challenge by deriving the final expressions, which do not involve the inverse operators, even though we operated with them in intermediate steps without considering their existence.

As opposed to the approach based on the HS transformation, Hubbard and Schofield introduced a consistent method for constructing the effective field Hamiltonian for simple fluids, which relies on an integral transformation similar to the Hubbard–Stratonovich one. However, Hubbard and Schofield only outlined a general framework for constructing the field-theoretical model of the fluid. In subsequent works [41–44], Brilliantov and coauthors further developed the Hubbard–Schofield approach and demonstrated that, in principle, all coefficients of the effective Hamiltonian for the one-component fluid can be expressed in terms of known quantities from the reference model of the fluid—specifically, the isothermal compressibility and its derivatives with respect to density. Let us discuss the details of this method. From a pedagogical perspective, we will be using the same techniques as in the references mentioned above.

Let's consider the Hamiltonian describing the interactions within the fluid:

$$U = \sum_{i<j} \phi(r_{ij}) - \sum_{i<j} v(r_{ij}) = U_R - U_A, \tag{2.302}$$

where $\phi(r)$ represents the repulsive part of the interaction potential, $-v(r)$ is the part responsible for particle attraction; $r_{ij} = |\mathbf{r}_i - \mathbf{r}_j|$, where $\mathbf{r}_i$ denotes the coordinates of the centers of mass of the fluid particles. It's important to note that the approach developed here is applicable to any interaction potential that can be represented as a sum of the repulsive and attractive parts. By utilizing the Fourier components of density fluctuations,

$$\hat{n}_{\mathbf{k}} = \frac{1}{\sqrt{V}} \sum_{j=1}^{N} e^{-i\mathbf{k}\mathbf{r}_j}, \tag{2.303}$$

and Fourier transform of the attraction potential,

$$v_{\mathbf{k}} = \int d\mathbf{r} e^{-i\mathbf{k}\mathbf{r}} v(r), \tag{2.304}$$

the attractive part of the Hamiltonian can be written as

$$U_A = -\frac{1}{2} \sum_{\mathbf{k}} v_{\mathbf{k}} \hat{n}_{\mathbf{k}} \hat{n}_{-\mathbf{k}} - U_{self}, \tag{2.305}$$

where $U_{self} = -Nv(0)/2$; here N is the number of particles and $V = L^3$ is the volume of the system. The summation in (2.305) is over a discrete set of wave vectors $\mathbf{k} = (k_x, k_y, k_z)$ satisfying the condition $k_i = 2\pi n_i / L$, where $i = x, y, z$ and n_i is an integer. It is also implied that the system is considered within the thermodynamic limit $L \to \infty$. Let's introduce the chemical potential μ for the entire system with a total interaction Hamiltonian (2.302), and μ_R as the chemical potential of the reference system with a Hamiltonian H_R in which only repulsive interactions are present. Then, the grand partition function Ξ can be represented through the corresponding partition function of the reference system Ξ_R as follows

$$\Xi = \Xi_R \left\langle \exp\left[\beta\mu' + \frac{\beta}{2} \sum_{\mathbf{k}} v_{\mathbf{k}} \hat{n}_{\mathbf{k}} \hat{n}_{-\mathbf{k}} \right] \right\rangle_R, \tag{2.306}$$

where

$$\mu' = \mu - \mu_R + \frac{1}{2}v(0). \tag{2.307}$$

Let's emphasize that the reference system with a purely repulsive interaction potential must not exhibit critical properties (such as the system with the Yukawa repulsive potential). An example of such a system could be the hard-sphere system with identical diameters. Consequently, the grand canonical partition function of the reference system behaves regularly over a wide range of the fluid's state parameters. Then, using the identity

$$e^{\frac{1}{2}a^2x^2} = \frac{1}{\sqrt{2\pi a^2}} \int_{-\infty}^{+\infty} dy e^{-\frac{1}{2}y^2/a^2+xy}, \tag{2.308}$$

After some straightforward transformations, we arrive at the ratio of the partition functions $Q = \Xi/\Xi_R$:

$$Q = \int \prod_{\mathbf{k}} \frac{d\varphi_{\mathbf{k}}}{\sqrt{2\pi v_{\mathbf{k}}}} \exp\left[\frac{\mu'}{v_0}\sqrt{V}\varphi_0 - \frac{1}{2\beta}\sum_{\mathbf{k}} \frac{\varphi_{\mathbf{k}}\varphi_{-\mathbf{k}}}{v_{\mathbf{k}}}\right] \left\langle \exp\left[\sum_{\mathbf{k}} \varphi_{\mathbf{k}}\hat{n}_{-\mathbf{k}}\right]\right\rangle_R . \tag{2.309}$$

Integration in (2.309) is carried out under the assumption that $\varphi_{-\mathbf{k}} = \varphi_{\mathbf{k}}$, where $*$ denotes complex conjugation. Then, by applying the cumulant expansion to $\langle \exp\left[\sum_{\mathbf{k}} \varphi_{\mathbf{k}}\hat{n} - \mathbf{k}\right]\rangle R$, let's rearrange (2.309) in the following form:

$$Q = \int \prod_{\mathbf{k}} \frac{d\varphi_{\mathbf{k}}}{\sqrt{2\pi v_{\mathbf{k}}}} \exp\left[-\beta\mathcal{H}[\{\varphi_{\mathbf{k}}\}]\right], \tag{2.310}$$

where

$$\beta\mathcal{H} = -h\sqrt{V}\varphi_0 + \sum_{l=2}^{\infty} V^{1-l/2} \sum_{\mathbf{k}_1,\ldots,\mathbf{k}_l} u_l(\mathbf{k}_1,\ldots,\mathbf{k}_l)\varphi_{\mathbf{k}_1}\cdots\varphi_{\mathbf{k}_l}, \tag{2.311}$$

$$h = v_0^{-1}(\mu - \mu_R + v(0)/2) + n, \tag{2.312}$$

$$u_2(\mathbf{k}_1,\mathbf{k}_2) = \frac{1}{2}\delta_{\mathbf{k}_1+\mathbf{k}_2,0}\left((\beta v_{\mathbf{k}_1})^{-1} - \langle \hat{n}_{\mathbf{k}_1}\hat{n}_{-\mathbf{k}_1}\rangle_{cR}\right), \tag{2.313}$$

$$u_l(\mathbf{k}_1,\ldots,\mathbf{k}_l) = \frac{V^{l/2-1}}{l!}\langle \hat{n}_{\mathbf{k}_1}\cdots\hat{n}_{\mathbf{k}_l}\rangle_{cR}, \tag{2.314}$$

here, $\langle(\ldots)\rangle_{cR}$ denotes he cumulant average, calculated for the reference system.

The coefficients of the obtained effective Hamiltonian depend on the correlation functions of the reference system with purely repulsive interactions. Utilizing the definitions of the equilibrium correlation functions of the fluid and the cumulant averages, we can express the coefficients at the degrees of the field variable φ directly through the Fourier transforms of the correlation functions of the reference system, $h_1, h_2, \ldots, h_l$, determined as

$$h_1(\mathbf{r}) = \delta(\mathbf{r}), \tag{2.315}$$

$$h_2(\mathbf{r}_1, \mathbf{r}_2) = g_2(\mathbf{r}_1, \mathbf{r}_2) - 1, \tag{2.316}$$

$$h_3(\mathbf{r}_1, \mathbf{r}_2, \mathbf{r}_3) = g_3(\mathbf{r}_1, \mathbf{r}_2, \mathbf{r}_3) - g_2(\mathbf{r}_1, \mathbf{r}_2) - g_2(\mathbf{r}_1, \mathbf{r}_3) - g_2(\mathbf{r}_2, \mathbf{r}_3) + 2, \ldots \tag{2.317}$$

Specifically, the first few coefficients take the form:

$$u_2 = \delta_{1,2} \left[\frac{k_B T}{v_0} - n(1 + n\tilde{h}_2(\mathbf{k}_1)) \right], \tag{2.318}$$

$$u_3 = -\delta_{1,2,3} n \left[1 + n(1 + n(\tilde{h}_2(\mathbf{k}_1) + \tilde{h}_2(\mathbf{k}_2) + \tilde{h}_2(\mathbf{k}_3)) + n^2 \tilde{h}_3(\mathbf{k}_1, \mathbf{k}_2) \right], \tag{2.319}$$

where the Fourier transforms of the correlation functions $\tilde{h}$ are introduced

$$\tilde{h}_l(\mathbf{k}_1, \ldots, \mathbf{k}_{l-1}) = \int d\mathbf{r}_1 \ldots \int d\mathbf{r}_{l-1} \exp\left[i \sum_{s=1}^{l-1} \mathbf{k}_s \mathbf{r}_s \right] h_l(\mathbf{r}_1, \ldots, \mathbf{r}_l) \tag{2.320}$$

and the following short-hand notation is adopted

$$\delta_{1,2,\ldots,l} = \frac{\delta_{\mathbf{k}_1 + \mathbf{k}_2 + \cdots + \mathbf{k}_l, 0}}{l!}. \tag{2.321}$$

We also introduce the average density of the reference system $n = n_R(\mu_R)$, which is a function of the chemical potential μ_R. Note that the number of arguments of the correlation function in direct space is one more than of their Fourier transforms in reciprocal space. This is due to the fact that in the thermodynamic limit the correlation functions $h_l(\mathbf{r}_1, \ldots, \mathbf{r}_l)$ depend only on the distances $|\mathbf{r}_i - \mathbf{r}_j|$, so one of the radius-vectors can be excluded due to the shift (we can put one of the l particles at the origin).

Representation (2.311) is formally exact, but in this form, the effective Hamiltonian is of little practical use. However, the effective Hamiltonian can be represented in a more convenient form for practical calculations using the mean field approximation. Namely, within the mean field approximation, only the long-wavelength modes need to be taken into account. Therefore, let's consider the expansion of the coefficients of the Hamiltonian at small $k = |\mathbf{k}|$. Following the idea of constructing the effective Landau–Ginsburg–Wilson-type Hamiltonian, it's necessary to keep only the zeroth-order terms in the expansion of the coefficients u_l at $l > 2$, and the coefficient u_2 has to be represented in the form:

$$u_2 = a + bk^2. \tag{2.322}$$

Utilizing the Ornstein–Zernike relation

$$\tilde{h}_2(\mathbf{k}) = \frac{\tilde{c}_2(\mathbf{k})}{1 - n\tilde{c}_2(\mathbf{k})} \tag{2.323}$$

and the representation of $\tilde{h}_2(0)$ in terms of the isothermal compressibility χ_R of the reference system

$$1 + n\tilde{h}_2(0) = nk_B T \chi_R, \quad \chi_R = n^{-1}\frac{\partial n}{\partial P_R}, \tag{2.324}$$

along with expansions

$$\tilde{c}_2(\mathbf{k}) = \tilde{c}_2(0) - \tilde{c}_2''(0)k^2 + \cdots . \tag{2.325}$$

$$v_k = v_0 - v_0''k^2 + \cdots ., \tag{2.326}$$

which are valid for the potentials decaying faster than r^{-5} in three-dimensional space, we derive the following relation for the coefficient u_2

$$u_2 = \delta_{1,2}\left[a_2 + b_2 k^2 + O(k^4)\right], \tag{2.327}$$

where

$$a_2 = (\beta v_0)^{-1} - nz_0, \tag{2.328}$$

$$b_2 = (\beta v_0)^{-1}\frac{v_0''}{v_0} + n^2 z_0^2 \tilde{c}_2''(0). \tag{2.329}$$

In a similar manner, we obtain the following for the coefficients of higher orders

$$u_3 = -\delta_{1,2,3}n\left(1 + 3n\tilde{h}_2(0) + n^2\tilde{h}_3(0)\right) = \delta_{1,2,3}u_3', \tag{2.330}$$

$$u_4 = -\delta_{1,2,3,4}n\left(1 + 7n\tilde{h}_2(0) + 6n^2\tilde{h}_3(0) + n^3\tilde{h}_4(0)\right) = \delta_{1,2,3,4}u_4', \ldots, \tag{2.331}$$

where we utilize the short-hand notation $\tilde{h}_l(0, 0, \ldots, 0) = \tilde{h}_l(0)$.

Now, let's apply the relation (2.117) for the correlation functions, which can be rewritten in a more expanded form as

$$\chi_R n^2 \frac{\partial}{\partial n}(n^l g_l(\mathbf{r}_1, \ldots, \mathbf{r}_l)) = \beta n^l \Big[l g_l(\mathbf{r}_1, \ldots, \mathbf{r}_l) + n \int d\mathbf{r}_{l+1} \left(g_{l+1}(\mathbf{r}_1, \ldots, \mathbf{r}_{l+1}) - g_l(\mathbf{r}_1, \ldots, \mathbf{r}_l)\right) \Big]. \tag{2.332}$$

For $l = 1$, considering that $g_1(\mathbf{r}_1) = 1$, we arrive at the previously derived expression

$$\chi_R n^2 = \beta n \left(1 + n \int d\mathbf{r}_2 \left(g_2(\mathbf{r}_1, \mathbf{r}_2) - 1\right)\right), \tag{2.333}$$

from which, knowing that $h_2(\mathbf{r}_1, \mathbf{r}_2) = g_2(\mathbf{r}_1, \mathbf{r}_2) - 1$, we obtain

$$\tilde{h}_2(0) = \frac{1}{n}\left(z_0 - 1\right). \tag{2.334}$$

For $l = 2$ we have

$$\chi_R n^2 \frac{\partial}{\partial n}(n^2 g_2(\mathbf{r}_1, \mathbf{r}_2)) = \beta n^2 (2g_2(\mathbf{r}_1, \mathbf{r}_2) + n \int d\mathbf{r}_3 \left(g_3(\mathbf{r}_1, \mathbf{r}_2, \mathbf{r}_3) - g_2(\mathbf{r}_1, \mathbf{r}_2)\right). \tag{2.335}$$

Taking into account that $g_3(\mathbf{r}_1, \ldots, \mathbf{r}_3) - g_2(\mathbf{r}_1, \mathbf{r}_2) = h_3(\mathbf{r}_1, \ldots, \mathbf{r}_3) + g_2(\mathbf{r}_1, \mathbf{r}_3) + g_2(\mathbf{r}_2, \mathbf{r}_3)$ and considering (2.333), we arrive at

$$\chi_R n^2 \frac{\partial}{\partial n}(n^2 h_2(\mathbf{r}_1, \mathbf{r}_2)) = \beta n^2 \left(2h_2(\mathbf{r}_1, \mathbf{r}_2) + n \int d\mathbf{r}_3 h_3(\mathbf{r}_1, \ldots, \mathbf{r}_3)\right), \tag{2.336}$$

which gives

$$\chi_R n^2 \frac{\partial}{\partial n}(n^2 \tilde{h}_2(0)) = 2\beta n^2 \tilde{h}_2(0) + \beta n^3 \tilde{h}_3(0). \tag{2.337}$$

After algebraic manipulations on the last expression, we obtain

$$\tilde{h}_3(0) = \frac{1}{n}\left(z_0^2 - 3z_0 + 2 + z_0 z_1\right), \tag{2.338}$$

where

$$z_1 = n\frac{\partial z_0}{\partial n}. \tag{2.339}$$

In this way we can arrive at the following recurrent relation

$$\chi_R n^2 \frac{\partial}{\partial n}(n^l \tilde{h}_l(0)) = \beta n^l \left[l\tilde{h}_l(0) + n\tilde{h}_{l+1}(0)\right], \tag{2.340}$$

from which we can get the following:

$$\tilde{h}_4(0) = \frac{1}{n^3}\left(z_0^3 - 6z_0^2 + 11z_0 - 6 + 4z_0^2 z_1 - 6z_0 z_1 + z_0 z_1^2 + z_0^2 z_2\right), \tag{2.341}$$

where

$$z_2 = n^2 \frac{\partial^2 z_0}{\partial n^2}. \tag{2.342}$$

Thus, within the formulated approach, it becomes evident that we can derive explicit relations for all coefficients of the effective Hamiltonian in the Landau–Ginzburg–Wilson form by utilizing the compressibility of the reference system χ_R and its derivatives with respect to density. Therefore, the task of constructing the effective Hamiltonian can be considered solved once the compressibility of the reference system is known. As known from liquid theory [19], in systems with exclusively repulsive interactions, one can adjust the particle diameter appropriately to fit the structural and thermodynamic properties of a hard sphere system to those of the original system. For non-singular repulsion, there exists a simple relation for the effective diameter of the hard sphere

$$d_{eff} = \int_0^{\sigma} dr \left(1 - e^{-\frac{\phi(r)}{k_B T}}\right), \tag{2.343}$$

corresponding to the repulsive potential $\phi(r)$, which is zero for $r \geq \sigma$. To describe hard sphere systems, the Carnahan–Starling equation of state is often employed due to its sufficient accuracy

$$\frac{\beta(F_R - F_{id})}{N} = \frac{4\eta - 3\eta^2}{(1-\eta)^2}, \tag{2.344}$$

$$\frac{\beta P_R}{n} = \frac{1 + \eta + \eta^2 - \eta^3}{(1-\eta)^3}, \tag{2.345}$$

where $\eta = \pi d^3 n/6$ is the packing fraction, and F_{id} denote the ideal-gas contributions to the free energy related to the translational degrees of freedom of the particles. For the system of hard spheres we obtain

$$z_0 = \frac{(1-\eta)^4}{1 + 4\eta + 4\eta^2 - 4\eta^3 + \eta^4}. \tag{2.346}$$

By employing the Wertheim–Thiele solution of the Percus–Yevick equation for the direct correlation function of the hard sphere system $c_2(r)$ [19]:

$$c_2(r) = \begin{cases} 0, & r > d \\ \frac{1}{(1-\eta)^4}\left(-(1+2\eta)^2 + 6\eta\left(1+\frac{\eta}{2}\right)^2 \frac{r}{d} - \frac{\eta}{2}(1+2\eta)^2 \left(\frac{r}{d}\right)^3\right), & r \leq d, \end{cases} \tag{2.347}$$

we obtain the following

$$\tilde{c}_2''(0) = \frac{1}{6}\int d\mathbf{r}\, r^2 c_2(r) = -\frac{\pi d^5}{120}\frac{16 - 11\eta + 4\eta^2}{(1-\eta)^4}. \tag{2.348}$$

Now, let's make a transition from the variables $\phi_{\mathbf{k}}$ to the field variable $\phi(\mathbf{r})$. This transition results in the integration over the variables $\phi_{\mathbf{k}}$ evolving into the functional integration over the scalar field $\phi(\mathbf{r})$ with the term $\sim k^2\phi_{\mathbf{k}}\phi_{-\mathbf{k}}$ transforming into the gradient term $\sim (\nabla\phi(\mathbf{r}))^2$. Consequently, we obtain

$$\beta\mathcal{H}[\phi] = \int d\mathbf{r}\left[\frac{1}{2}\kappa_0(\nabla\phi)^2 + W(\phi)\right], \tag{2.349}$$

where

$$W(\phi) = -h\phi(\mathbf{r}) + \frac{a_2}{2!}\phi^2(\mathbf{r}) + \frac{u_3'}{3!}\phi^3(\mathbf{r}) + \frac{u_4'}{4!}\phi^4(\mathbf{r}) + \cdots, \tag{2.350}$$

where we retain terms up to the fourth order in $\phi(\mathbf{r})$. The coefficient in the gradient term takes the form

$$\kappa_0 = \frac{3}{40\pi d}\left[\frac{\lambda_{\mathrm{eff}}^2}{\beta\epsilon_{\mathrm{eff}}} - B\right] \tag{2.351}$$

where $B = 4\eta^2(1-\eta)^4(16-11\eta+4\eta^2)/(1+4\eta+4\eta^2-4\eta^3+\eta^4)^2$ and constants ϵ_{eff} and λ_{eff} characterize the effective depth and width of the potential $v(r)$:

$$\epsilon_{\mathrm{eff}} = \frac{3}{4\pi d^3}\int d\mathbf{r}\, v(r), \tag{2.352}$$

$$\lambda_{\mathrm{eff}}^2 = \frac{5}{3v_0 d^2}\int d\mathbf{r}\, v(r) r^2. \tag{2.353}$$

The last step of our derivations involves eliminating the cubic term in the effective Hamiltonian, which is absent in the standard Landau–Ginzburg–Wilson form. To achieve this, we perform the transformation $\varphi \to \varphi + \bar{\varphi}$, where $\bar{\varphi}$ is chosen to make the coefficient at the cubic term vanish. This leads to the following effective Hamiltonian

$$\beta\mathcal{H}[\varphi] = \int d\mathbf{r}\left(-\bar{h}\varphi + \frac{\bar{a}_2}{2}\varphi^2 + \frac{u_4'}{4!}\varphi^4 + \frac{\kappa_0}{2}(\nabla\varphi)^2\right), \tag{2.354}$$

where

$$\bar{h} = h + \frac{a_2 u_3'}{u_4'} - \frac{u_3'^3}{3u_4^2}, \tag{2.355}$$

$$\bar{a}_2 = a_2 - \frac{u_3'^2}{2u_4'}. \tag{2.356}$$

The effective Hamiltonian (2.354) can be employed in the analysis of critical phenomena in simple liquids using standard methods such as self-consistent field theory and the renormalization group method [45, 46]. However, these topics are beyond the scope of this book. We will revisit the Hubbard–Schofield method in Chap. 3.

Chapter 3
Self-consistent Field Theory of Ionic Fluids. Canonical Ensemble Formulation

In previous chapter, we discussed theoretical approaches based on self-consistent field theory for describing the thermodynamic properties of fluids composed of electrically neutral particles interacting via short-range potentials. However, these approaches become inapplicable when the fluid particles possess nonzero electric charges and interact through Coulomb potentials, as in the case of the so-called ionic fluids. As it was mentioned above, in such systems the virial expansion cannot be used even at low concentrations due to the divergence of group integrals and virial coefficients. However, Mayer demonstrated within his approach (see, for example, [47]), based on diagrammatic framework, that summing certain divergent diagrammatic series for the simplest electrolyte model yields a finite expression for the free energy, identical to that obtained by Debye and Hückel using the Poisson–Boltzmann equation. However, Mayer's diagrammatic approach is cumbersome, especially when considering its generalization for electrolyte models considering internal electric structure of ions and their short-range specific interactions. An alternative approach based on statistical field theory [39, 48–58] seems to be more convenient, as it allows avoiding the explicit manipulation with divergent diagrammatic series and reduces the calculation of the free energy of ionic fluid to the approximate calculation of functional integrals on a Gaussian measure over one or several fluctuating fields. Within this approach, one can naturally derive the Poisson–Boltzmann equation and its modifications as the Euler–Lagrange equations of the field-theoretical Hamiltonian. Mayer's diagrammatic method is not covered in this monograph. Interested readers can refer to relevant literature [19, 47] for a detailed review. Instead, we proceed directly to the investigation of the statistical field theory of ionic fluids. In this chapter, we consider only formulation of the theory within the canonical ensemble. In Chap. 4, we will discuss the formulation within the grand canonical ensemble. Much of the following discussion will build upon the theory formulated in Chap. 2 for simple fluids.

Y. A. Budkov and N. N. Kalikin, *Statistical Field Theory of Ion-Molecular Fluids*,
https://doi.org/10.1007/978-3-031-68364-0_3

3.1 Electrostatic Free Energy of the Coulomb Gas and the Poisson–Boltzmann Equation

Let's consider the model of the Coulomb gas, which serves as the most basic model of the ionic fluids. We define an "Coulomb gas" as an electrically neutral system consisting of N_+ positively and N_- negatively charged point ions with charges $q_+ > 0$ and $q_- < 0$, respectively. The condition of electrical neutrality is satisfied by the relation $q_+N_+ + q_-N_- = 0$. We assume that these ions are confined within a volume of size V and are in thermodynamic equilibrium at temperature T. Additionally, we assume that the ions are immersed in a dielectric continuous medium with a constant dielectric permittivity ε. For the case when such model describes the high-temperature plasma, $\varepsilon = 1$, while for the case of diluted electrolyte solutions, $\varepsilon > 1$ (indicating the presence of a solvent). Furthermore, we consider the existence of fixed external charges with density $\rho_{ext}(\mathbf{r})$ alongside the ions within the volume of the system. These charges may, in principle, represent the charges of conducting electrodes or dielectric membranes. Thus, the Hamiltonian describing the interactions of the Coulomb gas, including the "self-action" of the charges, takes the form:

$$U_{el} = \frac{1}{2}\left(\hat{\rho}G_0\hat{\rho}\right) = \frac{1}{2}\int d\mathbf{r}\int d\mathbf{r}'\,\hat{\rho}(\mathbf{r})G_0(\mathbf{r}-\mathbf{r}')\hat{\rho}(\mathbf{r}'), \tag{3.1}$$

where

$$\hat{\rho}(\mathbf{r}) = \hat{\rho}_i(\mathbf{r}) + \rho_{ext}(\mathbf{r}) \tag{3.2}$$

denotes the local density of the total charge of the system; the microscopic ionic charge density is determined as

$$\hat{\rho}_i(\mathbf{r}) = q_+\sum_{k=1}^{N_+}\delta\left(\mathbf{r}-\mathbf{r}_k^{(+)}\right) + q_-\sum_{l=1}^{N_-}\delta\left(\mathbf{r}-\mathbf{r}_l^{(-)}\right), \tag{3.3}$$

where $\mathbf{r}_k^{(\pm)}$ are the positions of the ions in space. Moreover, we introduce the Green's function of the Poisson equation

$$G_0(\mathbf{r}-\mathbf{r}') = \frac{1}{\varepsilon|\mathbf{r}-\mathbf{r}'|}. \tag{3.4}$$

The total potential energy of interaction of ions with external fields has the form

$$U_{ext} = \sum_{k=1}^{N_+} u_+(\mathbf{r}_k^{(+)}) + \sum_{l=1}^{N_-} u_-(\mathbf{r}_l^{(-)}), \tag{3.5}$$

where $u_\pm(\mathbf{r})$ denote the potentials of interactions between the ions and the external fields. Thus, configuration integral of the system within the NVT-ensemble can be written as

$$Q = \int \cdots \int \prod_{k=1}^{N_+} \frac{d\mathbf{r}_k^{(+)}}{V} \prod_{l=1}^{N_-} \frac{d\mathbf{r}_l^{(-)}}{V} \exp\left[-\frac{\beta}{2}\left(\hat{\rho}G_0\hat{\rho}\right) - \beta U_{ext}\right]. \tag{3.6}$$

Then, applying the HS transformation

$$\exp\left[-\frac{\beta}{2}(\hat{\rho}G_0\hat{\rho})\right] = \int \frac{\mathcal{D}\varphi}{C_0} \exp\left[-\frac{\beta}{2}(\varphi G_0^{-1}\varphi) + i\beta(\hat{\rho}\varphi)\right], \tag{3.7}$$

where we introduce the following notations

$$(\varphi G_0^{-1}\varphi) = \int d\mathbf{r} \int d\mathbf{r}' \, \varphi(\mathbf{r}) G_0^{-1}(\mathbf{r}-\mathbf{r}')\varphi(\mathbf{r}'), \tag{3.8}$$

$$(\hat{\rho}\varphi) = \int d\mathbf{r} \, \hat{\rho}(\mathbf{r})\varphi(\mathbf{r}), \tag{3.9}$$

we obtain the configuration integral representation in the form of the following functional integral over the fluctuating electrostatic potential $\varphi(\mathbf{r})$

$$Q = \int \frac{\mathcal{D}\varphi}{C_0} \exp\left[-\frac{\beta}{2}(\varphi G_0^{-1}\varphi) + i\beta(\rho_{ext}\varphi)\right] Q_+^{N_+}[\varphi] Q_-^{N_-}[\varphi], \tag{3.10}$$

where we introduce the following one-particle configuration integrals

$$Q_\pm[\varphi] = \int \frac{d\mathbf{r}}{V} e^{i\beta q_\pm \varphi(\mathbf{r}) - \beta u_\pm(\mathbf{r})} \tag{3.11}$$

and the normalization constant of the Gaussian measure, $C_0 = \int \mathcal{D}\varphi \exp\left[-\frac{\beta}{2}(\varphi G_0^{-1}\varphi)\right]$, with the "inverse" Green's function

$$G_0^{-1}(\mathbf{r}, \mathbf{r}') = -\frac{\varepsilon}{4\pi}\nabla^2\delta(\mathbf{r}-\mathbf{r}'), \tag{3.12}$$

which satisfies the relation

$$\int d\mathbf{r}'' \, G_0(\mathbf{r}-\mathbf{r}'')G_0^{-1}(\mathbf{r}'', \mathbf{r}') = \delta(\mathbf{r}-\mathbf{r}'). \tag{3.13}$$

Now, considering the thermodynamic limit, $N_\pm \to \infty$, $V \to \infty$, and $N_\pm/V \to n_\pm$, where the following asymptotic equalities take place (see (2.141))

$$Q_\pm^{N_\pm}[\varphi] \simeq \exp\left[n_\pm \int d\mathbf{r} \left(e^{i\beta q_\pm \varphi(\mathbf{r}) - \beta u_\pm(\mathbf{r})} - 1\right)\right], \tag{3.14}$$

we arrive at the following expression for the configuration integral

$$Q = \int \frac{\mathcal{D}\varphi}{C_0} \exp\left[-S[\varphi]\right], \tag{3.15}$$

where we introduce the integrand functional

$$S[\varphi] = \frac{\beta}{2}(\varphi G_0^{-1}\varphi) - i\beta(\rho_{ext}\varphi) - W[\varphi] \tag{3.16}$$

with the auxiliary functional

$$W[\varphi] = n_+ \int d\mathbf{r} \left(e^{i\beta q_+\varphi(\mathbf{r}) - \beta u_+(\mathbf{r})} - 1\right) + n_- \int d\mathbf{r} \left(e^{i\beta q_-\varphi(\mathbf{r}) - \beta u_-(\mathbf{r})} - 1\right). \tag{3.17}$$

Let's formally evaluate the functional integral (3.15) within the Gaussian approximation. The Euler–Lagrange equitation has the form

$$\frac{\delta S[\varphi]}{\delta\varphi(\mathbf{r})} = 0. \tag{3.18}$$

Evaluating the variational derivative in (3.18) and introducing the self-consistent field potential $\varphi^{(MF)}(\mathbf{r}) = i\psi(\mathbf{r})$, we obtain the classical Poisson–Boltzmann (PB) equation

$$\Delta\psi(\mathbf{r}) = -\frac{4\pi}{\varepsilon}\rho_{ext}(\mathbf{r}) - \frac{4\pi}{\varepsilon}\left(q_- n_- e^{-\beta q_-\psi(\mathbf{r}) - \beta u_-(\mathbf{r})} + q_+ n_+ e^{-\beta q_+\psi(\mathbf{r}) - \beta u_+(\mathbf{r})}\right), \tag{3.19}$$

where $\Delta = \nabla^2$ denotes the Laplace operator.[1]

Thus, the self-consistent field approximation automatically leads us to the PB equation for the potential of the electrostatic field created by the ions and the external charges. By expanding the functional $S[\varphi]$ in expression (3.15) in a functional power series around the mean field configuration $\varphi^{(MF)}(\mathbf{r}) = i\psi(\mathbf{r})$ and neglecting terms beyond the second order, we obtain

$$Q \approx \exp\left[-S[i\psi]\right] \int \frac{\mathcal{D}\Lambda}{C_0} \exp\left[-\frac{\beta}{2}\left(\Lambda \mathcal{G}^{-1}\Lambda\right)\right], \tag{3.20}$$

where $\Lambda(\mathbf{r}) = \varphi(\mathbf{r}) - i\psi(\mathbf{r})$, and the renormalized inverse Green's function is determined by the expression

[1] For the first time the Poisson–Boltzmann equation was derived within the statistical field theory in grand canonical ensemble in [48].

$$\mathcal{G}^{-1}(\mathbf{r},\mathbf{r}'|\psi) = k_B T \frac{\delta^2 S[i\psi]}{\delta\varphi(\mathbf{r})\delta\varphi(\mathbf{r}')} = G_0^{-1}(\mathbf{r},\mathbf{r}')$$
$$+ \beta\left(q_+^2 n_+ e^{-\beta q_+ \psi(\mathbf{r}) - \beta u_+(\mathbf{r})} + q_-^2 n_- e^{-\beta q_- \psi(\mathbf{r}) - \beta u_-(\mathbf{r})}\right)\delta(\mathbf{r}-\mathbf{r}'). \quad (3.21)$$

Then, formally evaluating the Gaussian functional integral (see Chap. 1), we obtain the expression for the excess electrostatic free energy of the solution within the Gaussian approximation

$$F_{el} = -k_B T \ln Q = k_B T \left(S[i\psi] - \frac{1}{2}\left(tr \ln \mathcal{G} - tr \ln G_0\right)\right), \quad (3.22)$$

where the trace of the integral operator A, established by the kernel $A(\mathbf{r},\mathbf{r}')$, is as previously mentioned, determined by the following relation

$$tr A = \int d\mathbf{r} A(\mathbf{r},\mathbf{r}). \quad (3.23)$$

Applying the same approach as in Chap. 2 when deriving the expression (2.208), we arrive at the expression for the free energy of the Coulomb gas within the Gaussian approximation

$$\frac{F_{el}}{k_B T} = S[i\psi] + \int d\mathbf{r}\,\mathcal{I}(\mathbf{r}) \int\limits_0^1 d\tau\,\mathcal{G}(\mathbf{r},\mathbf{r};\tau), \quad (3.24)$$

where for brevity, we use the following notation for the Green's function, $\mathcal{G}(\mathbf{r},\mathbf{r}'|\psi) = \mathcal{G}(\mathbf{r},\mathbf{r}')$, which satisfies the equation following from (3.21)

$$\left(-\nabla^2 + \varkappa^2(\mathbf{r})\right)\mathcal{G}(\mathbf{r},\mathbf{r}') = \frac{4\pi}{\varepsilon}\delta(\mathbf{r}-\mathbf{r}'), \quad (3.25)$$

where

$$\varkappa^2(\mathbf{r}) = \frac{8\pi\beta}{\varepsilon}\mathcal{I}(\mathbf{r}), \quad (3.26)$$

$$\mathcal{I}(\mathbf{r}) = \frac{1}{2}\left(q_+^2 n_+ e^{-\beta q_+ \psi(\mathbf{r}) - \beta u_+(\mathbf{r})} + q_-^2 n_- e^{-\beta q_- \psi(\mathbf{r}) - \beta u_-(\mathbf{r})}\right). \quad (3.27)$$

Note, function $\mathcal{G}(\mathbf{r},\mathbf{r}';\tau)$ satisfies the following equation

$$\left(-\nabla^2 + \tau\varkappa^2(\mathbf{r})\right)\mathcal{G}(\mathbf{r},\mathbf{r}';\tau) = \frac{4\pi}{\varepsilon}\delta(\mathbf{r}-\mathbf{r}'). \quad (3.28)$$

3.2 Debye Screening and Debye–Hückel Limiting Law

Until now, we have focused on inhomogeneous ionic fluids in the thermodynamic limit within the mean field approximation, i.e. utilizing the Poisson–Boltzmann equation. Now, let us consider homogeneous fluids.

Utilizing the obtained PB equation, we can evaluate the potential of a point charge with charge q_0 immersed in the Coulomb gas within the linear approximation. In this case, the external charge density is determined by the density of the point charge, i.e., $\rho_{ext}(\mathbf{r}) = q_0\delta(\mathbf{r})$. In the region of small electrostatic interactions, $q_{\pm}\psi(\mathbf{r})/k_BT \ll 1$, the linearized PB equation takes the form

$$\Delta\psi(\mathbf{r}) - \kappa^2\psi(\mathbf{r}) = -\frac{4\pi q_0}{\varepsilon}\delta(\mathbf{r}), \tag{3.29}$$

where $\kappa^2 = 4\pi(q_+^2 n_+ + q_-^2 n_-)/(\varepsilon k_B T)$ is the value whose physical interpretation will be clarified later. Transitioning to the Fourier transform of the potential

$$\tilde{\psi}(\mathbf{k}) = \int d\mathbf{r}\, e^{-i\mathbf{k}\mathbf{r}}\psi(\mathbf{r}), \tag{3.30}$$

we can rewrite the Eq. (3.29) in the algebraic form

$$\tilde{\psi}(\mathbf{k}) = \frac{4\pi q_0}{\varepsilon(k^2 + \kappa^2)}, \tag{3.31}$$

which, upon inverse Fourier transformation, results in the Debye–Hückel potential

$$\psi(\mathbf{r}) = \frac{q_0}{\varepsilon r}\exp\left[-\kappa r\right]. \tag{3.32}$$

The expression (3.32) for the potential is straightforward to interpret. It illustrates that the presence of ions surrounding the point charge results in the screening of the Coulomb potential at distances of the order of κ^{-1}. In other words, at distances around $r_D = \kappa^{-1}$, known as the Debye radius, the surrounding ions are essentially unaffected by the electric field. Consequently, a "cloud" of predominantly oppositely charged ions forms around the test point charge, commonly referred to as the "ionic atmosphere". It's worth noting that the screening atmosphere is a characteristic feature of ionic systems, stemming from the presence of the long-range Coulomb interactions.

Let's now apply the formula (3.22) to evaluate the electrostatic free energy of the ideal gas when the external charges are absent, i.e., when $\rho_{ext}(\mathbf{r}) = 0$. In this case, $\psi(\mathbf{r}) = 0$, so the mean field contribution to the electrostatic free energy is zero, and it is solely determined by the thermal fluctuations of the electrostatic potential near zero. The renormalized inverse Green's function in such a case takes the following form:

$$\mathcal{G}^{-1}(\mathbf{r}, \mathbf{r}'|0) = G^{-1}(\mathbf{r} - \mathbf{r}') = -\frac{\varepsilon}{4\pi}(\Delta - \kappa^2)\delta(\mathbf{r} - \mathbf{r}'). \tag{3.33}$$

The electrostatic free energy of the Coulomb gas can then be calculated using the formula

$$F_{el} = \frac{V k_B T}{2} \int \frac{d\mathbf{k}}{(2\pi)^3} \ln \frac{G_0(\mathbf{k})}{G(\mathbf{k})}, \tag{3.34}$$

which can be easily obtained from the expression (1.198) of Chap. 1, where $G_0(\mathbf{k}) = 4\pi/(\varepsilon k^2)$ and $G(\mathbf{k}) = 4\pi/(\varepsilon(k^2 + \kappa^2))$ are the Fourier transforms of the Green's functions, which are translation invariant in the thermodynamic limit. Substituting them into relation (3.34), we obtain the following

$$\frac{F_{el}}{V k_B T} = \frac{1}{2} \int \frac{d\mathbf{k}}{(2\pi)^3} \left(\ln \left(1 + \frac{\kappa^2}{k^2} \right) - \frac{\kappa^2}{k^2} \right), \tag{3.35}$$

where we subtracted the self-energy of the ions in the medium with a dielectric permittivity ε from the final expression ε

$$U_{self} = \frac{V k_B T}{2} \int \frac{d\mathbf{k}}{(2\pi)^3} \frac{\kappa^2}{k^2}. \tag{3.36}$$

Evaluating the integral, we arrive at the well-known Debye–Hückel (DH) limiting law for the excess free energy density of the Coulomb gas [12]

$$\frac{F_{el}}{V k_B T} = -\frac{\kappa^3}{12\pi}. \tag{3.37}$$

Besides, utilizing the expression (2.21), we can easily obtain the equation of state of the Coulomb gas within the DH approximation. The result takes the form

$$\frac{P}{k_B T} = n_+ + n_- - \frac{\kappa^3}{24\pi}. \tag{3.38}$$

As observed, the contribution of the Coulomb interactions to the pressure (the third term in (3.38)) is negative. It is evident that the excess pressure is proportional to $n^{3/2}$, where n is the number density of the ions. The non-analyticity of the excess pressure with respect to concentration additionally indicates the impossibility of accounting for the electrostatic interactions within a finite number of terms in the virial expansion. It is noteworthy that the equation of state of the Coulomb gas obtained within the Gaussian approximation is valid only for sufficiently weak electrostatic interactions. To determine the applicability condition of the DH theory, we need to ensure that the third term in (3.38) is much smaller than the first two, leading to the straightforward condition $n_{\pm} r_D^3 \ll 1$.

3.3 Gouy–Chapman Theory of Electric Double Layer

As an illustration of the PB equation's application, let us examine the simplest theory of the electric double layer, which occurs at the metal-electrolyte solution interface—the Gouy–Chapman theory (GC). Within the GC theory, the electrolyte solution is described as the Coulomb gas model, inserted into a continuous dielectric medium (solvent) with a constant dielectric permittivity ε. The electrode is depicted as an infinitely charged flat surface with a constant surface charge density σ. We consider the most straightforward case of a $1:1$ electrolyte, meaning $q_+ = -q_- = q$. Thus, by placing the origin at the electrode and defining the z axis perpendicular to its surface, we can express the Poisson–Boltzmann equation for the electrostatic potential $\psi(z)$ of the self-consistent field as follows:

$$\frac{d^2\psi(z)}{dz^2} = \frac{8\pi nq}{\varepsilon}\sinh\left(\frac{q\psi(z)}{k_BT}\right), \tag{3.39}$$

where $n = n_\pm$ is the concentration of the ions away from the electrode. The boundary conditions for the Eq. (3.39) has the form

$$-\varepsilon\psi'(0) = 4\pi\sigma, \quad \psi'(\infty) = 0. \tag{3.40}$$

The first boundary condition fixes the value of the electric displacement at the surface of the electrode, while the second one guarantees the disappearance of the electric field far from the electrode (in the bulk solution). The first integral of Eq. (3.39) has the form:

$$2nk_BT\cosh(\beta q\psi) - \frac{\varepsilon\mathcal{E}^2}{8\pi} = 2nk_BT, \tag{3.41}$$

where $\mathcal{E} = -\psi'(z)$ is the electric field. Equation (3.41) has a straightforward physical meaning: it determines the condition of the mechanical equilibrium of the solution, i.e. the equality of the pressures in each of its layers. Indeed, the right-hand side of Eq. (3.41) defines the pressure of the Coulomb gas in the bulk solution (where $\psi = 0$), whereas the left-hand side represents the local pressure, consisting of the particles' pressure (the first term) and the pressure of the electrostatic field, arising from the violation of the local electric neutrality.[2] First-order differential Eq. (3.41) can be easily integrated by separating the variables, and the result takes the form:

$$\psi(z) = \frac{4k_BT}{q}\tanh^{-1}\left(\tau\exp\left(-\kappa z\right)\right), \tag{3.42}$$

[2] Left-hand side of the equation is nothing more than a component σ_{zz} of the stress tensor (see Chap. 3) with a negative sign.

where κ is the inverse Debye radius, introduced earlier, and $\tau = \tanh(\beta q\psi_0/4)$ is an auxiliary function of the surface potential $\psi_0 = \psi(0)$. Local ion concentrations, determined by the Boltzmann distribution, $n_\pm(z) = n\exp(\mp\beta q\psi(z))$, after substituting the expression (3.42), take the form:

$$n_\pm(z) = n\left(\frac{1 \mp \tau\exp[-\kappa z]}{1 \pm \tau\exp[-\kappa z]}\right)^2. \tag{3.43}$$

The fact that these are the ions concentrations can be verified by functional differentiation of the free energy of the solution within the mean field approximation with respect to the potentials of the external fields and equating them to zero, i.e.:

$$n_\pm(z) = k_B T\left.\frac{\delta S[i\psi]}{\delta u_\pm(z)}\right|_{u_\pm=0}. \tag{3.44}$$

Note that in the region of weak electrostatic interactions, when $\beta q\psi_0 \ll 1$, the formula for the potential simplifies to:

$$\psi(z) = \psi_0 e^{-\kappa z}. \tag{3.45}$$

Thus, as in the case of the point ion discussed in the previous section, an ionic atmosphere with a thickness of the order of the Debye radius κ^{-1}—a diffuse layer—is formed near the charged electrode. The diffuse layer of ions, along with the charge of the electrode, forms what is known as the diffuse electric double layer.

Thus, utilizing the first boundary condition from (3.50), we derive the expression that relates the surface charge density to the surface potential. This relation, known as the Grahame equation, was first introduced by Grahame [30, 59]

$$\sigma = \left(\frac{2n\varepsilon k_B T}{\pi}\right)^{1/2}\sinh\left(\frac{q\psi_0}{2k_B T}\right). \tag{3.46}$$

The expression (3.46) allows us to evaluate the differential electric capacitance of the electric double layer, which is an experimentally measurable property

$$C = \frac{d\sigma}{d\psi_0} = C_{GC}\cosh\left(\frac{q\psi_0}{2k_B T}\right), \tag{3.47}$$

where

$$C_{GC} = \left(\frac{nq^2\varepsilon}{2\pi k_B T}\right)^{1/2} = \frac{\varepsilon}{4\pi r_D} \tag{3.48}$$

is the differential electric capacitance at zero potential, known as the Gouy–Chapman capacitance. It's important to note that according to the GC theory, the differential electrical capacitance exhibits exponential growth at high electrode potentials. However, in real electrolyte solutions, the differential capacitance reaches a maximum at a certain potential value and then decreases monotonically [60]. The absence of such nonmonotonic behavior of the differential capacitance within the GC theory is attributed to the neglect of the effect of finite ion size. As we will demonstrate further, incorporating steric interactions of the ions allows for a description of the behavior of the differential capacitance at high surface potentials.

3.4 Electrolyte Solution Between Charged Surfaces

Now, let's consider the scenario of an electrolyte solution confined within a slit pore with charged walls, having a constant charge density, σ. Similar to the previous section, the electrostatic potential within the pore is governed by the Poisson–Boltzmann equation

$$\psi''(z) = \frac{8\pi nq}{\varepsilon} \sinh\left(\frac{q\psi(z)}{k_B T}\right), \tag{3.49}$$

with the boundary conditions at the walls, positioned at $z = 0$ and $z = H$

$$\psi'(0) = -\frac{4\pi\sigma}{\varepsilon}, \quad \psi'(H) = \frac{4\pi\sigma}{\varepsilon}. \tag{3.50}$$

The first integral of the PB equation yields

$$P = -\frac{\varepsilon\mathcal{E}^2(z)}{8\pi} + 2nk_B T \cosh\left(\frac{q\psi(z)}{k_B T}\right) = const, \tag{3.51}$$

which represents nothing more than the pressure inside the pore; $\mathcal{E}(z) = -\psi'(z)$. For practical purposes, one introduces the disjoining pressure as the excess osmotic pressure of the ions compared to the pressure in the bulk solution phase:

$$\Pi_{MF} = P - P_b = -\frac{\varepsilon\mathcal{E}^2(z)}{8\pi} + 2nk_B T\left(\cosh\left(\frac{q\psi(z)}{k_B T}\right) - 1\right), \tag{3.52}$$

where $P_b = 2nk_B T$ represents the osmotic pressure of ions in the bulk phase. Considering that $\mathcal{E}(H/2) = 0$ and defining the potential at the midpoint of the pore as $\psi(H/2) = \psi_m$, the disjoining pressure can be expressed as

$$\Pi_{MF} = 2nk_B T\left(\cosh\left(\frac{q\psi_m}{k_B T}\right) - 1\right). \tag{3.53}$$

The expression (3.53) was first heuristically proposed by Langmuir and derived by Derjaguin [61]. Thus, by solving the PB equation with specified boundary conditions and evaluating the potential at the midpoint of the pore, we can calculate the disjoining pressure in the charged slit pore as a function of its width, H.

Let us estimate the disjoining pressure under the condition of sufficiently weak electrostatics ($q\psi_m/k_BT \ll 1$), allowing us to linearize the Poisson–Boltzmann equation, i.e.

$$\psi'' - \kappa^2\psi = 0, \tag{3.54}$$

for which we arrive at the following solution

$$\psi(z) = \frac{4\pi\sigma}{\kappa\varepsilon}\left(\frac{1+\cosh\kappa H}{\sinh\kappa H}\cosh\kappa z - \sinh\kappa z\right). \tag{3.55}$$

Calculating the potential in the midpoint of the pore

$$\psi_m = \frac{4\pi\sigma}{\kappa\varepsilon\sinh\left(\frac{\kappa H}{2}\right)}, \tag{3.56}$$

we obtain the expression for the disjoining pressure

$$\Pi_{MF} = \frac{nq^2}{k_BT\varepsilon}\psi_m^2 = \frac{2\pi\sigma^2}{\varepsilon}\frac{1}{\sinh^2\left(\frac{\kappa H}{2}\right)}. \tag{3.57}$$

As observed in (3.57), the disjoining pressure exhibits a rapid increase at small distances ($\Pi \sim H^{-2}$) and an exponential decay at larger distances ($\Pi \sim e^{-\kappa H}$). The significant growth at small distances arises from the repulsion between walls bearing identical charges, where screening effects become negligible. Conversely, the exponential decay of the disjoining pressure at larger distances is attributed to the screening effect.

Now, let us calculate the correlation correction to the mean-field approximation of the disjoining pressure in the weak electrostatic limit. To do this, we will use the expression for the electrostatic free energy of a Coulomb gas that we derived earlier. We will write this expression down again for reference

$$\frac{F_{el}}{k_BT} = S[i\psi] + \int d\mathbf{r}\,\mathcal{I}(\mathbf{r})\int_0^1 d\tau\,\mathcal{G}(\mathbf{r},\mathbf{r};\tau), \tag{3.58}$$

where in this case

$$S[i\psi]/\mathcal{A} = \frac{\varepsilon}{8\pi}\int_0^H dz\,\psi(z)\psi''(z) + \sigma(\psi(0)+\psi(H)) - W[i\psi], \tag{3.59}$$

where we took into account that $\rho_{ext}(\mathbf{r}) = \sigma(\delta(z) + \delta(z - H))$ and introduced the following functional

$$W[i\psi] = 2n \int_0^H dz\,(\cosh(\beta q\psi(z)) - 1). \tag{3.60}$$

Considering that the Green's function is translation invariant in the tangent direction and performing a 2D Fourier transform,

$$\mathcal{G}(\mathbf{r}, \mathbf{r}'; \tau) = \mathcal{G}(\rho - \rho'; z, z'; \tau) = \int \frac{d^2\mathbf{p}}{(2\pi)^2} \mathcal{G}(\mathbf{p}|z, z'; \tau) e^{i\mathbf{p}(\rho-\rho')}, \tag{3.61}$$

we can rewrite Eq. (3.28) as follows

$$\left(-\frac{d^2}{dz^2} + \mathbf{p}^2 + \tau\varkappa^2(z)\right) \mathcal{G}(\mathbf{p}|z, z'; \tau) = \frac{4\pi}{\varepsilon}\delta(z - z'). \tag{3.62}$$

Thus, after integration by parts and using the boundary conditions (3.50), we eventually obtain

$$\frac{F_{el}}{k_B T \mathcal{A}} = \int_0^H dz\,\phi(\psi, \psi') + \int_0^H dz\,\mathcal{I}(z) \int_0^1 d\tau \int \frac{d^2\mathbf{p}}{(2\pi)^2} \mathcal{G}(\mathbf{p}|z, z; \tau), \tag{3.63}$$

where

$$\phi(\psi, \psi') = -\frac{\varepsilon\psi'^2(z)}{8\pi} - 2n(\cosh(\beta q\psi(z)) - 1), \tag{3.64}$$

$\mathcal{I}(z) = nq^2 \cosh(\beta q\psi(z))$. The electrostatic pressure can be calculated via thermodynamic relation

$$\Pi_{el} = -\frac{\partial}{\partial H}\left(\frac{F_{el}}{k_B T \mathcal{A}}\right) \tag{3.65}$$

Differentiation of first term gives

$$-\frac{\partial}{\partial H}\int_0^H dz\,\phi(\psi, \psi') = -\left(\phi - \frac{\partial\phi}{\partial\psi'}\right)\Bigg|_{z=H} = -\frac{\varepsilon\psi'^2(H)}{8\pi}$$

$$+ 2nk_B T\left(\cosh\left(\frac{q\psi(H)}{k_B T}\right) - 1\right) = 2n(\cosh(\beta q\psi_m) - 1) = \Pi_{MF}, \tag{3.66}$$

where we took into account the first integral (3.51) of the PB equation. We have used the following identity [62] for the functional

$$\mathcal{F}[\psi] = \int\limits_0^H dz\, \phi(\psi, \psi') \tag{3.67}$$

$$\frac{\partial \mathcal{F}}{\partial H} = \phi(\psi(H), \psi'(H)) - \frac{\partial \phi(\psi(H), \psi'(H))}{\partial \psi'(H)} \psi'(H). \tag{3.68}$$

Differentiation of the second term is not so trivial. Let us calculate the fluctuation correction to the mean-field approximation of the disjoining pressure in the limits of weak electrostatics, where $\varkappa^2(z) \approx \kappa^2 = 8\pi nq^2/(\varepsilon k_B T), \mathcal{I}(z) \approx nq^2$. The Green's function satisfies the following equation

$$\left(-\frac{d^2}{dz^2} + \mathbf{p}^2 + \tau\kappa^2\right) \mathcal{G}(\mathbf{p}|z, z'; \tau) = \frac{4\pi}{\varepsilon}\delta(z - z'). \tag{3.69}$$

Let us consider here only the case of metal walls. The case of dielectric walls will be considered in Chap. 4 within different approach. For metal walls we can use the following boundary conditions for $\mathcal{G}(\mathbf{p}|z, z'; \tau)$

$$\begin{gathered} \mathcal{G}(\mathbf{p}|0, z'; \tau) = \mathcal{G}(\mathbf{p}|H, z'; \tau) = 0, \\ \frac{\varepsilon}{4\pi}\left(\frac{\partial \mathcal{G}(\mathbf{p}|z' - 0, z'; \tau)}{\partial z} - \frac{\partial \mathcal{G}(\mathbf{p}|z' + 0, z'; \tau)}{\partial z}\right) = 1, \\ \mathcal{G}(\mathbf{p}|z' - 0, z'; \tau) = \mathcal{G}(\mathbf{p}|z' + 0, z'; \tau). \end{gathered} \tag{3.70}$$

Such boundary conditions can be motivated by the fact that the Green's function $\mathcal{G}(\mathbf{r}, \mathbf{r}')$ can be interpreted as the electrostatic potential created by unit trial charge placed at point $\mathbf{r}'$.

The solution is

$$\mathcal{G}(\mathbf{p}|z, z'; \tau) = \begin{cases} \frac{4\pi \sinh \kappa_p(\tau) z \left(\cosh \kappa_p(\tau) z' - \coth \kappa_p(\tau) H \sinh \kappa_p(\tau) z'\right)}{\varepsilon \kappa_p(\tau)}, & z < z', \\ \frac{4\pi \sinh \kappa_p(\tau) z' \left(\cosh \kappa_p(\tau) z - \coth \kappa_p(\tau) H \sinh \kappa_p(\tau) z\right)}{\varepsilon \kappa_p(\tau)}, & z > z', \end{cases} \tag{3.71}$$

where $\kappa_p(\tau) = \sqrt{\mathbf{p}^2 + \kappa^2 \tau}$. Thus, we have

$$\begin{aligned} nq^2 \int\limits_0^H dz \int\limits_0^1 d\tau \int \frac{d^2\mathbf{p}}{(2\pi)^2} \mathcal{G}(\mathbf{p}|z, z; \tau) &= \frac{2\pi nq^2 H}{\varepsilon} \int\limits_0^1 d\tau \int \frac{d^2\mathbf{p}}{(2\pi)^2} \frac{\coth \kappa_p(\tau)}{\kappa_p(\tau)} \\ &= \frac{k_B T}{2} \int \frac{d^2\mathbf{p}}{(2\pi)^2} \ln \frac{\sinh \kappa_p H}{\sinh pH}, \end{aligned} \tag{3.72}$$

where $p = |\mathbf{p}|$. Therefore, we obtain

$$-\frac{\partial}{\partial H}\left(\frac{k_BT}{2}\int\frac{d^2\mathbf{p}}{(2\pi)^2}\ln\frac{\sinh\kappa_pH}{\sinh pH}\right)=-\frac{k_BT}{2}\int\frac{d^2\mathbf{p}}{(2\pi)^2}\left(\kappa_p\coth\kappa_pH-p\coth pH\right). \tag{3.73}$$

Subtracting the constant terms occurring at $H\to\infty$, we obtain the following fluctuation correction to the mean-field approximation of the disjoining pressure

$$\Pi_{GA}=-\frac{k_BT}{2}\int\frac{d^2\mathbf{p}}{(2\pi)^2}\kappa_p\left(\coth\kappa_pH-1\right)+\frac{k_BT}{2}\int\frac{d^2\mathbf{p}}{(2\pi)^2}p\left(\coth pH-1\right). \tag{3.74}$$

The second term in the right hand side of Eq. (3.74) can be omitted since it has to be cancelled with a term corresponding to zero Matsubara mode of the van der Waals contribution to the disjoining pressure [63, 64]

$$\Pi_{vdW}=-\frac{k_BT}{2}\int\frac{d^2\mathbf{p}}{(2\pi)^2}p\left(\coth pH-1\right). \tag{3.75}$$

Thus, the fluctuation correction to the disjoining pressure in the linear approximation, taking into account classical van der Waals interactions between metals, is

$$\Pi_{GA}=-\frac{k_BT}{2}\int\frac{d^2\mathbf{p}}{(2\pi)^2}\kappa_p\left(\coth\kappa_pH-1\right). \tag{3.76}$$

The latter expression yields the following limiting cases

$$\Pi_{GA}\simeq\begin{cases}-\frac{k_BT}{8\pi H^3}\zeta(3), & \kappa H\ll 1\\ -\frac{k_BT\kappa^2}{4\pi H}e^{-2\kappa H}, & \kappa H\gg 1,\end{cases} \tag{3.77}$$

where $\zeta(x)$ is the Riemann zeta function, which have been obtained for the first time by Ninham and Yaminsky [63].

3.5 Corrections to the Debye–Hückel Limiting Law

Now, let's explore how we can systematically evaluate the corrections to the Gaussian approximation, i.e. to the Debye–Hückel approximation. To do this, we will apply the Gaussian equivalent representation method, as formulated in Chap. 2, which was initially applied to simple fluids with repulsive pairwise potentials. Let's focus only on the case of a spatially-homogeneous Coulomb gas with symmetric ionic charges $q_+=-q_-=q$, so $N_+=N_-=N$. We have:

$$Q=\frac{1}{V^{N_++N_-}}\int\frac{\mathcal{D}\varphi}{C_0}e^{-\frac{\beta}{2}(\varphi G_0^{-1}\varphi)}Q_+^{N_+}[\varphi]Q_-^{N_-}[\varphi], \tag{3.78}$$

where

$$Q_{\pm}[\varphi] = \int d\mathbf{r} : e^{\pm i\beta q\varphi(\mathbf{r})} :_{G_0} = \int d\mathbf{r}\, e^{\pm i\beta q\varphi(\mathbf{r})} e^{\frac{\beta q^2 G_0(0)}{2}}. \tag{3.79}$$

The symbol $: (\dots) :_{G_0}$ denotes the normal form of the functional (see Chap. 1).

Let's utilize the Gaussian equivalent representation method to compute the configuration integral. Using the identity

$$Q^N = \frac{N!}{2\pi i} \oint\limits_C \frac{e^{Qz}}{z^{N+1}} dz, \tag{3.80}$$

we can represent the functional integral in the form

$$Q = \frac{(N!)^2}{(2\pi i)^2 V^{2N}} \oint\limits_{C_1} \frac{dz_1}{z_1^{N+1}} \oint\limits_{C_2} \frac{dz_2}{z_2^{N+1}} I(z_1, z_2),$$

where we introduce the function of complex variables

$$I(z_1, z_2) = \int \frac{\mathcal{D}\varphi}{C_0} e^{-\frac{\beta}{2}(\varphi G_0^{-1}\varphi) + z_1 Q_+[\varphi] + z_2 Q_-[\varphi]}. \tag{3.81}$$

Integration is performed over closed contours in the complex planes around the points $z_{1,2} = 0$.

As we consider the case of a spatially-homogeneous Coulomb gas, where the local electrical neutrality condition must be met, there is no need for a variable shift. Instead, we simply transition to a new Gaussian measure with a Green's function $G(\mathbf{x} - \mathbf{y})$:

$$\begin{aligned} I(z_1, z_2) &= \frac{C}{C_0} \int \frac{\mathcal{D}\varphi}{C} \exp\left[-\frac{\beta}{2}(\varphi G^{-1}\varphi)\right] \\ &\times \exp\left[-\frac{1}{2}(\varphi[G_0^{-1} - G^{-1}]\varphi) + z_1 A \int_V d\mathbf{x} : e^{i\beta q\varphi(\mathbf{x})} :_G + z_2 A \int_V d\mathbf{x} : e^{-i\beta q\varphi(\mathbf{x})} :_G\right] \\ &= \exp\left[V R_0(z_1, z_2)\right] \int \frac{\mathcal{D}\varphi}{C} e^{-\frac{\beta}{2}(\varphi G^{-1}\varphi) + \tilde{W}[\varphi]}, \end{aligned} \tag{3.82}$$

where

$$R_0(z_1, z_2) = \frac{1}{V} \ln \frac{C}{C_0} - \frac{1}{2V} tr\left(G[G_0^{-1} - G^{-1}]\right) + (z_1 + z_2)A, \tag{3.83}$$

$$A = \exp\left[\frac{\beta q^2 \left(G_0(0) - G(0)\right)}{2}\right], \tag{3.84}$$

$$\tilde{W}[\varphi] = z_1 A \int d\mathbf{x} : e_2^{i\beta q\varphi(\mathbf{x})} :_G + z_2 A \int_V d\mathbf{x} : e_2^{-i\beta q\varphi(\mathbf{x})} :_G, \tag{3.85}$$

$$e_2^x = e^x - 1 - x - \frac{x^2}{2}. \tag{3.86}$$

The condition of absence of the quadratic terms in the exponent leads to the following self-consistency equation:

$$\int d\mathbf{x} \int d\mathbf{y}\, \varphi(\mathbf{x})[G^{-1}(\mathbf{x}-\mathbf{y}) - G_0^{-1}(\mathbf{x}-\mathbf{y})]\varphi(\mathbf{y}) = \beta q^2(z_1+z_2)A \int d\mathbf{x}\, \varphi^2(\mathbf{x}). \tag{3.87}$$

Turning to the Fourier representation:

$$\varphi(\mathbf{x}) = \int \frac{d\mathbf{k}}{(2\pi)^3} \widetilde{\varphi}(\mathbf{k}) e^{i\mathbf{k}\mathbf{x}}, \tag{3.88}$$

$$G_0(\mathbf{x}) = \int \frac{d\mathbf{k}}{(2\pi)^3} G_0(\mathbf{k}) e^{i\mathbf{k}\mathbf{x}}, \tag{3.89}$$

$$G(\mathbf{x}) = \int \frac{d\mathbf{k}}{(2\pi)^3} G(\mathbf{k}) e^{i\mathbf{k}\mathbf{x}}, \tag{3.90}$$

$$\tilde{G}_0(\mathbf{k}) = \frac{4\pi}{\varepsilon \mathbf{k}^2}, \tag{3.91}$$

we obtain:

$$\int \frac{d\mathbf{k}}{(2\pi)^3} \left[\frac{1}{G(\mathbf{k})} - \frac{1}{G_0(\mathbf{k})} - \beta q^2(z_1+z_2)A\right] |\widetilde{\varphi}(\mathbf{k})|^2 = 0. \tag{3.92}$$

As this identity is valid for the arbitrary functions $\widetilde{\varphi}(\mathbf{k})$, we get:

$$G(\mathbf{k}) = \frac{G_0(\mathbf{k})}{1 + \beta q^2(z_1+z_2)AG_0(\mathbf{k})}. \tag{3.93}$$

In the zeroth approximation:

$$I(z_1, z_2) \approx e^{VR_0(z_1,z_2)}, \tag{3.94}$$

where

$$R_0(z_1, z_2) = \frac{1}{2} \int \frac{d\mathbf{k}}{(2\pi)^3} \left[\ln \frac{G(\mathbf{k})}{G_0(\mathbf{k})} - \frac{G(\mathbf{k})}{G_0(\mathbf{k})} + 1\right] + (z_1+z_2)A. \tag{3.95}$$

Therefore,

$$Q \approx \frac{(N!)^2}{(2\pi i)^2 V^{2N}} \oint_{C_1} \frac{dz_1}{z_1^{N+1}} \oint_{C_2} \frac{dz_2}{z_1^{N+1}} e^{VR_0(z_1,z_2)}. \tag{3.96}$$

We can evaluate the obtained integral using the steepest descent method. It is convenient to introduce new integration variables, $c_{1,2} = z_{1,2}\beta q^2 A$, by which the configuration integral will take the form:

$$Q \approx \frac{1}{(2\pi i)^2} \oint\limits_{C_1'} \frac{dc_1}{c_1} \oint\limits_{C_2'} \frac{dc_2}{c_2} e^{V\mathcal{F}(c_1,c_2)},$$

where

$$\mathcal{F}(c_1, c_2) = \frac{1}{2} \int \frac{d\mathbf{k}}{(2\pi)^3} \left[\frac{(c_1 + c_2)G_0(\mathbf{k})}{1 + (c_1 + c_2)G_0(\mathbf{k})} + \frac{2n(c_1 + c_2)G_0^2(\mathbf{k})}{1 + (c_1 + c_2)G_0(\mathbf{k})} \right.$$
$$\left. - \ln\left(1 + (c_1 + c_2)G_0(\mathbf{k})\right) \right] + \frac{c_1 + c_2}{\beta q^2} + n \left(\ln \frac{n\beta q^2}{c_1} + \ln \frac{n\beta q^2}{c_2} \right) - 2n. \quad (3.97)$$

Equations for the saddle points are

$$\frac{\partial \mathcal{F}}{\partial c_1} = 0, \quad \frac{\partial \mathcal{F}}{\partial c_2} = 0. \quad (3.98)$$

From the symmetry of the problem, it follows that $c_1 = c_2 = \beta n q^2$. Substituting the expression $c = \beta n q^2$ in (3.97) after the evaluation of the integrals, we arrive at the well-known Debye–Hückel expression

$$\frac{\beta F_{ex}^{(0)}}{V} = -\frac{\kappa^3}{12\pi}. \quad (3.99)$$

Fourier transform of the Green's function is

$$G(\mathbf{k}) = \frac{4\pi}{\varepsilon(\mathbf{k}^2 + \kappa^2)}, \quad (3.100)$$

and the Green's function is given by

$$G(\mathbf{x} - \mathbf{y}) = \frac{e^{-\kappa|\mathbf{x}-\mathbf{y}|}}{\varepsilon|\mathbf{x} - \mathbf{y}|}, \quad (3.101)$$

where $\kappa = (8\pi q^2 n/\varepsilon k_B T)^{1/2}$.

Thus, we obtain the following representation for the configuration integral

$$Q = e^{-\beta F_{ex}^{(0)}} \int \frac{\mathcal{D}\varphi}{C} e^{-\frac{\beta}{2}(\varphi G^{-1}\varphi) + \tilde{W}[\varphi]}, \quad (3.102)$$

where

$$\tilde{W}[\varphi] = 2n \int d\mathbf{x} \left(: \cos\left(\beta q \varphi(\mathbf{x})\right) - 1 + \frac{\beta^2 q^2 \varphi^2(\mathbf{x})}{2} :_G \right). \tag{3.103}$$

To evaluate the corrections to the lowest approximation (3.99), we need to use the cumulant expansion (see Chap. 1) of the functional $\tilde{W}[\varphi]$, i.e.,

$$Q = e^{-\beta F_{ex}^{(0)}} \exp\left(\sum_{k=1}^{\infty} \frac{1}{k!} \left\langle \tilde{W}^k \right\rangle_c \right), \tag{3.104}$$

where

$$\left\langle \tilde{W}[\varphi] \right\rangle_c = \left\langle \tilde{W}[\varphi] \right\rangle,$$

$$\left\langle \tilde{W}^2[\varphi] \right\rangle_c = \left\langle \tilde{W}^2[\varphi] \right\rangle - \left\langle \tilde{W}[\varphi] \right\rangle^2,$$

$$\left\langle \tilde{W}^3[\varphi] \right\rangle_c = \left\langle \tilde{W}^3[\varphi] \right\rangle - 3\left\langle \tilde{W}^2[\varphi] \right\rangle \left\langle \tilde{W}[\varphi] \right\rangle + 2\left\langle \tilde{W}[\varphi] \right\rangle^3$$

etc. The symbol $\langle (\ldots) \rangle$ denotes the average over the Gaussian measure with a new Green's function $G(\mathbf{x} - \mathbf{y})$.

The first correction is identically zero due to the determination of the normal form of the functional

$$\left\langle \tilde{W}[\varphi] \right\rangle_c = \left\langle \tilde{W}[\varphi] \right\rangle \equiv 0. \tag{3.105}$$

To evaluate the second correction, we need the following auxiliary relations, which can be obtained after calculating the Gaussian functional integrals and applying Wick's theorem (see Chap. 1):

$$\left\langle : \varphi^2(\mathbf{x}_1) :_G : \varphi^2(\mathbf{x}_2) :_G \right\rangle = 2G^2(\mathbf{x}_1 - \mathbf{x}_2), \tag{3.106}$$

$$\left\langle : \varphi^2(\mathbf{x}_1) :_G : \cos(\beta q \varphi(\mathbf{x}_2)) :_G \right\rangle = -\beta q^2 G^2(\mathbf{x}_1 - \mathbf{x}_2), \tag{3.107}$$

$$\langle : \cos(\beta q \varphi(\mathbf{x}_1)) :_G : \cos(\beta q \varphi(\mathbf{x}_2)) :_G \rangle = \cosh\left(\beta q^2 G(\mathbf{x}_1 - \mathbf{x}_2) \right). \tag{3.108}$$

Thus, we have the following:

$$\begin{aligned}
\left\langle \tilde{W}^2[\varphi] \right\rangle_c &= \left\langle \tilde{W}^2[\varphi] \right\rangle - \left\langle \tilde{W}[\varphi] \right\rangle^2 = \left\langle \tilde{W}^2[\varphi] \right\rangle \\
&= 4n^2 \iint d\mathbf{x}_1 d\mathbf{x}_2 \left(: \cos\left(\beta q \varphi(\mathbf{x}_1)\right) - 1 + \frac{\beta^2 q^2 \varphi^2(\mathbf{x}_1)}{2} :_G \right) \\
&\times \left(: \cos\left(\beta q \varphi(\mathbf{x}_2)\right) - 1 + \frac{\beta^2 q^2 \varphi^2(\mathbf{x}_2)}{2} :_G \right) \\
&= 4n^2 V \int d\mathbf{x} \left(\cosh\left(\beta q^2 G(\mathbf{x})\right) - 1 - \frac{\beta^2 q^4 G^2(\mathbf{x})}{2} \right).
\end{aligned} \tag{3.109}$$

Therefore, the correction to the zeroth approximation of the Gaussian equivalent representation has the form

$$\frac{\beta F_{ex}^{(2)}}{V} = -8\pi n^2 \int_0^\infty drr^2 \left(\cosh\left(\beta q^2 G(r)\right) - 1 - \frac{\beta^2 q^4 G^2(r)}{2} \right), \tag{3.110}$$

where

$$G(r) = \frac{e^{-\kappa r}}{\varepsilon r}. \tag{3.111}$$

As easily seen, the integral (3.110) diverges at the lower limit. It can be shown that the subsequent corrections, which can be evaluated within the cumulant expansion, also diverge. The divergence of the integral is related to the fact that the Coulomb gas model represents ions as point charges. To give meaning to the expression (3.110), we need to introduce a "cut-off" at the lower limit of the integral by some value a, which determines the characteristic size of the ions, i.e.

$$\frac{\beta F_{ex}^{(2)}}{V} = -8\pi n^2 \int_a^\infty dr\, r^2 \left(\cosh\left(\beta q^2 G(r)\right) - 1 - \frac{\beta^2 q^4 G^2(r)}{2} \right). \tag{3.112}$$

The divergence at the lower limit arises because the Coulomb gas model treats ions as point charges, causing the electrostatic energy of the ion pair to tend toward infinity. Thus, we can choose a cut-off parameter as the sum of the ionic radii, i.e. $a = a_+ + a_-$, which represents the distance between the centres of the ions in an ion pair. A similar approach will be taken in Chap. 4, when considering the problem of surface tension in electrolyte solutions. To self-consistently resolve these divergences, it's necessary to consider the finite size of the ions (hard core) within the theoretical framework. A field-theoretic approach, which is capable to systematically incorporate the excluded volume of ions, thereby eliminating the divergence of corrections to the Gaussian approximation, is proposed in Ref. [65].

3.6 One-Component Plasma. Gaussian Approximation and Further Corrections

Let's explore the one-component plasma (OCP) model, widely used in condensed matter physics. OCP consists of N point charged particles (ions) placed in a structureless compensating background with a constant charge density $\rho = -qN/V$, where V is the volume of the system. The total interaction potential energy is given by:

$$U = \frac{1}{2}\left(\hat{\rho} G_0 \hat{\rho}\right) - U_{self}, \tag{3.113}$$

where the microscopic charge density is

$$\hat{\rho}(\mathbf{r}) = q\sum_{j=1}^{N}\delta\left(\mathbf{r}-\mathbf{r}_j\right)+\rho, \tag{3.114}$$

so

$$\int d\mathbf{r}\,\hat{\rho}(\mathbf{r}) = 0. \tag{3.115}$$

After applying the HS transformation, the configuration integral of the OCP can be expressed as follows:

$$Q = \frac{1}{V^N}\int\frac{\mathcal{D}\varphi}{C_0}e^{-\frac{\beta}{2}(\varphi G_0^{-1}\varphi)}Q_{ion}^N[\varphi], \tag{3.116}$$

where

$$Q_{ion}[\varphi] = \int d\mathbf{r} : e^{\pm i\beta q\varphi(\mathbf{r})} :_{G_0} . \tag{3.117}$$

As before, the symbol $:(\ldots):_{G_0}$ denotes the normal form of the functional. Similarly to the previous section concerning the Coulomb gas, we utilize the technique of Gaussian equivalent representation to compute the configuration integral. Utilizing the identity

$$Q_{ion}^N = \frac{N!}{2\pi i}\oint_C\frac{e^{Q_{ion}z}}{z^{N+1}}dz, \tag{3.118}$$

we represent the configuration integral in the form

$$Q = \frac{N!}{2\pi i V^N}\oint_C\frac{dz}{z^{N+1}}I(z), \tag{3.119}$$

where we introduce the function of the complex variable

$$I(z) = \int\frac{\mathcal{D}\varphi}{C_0}e^{-\frac{\beta}{2}(\varphi G_0^{-1}\varphi)+zQ_{ion}[\varphi]}. \tag{3.120}$$

The integration is performed over the closed contour in the complex plane around the point $z = 0$. Additionally, we consider that the fluctuation potential satisfies the relation $\int d\mathbf{r}\,\varphi(\mathbf{r}) = 0$.

After evaluations similar to those in the previous section, we obtain:

$$I(z) = \exp\left[V R_0(z)\right]\int\frac{\mathcal{D}\varphi}{C}e^{-\frac{\beta}{2}(\varphi G^{-1}\varphi)+\tilde{W}[\varphi]}, \tag{3.121}$$

where

$$R_0(z) = \frac{1}{V}\ln\frac{C}{C_0} - \frac{1}{2V}tr\left(G[G_0^{-1} - G^{-1}]\right) + zA, \tag{3.122}$$

$$A = \exp\left[\frac{\beta q^2\left(G_0(0) - G(0)\right)}{2}\right], \tag{3.123}$$

$$\tilde{W}[\varphi] = zA\int d\mathbf{x} : e_2^{i\beta q\varphi(\mathbf{x})} :_G, \tag{3.124}$$

$$e_2^x = e^x - 1 - x - \frac{x^2}{2}. \tag{3.125}$$

Thus, we arrive at the expression for the zeroth approximation within the GER

$$Q \approx \frac{N!}{2\pi i V^N}\oint_C \frac{dz}{z^{N+1}} e^{VR_0(z)}. \tag{3.126}$$

As before, we evaluate the integral using the saddle-point method. It is convenient to introduce new integration variable $c = z\beta q^2 A$, which leads to the following form of the functional integral:

$$Q \approx \frac{1}{2\pi i}\oint_C \frac{dc}{c} e^{V\mathcal{F}(c)},$$

where

$$\mathcal{F}(c) = \frac{1}{2}\int \frac{d\mathbf{k}}{(2\pi)^3}\left[\frac{cG_0(\mathbf{k})}{1 + cG_0(\mathbf{k})} + \frac{2ncG_0^2(\mathbf{k})}{1 + cG_0(\mathbf{k})} - \ln\left(1 + cG_0(\mathbf{k})\right)\right] + \frac{c}{\beta q^2} + n\ln\frac{n\beta q^2}{c} - n. \tag{3.127}$$

The equation for the saddle point, $\partial\mathcal{F}/\partial c = 0$, yields $c = \beta n q^2$. Substituting this expression into (3.127) after evaluating the integral, we obtain the Debye–Hückel expression

$$\frac{\beta F_{ex}^{(0)}}{V} = \frac{1}{2}\int \frac{d\mathbf{k}}{(2\pi)^3}\left(\ln\left(1 + \frac{\kappa^2}{k^2}\right) - \frac{\kappa^2}{k^2}\right) = -\frac{\kappa^3}{12\pi}, \tag{3.128}$$

where $\kappa = r_D^{-1} = (4\pi q^2 n/k_B T\varepsilon)^{1/2}$ is the inverse Debye radius, related to the presence of the mobile ions in OCP.

To evaluate the corrections to the zeroth approximation (3.128), we can make use of the following representation of the configuration integral

$$Q = e^{-\beta F_{ex}^{(0)}}\int \frac{\mathcal{D}\varphi}{C} e^{-\frac{\beta}{2}(\varphi G^{-1}\varphi) + \tilde{W}[\varphi]}, \tag{3.129}$$

where

$$\tilde{W}[\varphi] = n \int d\mathbf{x} : e_2^{i\beta q\varphi(\mathbf{x})} :_G, \tag{3.130}$$

As in the case of the Coulomb gas, the first correction is identically zero. Utilizing Wick's theorem, we obtain the correction to the zeroth approximation within the Gaussian equivalent representation for OCP

$$\frac{\beta F_{ex}^{(2)}}{V} = -2\pi n^2 \int_0^\infty drr^2 \left(e^{-\beta q^2 G(r)} - 1 + \beta q^2 G(r) - \frac{\beta^2 q^4 G^2(r)}{2} \right), \tag{3.131}$$

where

$$G(r) = \frac{e^{-\kappa r}}{\varepsilon r}. \tag{3.132}$$

As easily understood, the integral (3.131), unlike the correction (3.110) for the Coulomb gas, converges without the need for a cutoff at the lower limit of the integration. It can be shown that further corrections within the cumulant expansion also converge. The convergence of the integral is due to the fact that OCP considers only one type of ions in the compensating background, unlike the Coulomb gas model, where point ions of both charges are present.

It should be noted that a diagrammatic field-theoretic approach for the systematic evaluation of the corrections to the Debye–Hückel approximation for OCP model, based on the Hubbard-Schofield transformation, was proposed in Ref. [66].

3.7 Equation of State of the One-Component Plasma Beyond the Gaussian Approximation. Modified Gaussian Approximation

Now let's focus on the Gaussian approximation for the excess free energy of the OCP

$$\frac{\beta F_{ex}}{V} = \frac{1}{2} \int \frac{d\mathbf{k}}{(2\pi)^3} \left(\ln\left(1 + \frac{\kappa^2}{k^2}\right) - \frac{\kappa^2}{k^2} \right). \tag{3.133}$$

As demonstrated earlier, integrating over the infinite reciprocal space yields the Debye–Hückel expression, applicable under condition of sufficiently weak electrostatic interactions between the ions (weak-coupling regime), i.e., when $n\kappa^{-3} \ll 1$. Equation (3.133) serves as a foundation to derive an interpolation formula for the OCP free energy, extending its validity beyond the weak Coulombic interaction regime. To achieve this, we need to introduce an ultraviolet cutoff in the integration over the reciprocal space wave-vectors $\mathbf{k}$ in (3.133), akin to the cutoff in Debye's quantum theory of crystal heat capacity [12]. Such a cutoff was proposed by Brilliantov [67], based on the condition that the total number of modes contributing to the excess free energy of the OCP equals the number of translational degrees of freedom of the ions, i.e.

$$2V \int\limits_{|\mathbf{k}|<\Lambda} \frac{d\mathbf{k}}{(2\pi)^3} = 3N, \tag{3.134}$$

from where we obtain

$$\Lambda = (9\pi^2 n)^{1/3}. \tag{3.135}$$

Note that the coefficient 2 in (3.134) accounts for the fact that the number of modes is twice the number of wave vectors, as each **k** corresponds to both sinusoidal and cosinusoidal modes (the amplitude of **k**th mode is a complex number). This variant of the Gaussian approximation, with an ultraviolet cutoff, we will refer to as the "modified Gaussian approximation".

Applying this cutoff, we obtain

$$-\frac{\beta F_{ex}}{N} = \frac{V}{4\pi^2 N} \int\limits_0^\Lambda dk\, k^2 \left(\ln\left(1 + \frac{\kappa^2}{k^2}\right) - \frac{\kappa^2}{k^2} \right) = \frac{9}{4} \int\limits_0^1 dx\, x^2 \left(\ln\left(1 + \frac{b\Gamma}{x^2}\right) - \frac{b\Gamma}{x^2} \right), \tag{3.136}$$

where we introduce a numerical factor $b = 2/3(2/\pi^2)^{1/3}$, the so-called plasma-parameter $\Gamma = q^2/(\varepsilon k_B T a_c)$ and the length scale of the order of the average distance between ions in the OCP $a_c = (3/4\pi n)^{1/3}$.

Evaluating the integral, we arrive at the final interpolation formula

$$\frac{F_{ex}}{N k_B T} = \frac{3}{4} \left(\ln\left(1 + b\Gamma\right) - b\Gamma \right) - \frac{3}{2} (b\Gamma)^{3/2} \arctan\left(\frac{1}{\sqrt{b\Gamma}} \right), \tag{3.137}$$

from which we obtain the following asymptotics

$$\frac{F_{ex}}{N k_B T} = \begin{cases} -\frac{1}{\sqrt{3}} \Gamma^{3/2}, & \Gamma \ll 1 \\ -C\Gamma, & \Gamma \gg 1, \end{cases} \tag{3.138}$$

where $C = 9/4b \approx 0.881\ldots$ is a numeric factor. The first regime describes the Debye–Hückel theory for the case of sufficiently dilute OCP, whereas the second one reproduces with good accuracy (the true coefficient, obtained within the Monte Carlo simulation, is $C = 0.899\ldots$) the expression for the excess free energy of the OCP in the strong coupling regime. As shown in Ref. [67], the interpolation formula (3.137) is in good agreement with the Monte Carlo data not only in the limiting regimes but also for intermediate values of the plasma parameter ($\Gamma \sim 1 - 10$), where the Gaussian approximation is no longer applicable (Fig. 3.1).

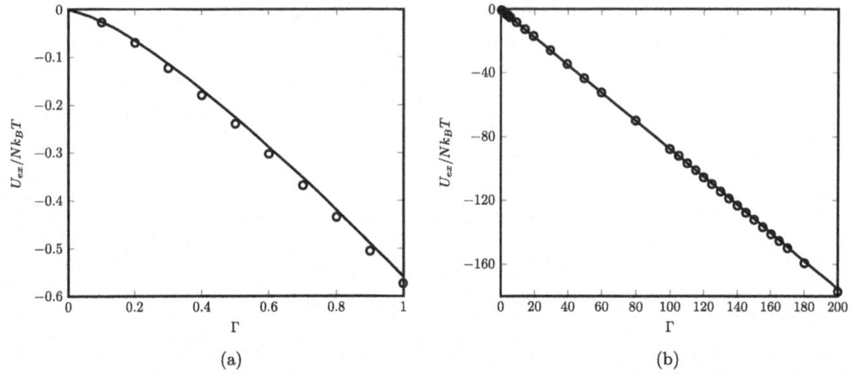

Fig. 3.1 Dependence of the excess internal energy of the OCP, U_{ex}/Nk_BT, on the plasma parameter Γ; points represent the Monte Carlo data (Refs. [68, 69] for $0.1 \leq \Gamma \leq 1$ a, and Refs. [70, 71] for $\Gamma > 1$ b); lines represent the theoretical result

We should note that an earlier study by Kaklugin [72, 73] applied a different criterion for the ultraviolet cutoff to estimate the excess free energy of the OCP. However, this criterion led to similar interpolation formula (differing by a factor of order of unity). In other words, similar to the described above cutoff procedure, the modes corresponding to distances between the particles smaller than the average distance $n^{-1/3}$, were excluded from the integration. It is important to emphasize that both approaches provide an estimation for the excess free energy of the OCP, and there is no strict basis for favoring one cutoff criterion over the other. Thus, the modified Gaussian approximation is a heuristic method, that allows for the extension of the applicability region of the standard Gaussian approximation.

The modified Gaussian approximation is applied to describe the thermodynamic properties of various ionic systems [74–79].

3.8 Equation of State of the Hard-Core One-Component Plasma

A model similar to the OCP, known as the hard-core one-component plasma (HCOCP), incorporates the repulsion of hard cores of particles to provide a more satisfactory description of electrostatic interactions at short distances. HCOCP is of independent interest as it belongs to the class of so-called primitive models, used as a basis for the theoretical description of various physico-chemical systems. In this section, we obtain the equation of state of the HCOCP within the Hubbard-Schofield-Brilliantov method, formulated in Chap. 2, applying the modified Gaussian approximation formulated in the previous section.

Effective Hamiltonian of HCOCP. Let us examine the Hamiltonian of HCOCP interactions:

$$H = U_{hc} + \frac{1}{2}{\sum_{\mathbf{k}}}' \nu_{\mathbf{k}}(\hat{n}_{\mathbf{k}}\hat{n}_{-\mathbf{k}} - \rho). \tag{3.139}$$

where

$$\hat{n}_{\mathbf{k}} = \frac{1}{\sqrt{V}}\sum_{j=1}^{N} e^{-i\mathbf{k}\mathbf{r}_j}, \tag{3.140}$$

$\mathbf{r}j$ denotes the radius-vector of the jth particle, $\nu_{\mathbf{k}} = 4\pi q^2/\varepsilon k^2$, and U_{hc} corresponds to the the hard cores (spheres) interaction. Here, $L^3 = V$, $\mathbf{k} = (k_x, k_y, k_z)$, $k_i = 2\pi l_i/L$. The prime at the sum symbol denotes the absence of the term with $\mathbf{k} = 0$ in a summation over $\mathbf{k}$, which cancels with the terms describing the self-action of the compensating background and its interaction with the particles. U_{hc} is the Hamiltonian of the interactions among the hard spheres.

Configuration integral in NVT-ensemble can be written as

$$Q = Q_R \left\langle \exp\left[-\frac{\beta}{2}{\sum_{\mathbf{k}}}' \nu_{\mathbf{k}}(\hat{n}_{\mathbf{k}}\hat{n}_{-\mathbf{k}} - \rho)\right]\right\rangle_R , \tag{3.141}$$

where Q_R is the configuration integral of the reference hard spheres system. Following the Hubbard-Schofield scheme discussed in Chap. 2, we can transform the integral to the form

$$Q = Q_R \int {\prod_{\mathbf{k}}}' a_{\mathbf{k}} d\varphi_{\mathbf{k}} \exp\left[-\frac{\beta}{2}{\sum_{\mathbf{k}}}' \nu_{\mathbf{k}}^{-1}\varphi_{\mathbf{k}}\varphi_{-\mathbf{k}}\right]\left\langle \exp\left[i\beta{\sum_{\mathbf{k}}}' \hat{n}_{\mathbf{k}}\varphi_{-\mathbf{k}}\right]\right\rangle_R , \tag{3.142}$$

where $a_{\mathbf{k}} = (2\pi\beta\nu_{\mathbf{k}})^{-1/2}\exp[\beta\rho\nu_{\mathbf{k}}/2]$, and the integration is performed under the condition $\varphi_{-\mathbf{k}} = \varphi^*_{\mathbf{k}}$. Applying the cumulant expansion, we arrive at the following expression for the configuration integral

$$Q = Q_R \int {\prod_{\mathbf{k}}}' a_{\mathbf{k}} d\varphi_{\mathbf{k}} \exp[-\beta\mathcal{H}[\{\varphi_{\mathbf{k}}\}]] \tag{3.143}$$

with the effective Hamiltonian

$$\beta\mathcal{H} = \sum_{l=2}^{\infty} V^{1-l/2} {\sum_{\mathbf{k}_1,\ldots,\mathbf{k}_l}}' u_l(\mathbf{k}_1, \ldots, \mathbf{k}_l)\varphi_{\mathbf{k}_1}\ldots\varphi_{\mathbf{k}_l}, \tag{3.144}$$

where

$$u_2(\mathbf{k}_1, \mathbf{k}_2) = \frac{\beta}{2}\delta_{\mathbf{k}_1+\mathbf{k}_2,0}\left(\frac{\varepsilon k_1^2}{4\pi q^2} + \beta\langle\hat{n}_{\mathbf{k}_1}\hat{n}_{\mathbf{k}_2}\rangle_{cR}\right), \tag{3.145}$$

$$u_l(\mathbf{k}_1, \ldots, \mathbf{k}_l) = -i^l \beta^l \frac{V^{l/2-1}}{l!} \delta_{\mathbf{k}_1+\cdots+\mathbf{k}_l,0} \left\langle \hat{n}_{\mathbf{k}_1} \ldots \hat{n}_{\mathbf{k}_l} \right\rangle_{cR}, \; l > 2. \tag{3.146}$$

As demonstrated in Chap. 2, all coefficients of the effective Hamiltonian can be expressed through the correlation functions of the reference system. Specifically, the second-order cumulant takes the form:

$$\left\langle \hat{n}_{\mathbf{k}} \hat{n}_{-\mathbf{k}} \right\rangle_{cR} = n \left(1 + n\tilde{h}_2(\mathbf{k}) \right). \tag{3.147}$$

Gaussian approximation for the effective Hamiltonian. To evaluate the excess free energy of the HCOCP within the Gaussian approximation for the effective Hamiltonian, let's omit the terms higher the second order with respect to the field variable φ. In this case, we obtain

$$Q = Q_R \int \prod_{\mathbf{k}}{}' a_{\mathbf{k}} d\varphi_{\mathbf{k}} \exp\left[-\frac{\beta}{2} \sum_{\mathbf{k}}{}' \left(\nu_{\mathbf{k}}^{-1} + \frac{\beta n}{1 - n\tilde{c}_2(\mathbf{k})} \right) \varphi_{\mathbf{k}} \varphi_{-\mathbf{k}} \right], \tag{3.148}$$

where we use the Ornstein–Zernike equation

$$\tilde{h}_2(\mathbf{k}) = \frac{\tilde{c}_2(\mathbf{k})}{1 - n\tilde{c}_2(\mathbf{k})}. \tag{3.149}$$

After evaluating the Gaussian integrals of $\varphi_{\mathbf{k}}$, and performing simple computations, we arrive at the following expression:

$$Q = Q_R \prod_{k_z>0} \frac{\exp\left[\beta n \nu_{\mathbf{k}}\right]}{\nu_{\mathbf{k}}} \left(\frac{1}{\nu_{\mathbf{k}}} + \frac{\beta n}{1 - n\tilde{c}_2(\mathbf{k})} \right)^{-1}. \tag{3.150}$$

Due to the long-range character of the Coulomb interaction, the main contribution to the free energy comes from the long-wave modes of the density fluctuations, corresponding to small values of the wave vector $|\mathbf{k}|$. Therefore, we can utilize the expansion of $\tilde{c}_2(\mathbf{k})$ around zero. The configuration integral of the system then takes the form:

$$Q = Q_R \prod_{k_z>0} \frac{\exp\left[\beta n \nu_{\mathbf{k}}\right]}{\nu_{\mathbf{k}}} \left(\frac{1}{\nu_{\mathbf{k}}} + \frac{\beta n}{1 - n\tilde{c}_2(0)} - k^2 \frac{n^2 \tilde{c}_2''(0)}{(1 - n\tilde{c}_2(0))^2} \right)^{-1}. \tag{3.151}$$

Utilizing the relation for $\tilde{c}_2''(0)$ and $\tilde{c}_2(0) = (z_0 - 1)/(nz_0)$, as represented in Chap. 2, we arrive at the following expression for the configuration integral:

$$Q = Q_R \prod_{k_z>0} \exp\left[\beta n \nu_{\mathbf{k}}\right] \left(\Theta + n z_0 \nu_{\mathbf{k}}\right)^{-1}, \tag{3.152}$$

where

$$\Theta = 1 + \frac{6q^2}{5\varepsilon d k_B T} \frac{\eta^2(1-\eta)^4(16-11\eta+4\eta^2)}{(1+4\eta+4\eta^2-4\eta^3+\eta^4)^2}. \tag{3.153}$$

Thus, the excess free energy of the HCOCP has the form

$$-\beta F_{ex} = \ln Q = -\beta F_{hc} + \frac{1}{2}\sum_{\mathbf{k}}{}' (\beta n \nu_{\mathbf{k}} - \ln(\Theta + n z_0 \nu_{\mathbf{k}})). \tag{3.154}$$

The first term on the right-hand side of (3.154) is related to the excess free energy of the hard spheres system, for which we can utilize the rather accurate Carnahan-Starling approximation, while the second term corresponds to the contribution of the electrostatic interactions. Similar to the previous section, we apply an ultraviolet cutoff with a cutoff parameter $\Lambda = (9\pi^2 n)^{1/3}$, determined from the relation (3.134), when evaluating the electrostatic contribution to the free energy. Thus, we obtain:

$$\begin{aligned} \frac{\beta F}{N} &= \frac{\beta F_{hc}}{N} + \frac{1}{2}\frac{V}{8\pi^3}\frac{4\pi}{N}\int_0^{\Lambda} dk\, k^2 \left(\ln(\Theta + \beta z_0 \nu_{\mathbf{k}}) - \beta n \nu_{\mathbf{k}}\right) \\ &= \frac{\beta F_{hc}}{N} + \frac{9}{4}\int_0^1 dx\, x^2 \left[\ln\left(\Theta + \frac{b\Gamma z_0}{x^2}\right) - \frac{b\Gamma}{x^2}\right], \end{aligned} \tag{3.155}$$

where we replace the sum by the integral in the thermodynamic limit according to the rule

$$\sum_{\mathbf{k}}{}'(\ldots) = V\int \frac{d\mathbf{k}}{(2\pi)^3}(\ldots). \tag{3.156}$$

After the evaluation of the integral we arrive at the following expression

$$\frac{F_{ex}}{N k_B T} = \frac{4\eta - 3\eta^2}{(1-\eta)^2} + \frac{3}{4}\left[\ln(\Theta + b\Gamma z_0) - b\Gamma\left(3 - \frac{2z_0}{\Theta}\right)\right] - \frac{3}{2}\left(\frac{b\Gamma z_0}{\Theta}\right)^{3/2} \arctan\left(\sqrt{\frac{\Theta}{b\Gamma z_0}}\right), \tag{3.157}$$

where

$$z_0 = \frac{(1-\eta)^4}{1+4\eta+4\eta^2-4\eta^3+\eta^4}. \tag{3.158}$$

At $\eta \to 0$, the expression (3.157) converges to the previously obtained expression (3.137). As stated in Ref. [80], the expression (3.157) accurately describes the thermodynamic properties of the HCOCP over a wide range of plasma parameter and packing fraction.

Chapter 4
Self-consistent Field Theory of Ionic Fluids. Grand Canonical Ensemble Formulation

In this chapter, we continue the formulation of statistical field theory for ionic fluids using the grand canonical ensemble approach. This approach allows us to take into account the excluded volume of ions and to derive the modified Poisson-Boltzmann equations that describe the system more accurately. Additionally, the grand canonical framework makes it easier to formulate a variational field theory that takes into account ionic correlations beyond the mean-field approximation, which was introduced in Chap. 1.

4.1 A Self-consistent Field Theory of Ionic Fluids Accounting for Steric and Specific Interactions of the Ions

In Chap. 3, we have presented the self-consistent field theory of the Coulomb gas, which considers only the Coulomb interactions. In this section, we explore a more comprehensive theory [81], which incorporates both steric ion interactions (finite size), characterized by purely repulsive potentials, and short-range specific interactions associated with the chemical nature of the ions, which can be reduced to both attraction and repulsion. Let us assume that the potentials of the non-electrostatic interactions between the ions can be represented as

$$\mathcal{U}_{\alpha\gamma}(\mathbf{r}) = U_{\alpha\gamma}(\mathbf{r}) + V_{\alpha\gamma}(\mathbf{r}), \tag{4.1}$$

where the first term represents purely repulsive short-range interactions, while the second term accounts for contributions from the short-range specific interactions, where $\alpha, \gamma = \pm$. To construct the theory, it is more convenient to work in the Grand canonical (μVT) ensemble.

Y. A. Budkov and N. N. Kalikin, *Statistical Field Theory of Ion-Molecular Fluids*,
https://doi.org/10.1007/978-3-031-68364-0_4

Let us express the grand canonical partition function of the ionic fluid as follows

$$\Xi = \sum_{N_+=0}^{\infty}\sum_{N_-=0}^{\infty}\frac{z_+^{N_+}z_-^{N_-}}{N_+!N_-!}Q_{N_+N_-}, \tag{4.2}$$

where $z_{\pm} = \Lambda_{\mp}^{-3}e^{\beta\mu_{\pm}}\theta_{\pm}$ represents the activities of the ions with chemical potentials $\mu_{\pm}$, and the summation is performed over all possible numbers of ions; $\theta_{\pm}$—internal partition functions of the ions. The configuration integral takes the form

$$Q_{N_+N_-} = \int\prod_{j=1}^{N_+}d\mathbf{r}_j^{(+)}\int\prod_{k=1}^{N_-}d\mathbf{r}_k^{(-)}\exp\left[-\beta U_{N_+N_-}\right], \tag{4.3}$$

where the Hamiltonian of the interaction is expressed as follows

$$U_{N_+N_-} = U_{N_+N_-}^{(el)} + U_{N_+N_-}^{(R)} + U_{N_+N_-}^{(spc)} + U_{N_+N_-}^{(ext)} - U_{self} \tag{4.4}$$

where

$$U_{N_+N_-}^{(el)} = \frac{1}{2}\left(\hat{\rho}G_0\hat{\rho}\right) \tag{4.5}$$

denotes the Hamiltonian of the electrostatic interactions,

$$U_{N_+N_-}^{(R)} = \frac{1}{2}\sum_{\alpha\gamma}\left(\hat{n}_{\alpha}U_{\alpha\gamma}\hat{n}_{\gamma}\right), \tag{4.6}$$

defines the contribution of steric interactions of the ions and

$$U_{N_+N_-}^{(spc)} = \frac{1}{2}\sum_{\alpha\gamma}\left(\hat{n}_{\alpha}V_{\alpha\gamma}\hat{n}_{\gamma}\right) \tag{4.7}$$

is a contribution of the specific interactions of the ions. The Hamiltonian governing the interactions of the ions with external fields, characterized by potential energies $u_{\pm}(\mathbf{r})$, is given by

$$U_{N_+N_-}^{(ext)} = \sum_{\alpha}\left(\hat{n}_{\alpha}u_{\alpha}\right) \tag{4.8}$$

Besides, we introduce the local charge density of the system, $\hat{\rho}(\mathbf{r}) = \hat{\rho}_i(\mathbf{r}) + \rho_{ext}(\mathbf{r})$, which includes the charge density of the ions

$$\hat{\rho}_i(\mathbf{r}) = \sum_{\alpha} q_{\alpha}\hat{n}_{\alpha}(\mathbf{r}) \tag{4.9}$$

and density of the external charges $\rho_{ext}(\mathbf{r})$; $\hat{n}_\alpha(\mathbf{r}) = \sum_{j=1}^{N_\alpha} \delta\left(\mathbf{r} - \mathbf{r}_j^{(\alpha)}\right)$ denotes a local concentration of the ions of α-th species ($\alpha = \pm$). Intrinsic free energy of the particles is determined by the expression

$$U_{self} = \left(\frac{N_+ q_+^2}{2} + \frac{N_- q_-^2}{2}\right) G_0(0) + \frac{N_+ \mathcal{U}_{++}(0)}{2} + \frac{N_- \mathcal{U}_{--}(0)}{2}. \tag{4.10}$$

Applying the HS transformation, we arrive at the following representation of the grand canonical partition function

$$\Xi = \int \frac{\mathcal{D}\varphi}{C_0} \int \frac{\mathcal{D}\Psi}{\mathcal{N}_V} \exp\left[-\frac{\beta}{2}\left(\varphi G_0^{-1} \varphi\right) - \frac{\beta}{2} \sum_{\alpha\gamma} (\Psi_\alpha V_{\alpha\gamma}^{-1} \Psi_\gamma) + i\beta(\rho_{ext}\varphi)\right] \Xi_R[\mu_+ + i\chi_+; \mu_- + i\chi_-], \tag{4.11}$$

where

$$\Xi_R = \int \frac{\mathcal{D}\Phi}{\mathcal{N}_U} e^{-\frac{\beta}{2} \sum_{\alpha\gamma} (\Phi_\alpha U_{\alpha\gamma}^{-1} \Phi_\gamma) + W[\Phi;\chi]} \tag{4.12}$$

with an auxiliary functional

$$W[\Phi; \chi] = \sum_\alpha \bar{z}_\alpha \int d\mathbf{r}\, e^{i\beta\Phi_\alpha + i\beta\chi_\alpha}, \tag{4.13}$$

We also introduce the following short-hand notations

$$\chi_\alpha = q_\alpha \varphi + \Psi_\alpha + i u_\alpha, \tag{4.14}$$

and $\bar{z}_\pm = e^{\frac{\beta \mathcal{U}_{\pm\pm}(0)}{2} + \frac{\beta q_\pm^2 G_0(0)}{2}} z_\pm \theta_\pm$ are the "renormalized" activities. The inverse matrix operator $\mathbf{U}^{-1}$ is determined by

$$\int d\mathbf{r}'' \sum_\lambda U_{\alpha\lambda}^{-1}(\mathbf{r} - \mathbf{r}'') U_{\lambda\gamma}(\mathbf{r}'' - \mathbf{r}') = \delta_{\alpha\gamma} \delta(\mathbf{r} - \mathbf{r}'), \tag{4.15}$$

where $\delta_{\alpha\gamma}$ is the Kronecker delta. The partition function of the reference system can be represented using the cumulant expansion (see Chap. 2)

$$\Xi_R = \exp\left[\sum_{n=1}^{\infty} \frac{\langle W^n[\Phi; \chi] \rangle_c}{n!}\right]. \tag{4.16}$$

The cumulants are determined as

$$\langle W[\Phi; \chi] \rangle_c = \langle W[\Phi; \chi] \rangle, \tag{4.17}$$

$$\left\langle W^2[\Phi;\chi]\right\rangle_c = \left\langle W^2[\Phi;\chi]\right\rangle - \left\langle W[\Phi;\chi]\right\rangle^2, \tag{4.18}$$

$$\left\langle W^3[\Phi;\chi]\right\rangle_c = \left\langle W^3[\Phi;\chi]\right\rangle - 3\left\langle W^2[\Phi;\chi]\right\rangle\left\langle W[\Phi;\chi]\right\rangle + 2\left\langle W[\Phi;\chi]\right\rangle^3, \ldots, \tag{4.19}$$

where we introduce the following notation

$$\langle(\ldots)\rangle = \int \frac{\mathcal{D}\Phi}{\mathcal{N}_U} \exp\left[-\frac{\beta}{2}\sum_{\alpha\gamma}(\Phi_\alpha U^{-1}_{\alpha\gamma}\Phi_\gamma)\right](\ldots). \tag{4.20}$$

After evaluating the Gaussian integrals, we obtain

$$\left\langle e^{i\beta\sum\limits_{s=1}^{k}\Phi_+(\mathbf{r}_s^{(+)})+i\beta\sum\limits_{l=1}^{n-k}\Phi_-(\mathbf{r}_s^{(-)})}\right\rangle_c = e^{\frac{\beta k U_{++}(0)}{2}+\frac{\beta(n-k)U_{--}(0)}{2}} B_R^{nk}\left(\mathbf{r}_1^{(+)},\ldots,\mathbf{r}_k^{(+)};\mathbf{r}_1^{(-)},\ldots,\mathbf{r}_{n-k}^{(-)}\right), \tag{4.21}$$

where we use the following cluster functions

$$B_R^{11}\left(\mathbf{r}^{(+)}\right) = B_R^{10}\left(\mathbf{r}^{(-)}\right) = 1, \tag{4.22}$$

$$B_R^{22}\left(\mathbf{r}_1^{(+)},\mathbf{r}_2^{(+)}\right) = e^{-\beta U_{++}\left(\mathbf{r}_1^{(+)}-\mathbf{r}_2^{(+)}\right)} - 1,\ B_R^{21}\left(\mathbf{r}_1^{(+)};\mathbf{r}_1^{(-)}\right) = e^{-\beta U_{+-}\left(\mathbf{r}_1^{(+)}-\mathbf{r}_1^{(-)}\right)} - 1,$$

$$B_R^{20}\left(\mathbf{r}_1^{(-)},\mathbf{r}_2^{(-)}\right) = e^{-\beta U_{--}\left(\mathbf{r}_1^{(-)}-\mathbf{r}_2^{(-)}\right)} - 1, \tag{4.23}$$

$$B_R^{33}\left(\mathbf{r}_1^{(+)},\mathbf{r}_2^{(+)},\mathbf{r}_3^{(+)}\right) = e^{-\beta\left(U_{++}\left(\mathbf{r}_1^{(+)}-\mathbf{r}_2^{(+)}\right)+U_{++}\left(\mathbf{r}_1^{(+)}-\mathbf{r}_3^{(+)}\right)+U_{++}\left(\mathbf{r}_2^{(+)}-\mathbf{r}_3^{(+)}\right)\right)}$$

$$- e^{-\beta U_{++}\left(\mathbf{r}_1^{(+)}-\mathbf{r}_2^{(+)}\right)} - e^{-\beta U_{++}\left(\mathbf{r}_1^{(+)}-\mathbf{r}_3^{(+)}\right)} - e^{-\beta U_{++}\left(\mathbf{r}_2^{(+)}-\mathbf{r}_3^{(+)}\right)} + 2,$$

$$B_R^{32}\left(\mathbf{r}_1^{(+)},\mathbf{r}_2^{(+)};\mathbf{r}_1^{(-)}\right) = e^{-\beta\left(U_{++}\left(\mathbf{r}_1^{(+)}-\mathbf{r}_2^{(+)}\right)+U_{+-}\left(\mathbf{r}_1^{(+)}-\mathbf{r}_1^{(-)}\right)+U_{+-}\left(\mathbf{r}_2^{(+)}-\mathbf{r}_1^{(-)}\right)\right)}$$

$$- e^{-\beta U_{++}\left(\mathbf{r}_1^{(+)}-\mathbf{r}_2^{(+)}\right)} - e^{-\beta U_{+-}\left(\mathbf{r}_1^{(+)}-\mathbf{r}_1^{(-)}\right)} - e^{-\beta U_{+-}\left(\mathbf{r}_2^{(+)}-\mathbf{r}_1^{(-)}\right)} + 2,$$

$$B_R^{31}\left(\mathbf{r}_1^{(+)};\mathbf{r}_1^{(-)},\mathbf{r}_2^{(-)}\right) = e^{-\beta\left(U_{+-}\left(\mathbf{r}_1^{(+)}-\mathbf{r}_1^{(-)}\right)+U_{+-}\left(\mathbf{r}_1^{(+)}-\mathbf{r}_2^{(-)}\right)+U_{--}\left(\mathbf{r}_1^{(-)}-\mathbf{r}_2^{(-)}\right)\right)}$$

$$- e^{-\beta U_{+-}\left(\mathbf{r}_1^{(+)}-\mathbf{r}_1^{(-)}\right)} - e^{-\beta U_{+-}\left(\mathbf{r}_1^{(+)}-\mathbf{r}_2^{(-)}\right)} - e^{-\beta U_{--}\left(\mathbf{r}_1^{(-)}-\mathbf{r}_2^{(-)}\right)} + 2,$$

$$B_R^{30}\left(\mathbf{r}_1^{(-)},\mathbf{r}_2^{(-)},\mathbf{r}_3^{(-)}\right) = e^{-\beta\left(U_{--}\left(\mathbf{r}_1^{(-)}-\mathbf{r}_2^{(-)}\right)+U_{--}\left(\mathbf{r}_2^{(-)}-\mathbf{r}_3^{(-)}\right)+U_{--}\left(\mathbf{r}_1^{(-)}-\mathbf{r}_3^{(-)}\right)\right)}$$

$$- e^{-\beta U_{--}\left(\mathbf{r}_1^{(-)}-\mathbf{r}_2^{(-)}\right)} - e^{-\beta U_{--}\left(\mathbf{r}_2^{(-)}-\mathbf{r}_3^{(-)}\right)} - e^{-\beta U_{--}\left(\mathbf{r}_1^{(-)}-\mathbf{r}_3^{(-)}\right)} + 2, \ldots \tag{4.24}$$

The index k denotes the number of positively charged ions in the group of n ions. Using the same arguments as those used to derive expression (2.266), we arrive at

$$\sum_{n=1}^{\infty}\frac{\langle W^n[\Phi;\chi]\rangle_c}{n!}=\sum_{n=1}^{\infty}\sum_{k=0}^{n}\frac{\tilde{z}_+^k\tilde{z}_-^{n-k}}{k!(n-k)!}\int d\mathbf{r}_1^{(+)}\dots\int d\mathbf{r}_k^{(+)}\int d\mathbf{r}_1^{(-)}\dots\int d\mathbf{r}_{n-k}^{(-)}$$

$$\times B_R^{nk}\left(\mathbf{r}_1^{(+)},\dots,\mathbf{r}_k^{(+)};\mathbf{r}_1^{(-)},\dots,\mathbf{r}_{n-k}^{(-)}\right)e^{i\beta\sum_{s=1}^{k}\chi_+\left(\mathbf{r}_s^{(+)}\right)+i\beta\sum_{l=1}^{n-k}\chi_-\left(\mathbf{r}_l^{(-)}\right)}$$

$$\approx\int d\mathbf{r}\sum_{n=1}^{\infty}\sum_{k=0}^{n}\frac{\tilde{z}_+^k\tilde{z}_-^{n-k}}{k!(n-k)!}e^{i\beta k\chi_+(\mathbf{r})+i\beta(n-k)\chi_-(\mathbf{r})}$$

$$\times\int d\mathbf{r}_2^{(+)}\dots\int d\mathbf{r}_k^{(+)}\int d\mathbf{r}_1^{(-)}\dots\int d\mathbf{r}_{n-k}^{(-)}B_R^{nk}\left(\mathbf{r}_1^{(+)},\dots,\mathbf{r}_k^{(+)};\mathbf{r}_1^{(-)},\dots,\mathbf{r}_{n-k}^{(-)}\right)$$

$$=\int d\mathbf{r}\sum_{n=1}^{\infty}\sum_{k=0}^{n}b_{nk}\tilde{z}_+^k\tilde{z}_-^{n-k}e^{i\beta k\chi_+(\mathbf{r})+i\beta(n-k)\chi_-(\mathbf{r})},$$

where $\tilde{z}_\pm=z_\pm e^{\beta V_{\pm\pm}(0)/2+\beta q_\pm^2 G_0(0)/2}\theta_\pm$ and

$$b_{nk}=\frac{1}{k!(n-k)!}\int d\mathbf{r}_2^{(+)}\dots\int d\mathbf{r}_k^{(+)}\int d\mathbf{r}_1^{(-)}\dots\int d\mathbf{r}_{n-k}^{(-)}B_R^{nk}\left(\mathbf{r}_1^{(+)},\dots,\mathbf{r}_k^{(+)};\mathbf{r}_1^{(-)},\dots,\mathbf{r}_{n-k}^{(-)}\right)$$

are group integrals. In other words, within the local approximation, the cluster functions can be represented as

$$B_R^{nk}\left(\mathbf{r}_1^{(+)},\dots,\mathbf{r}_k^{(+)};\mathbf{r}_1^{(-)},\dots,\mathbf{r}_{n-k}^{(-)}\right)=k!(n-k)!b_{nk}\prod_{j=2}^{k}\delta\left(\mathbf{r}_j^{(+)}-\mathbf{r}_1^{(+)}\right)\prod_{l=1}^{n-k}\delta\left(\mathbf{r}_l^{(-)}-\mathbf{r}_1^{(+)}\right).\tag{4.25}$$

Thus, within this approximation, the cumulant expansion becomes a cluster expansion for the fluid within the volume, where the pressure is a function of shifted chemical potentials, $\bar{\mu}_\pm=\mu_\pm+i\chi_\pm+q_\pm^2G_0(0)/2+\beta V_{\pm\pm}(0)/2+k_BT\ln\theta_\pm$, i.e.

$$\sum_{n=1}^{\infty}\frac{\langle W^n[\Phi;\chi]\rangle_c}{n!}\approx\beta\int d\mathbf{r}\,P_R(\bar{\mu}_+,\bar{\mu}_-).\tag{4.26}$$

Redefining the chemical potentials, $\bar{\mu}_\pm\to\bar{\mu}_\pm-q_\pm^2G_0(0)/2-\beta V_{\pm\pm}(0)/2-k_BT\ln\theta_\pm$, we obtain the following local approximation

$$\Xi_R=\exp\left[\beta\int d\mathbf{r}\,P_R(\bar{\mu}_+,\bar{\mu}_-)\right],\tag{4.27}$$

where $P_R(\bar{\mu}_+,\bar{\mu}_-)$ represents the pressure of the reference system as a function of the intrinsic chemical potentials of the components, $\bar{\mu}_\pm=\mu_\pm+iq_\pm\varphi+i\Psi_\pm-u_\pm$, and temperature.

Similarly, we can derive the expression for Ξ_R for a system consisting of an arbitrary number, s, of ion species

$$\Xi_R = \exp\left[\beta \int d\mathbf{r} P_R(\{\bar{\mu}_\alpha\})\right], \quad \alpha = 1, 2, \ldots, s. \tag{4.28}$$

Thus, we obtain the following

$$\Xi = \int \frac{\mathcal{D}\varphi}{C_0} \int \frac{\mathcal{D}\Psi}{\mathcal{N}_V} \exp\left[-\beta\Omega[i\varphi, i\Psi]\right], \tag{4.29}$$

where

$$\Omega[i\varphi, i\Psi] = \frac{1}{2}\left(\varphi G_0^{-1}\varphi\right) - i(\rho_{ext}\varphi) + \frac{1}{2}\sum_{\alpha\gamma}(\Psi_\alpha V_{\alpha\gamma}^{-1}\Psi_\gamma) - \int d\mathbf{r} P_R\left(\bar{\mu}_+, \bar{\mu}_-\right), \tag{4.30}$$

where $P_R(\mu_+, \mu_-)$ represents the dependence of pressure on the chemical potential of the reference system with the repulsive interaction potentials. The integral matrix operator $\mathbf{V}^{-1}$ is denoted by

$$\int d\mathbf{r}'' \sum_\lambda V_{\alpha\lambda}^{-1}(\mathbf{r} - \mathbf{r}'') V_{\lambda\gamma}(\mathbf{r}'' - \mathbf{r}') = \delta_{\alpha\gamma}\delta(\mathbf{r} - \mathbf{r}'). \tag{4.31}$$

Then, applying the mean-field approximation to the functional integral (4.30), i.e. equating the variational derivatives with respect to the field variables of the integrand functional to zero and introducing the potentials of the self-consistent fields, $\phi^{(MF)} = i\psi$ and $\Psi_\alpha^{(MF)} = i\phi_\alpha$, we derive the following system of equations:

$$\nabla^2\psi(\mathbf{r}) = -\frac{4\pi}{\varepsilon}\left(\rho_{ext}(\mathbf{r}) + q_+\bar{n}_+(\mathbf{r}) + q_-\bar{n}_-(\mathbf{r})\right) \tag{4.32}$$

$$\phi_\alpha(\mathbf{r}) = \int d\mathbf{r}' \sum_\gamma V_{\alpha\gamma}(\mathbf{r} - \mathbf{r}')\bar{n}_\gamma(\mathbf{r}'), \tag{4.33}$$

where

$$\bar{n}_\pm = \frac{\delta\Omega}{\delta u_\pm} = \frac{\partial P_R}{\partial\bar{\mu}_\pm} \tag{4.34}$$

are the equilibrium concentrations of the ions. The local chemical potentials can be written as

$$\bar{\mu}_\pm = \mu_\pm - q_\pm\psi - \phi_\pm - u_\pm. \tag{4.35}$$

In the case of short-range potentials describing specific interactions, $V_{\alpha\gamma}(\mathbf{r}) = A_{\alpha\gamma}\delta(\mathbf{r} - \mathbf{r}')$, where $A_{\alpha\gamma}$ represents the matrix of parameters of specific interactions, the system of self-consistent field equations takes the form

$$\nabla^2\psi(\mathbf{r}) = -\frac{4\pi}{\varepsilon}\left(\rho_{ext}(\mathbf{r}) + q_+\bar{n}_+(\mathbf{r}) + q_-\bar{n}_-(\mathbf{r})\right) \tag{4.36}$$

$$\phi_\alpha(\mathbf{r}) = \sum_\gamma A_{\alpha\gamma}\bar{n}_\gamma(\mathbf{r}). \tag{4.37}$$

In this case, to determine the potentials ψ and ϕ_α, we need to solve the differential Poisson equation (4.36) with appropriate boundary conditions along with the system of transcendental equations. The grand thermodynamic potential within the self-consistent field approximation is expressed as:

$$\Omega = \int d\mathbf{r}\left(-\frac{\varepsilon(\nabla\psi)^2}{8\pi} - \frac{1}{2}\sum_{\alpha\gamma} g_{\alpha\gamma}\phi_\alpha\phi_\gamma - P_R(\bar{\mu}_+, \bar{\mu}_-)\right), \tag{4.38}$$

where $g_{\alpha\gamma} = A_{\alpha\gamma}^{-1}$ is a matrix that is the inverse of the matrix of parameters for specific interactions. It is worth noting that we can omit the term $\rho_{ext}\psi$ in the integrand of (4.30), as the external charge is typically surface-bound (the charge on the surface of the conductor), and we account for it through the boundary conditions of the self-consistent field equation.

The theory outlined above can also be approached through a different framework [82–85]. Let us write down the grand thermodynamic potential of the ionic fluid with short-range specific interactions as follows:

$$\Omega = \int d\mathbf{r}\left(-\frac{\varepsilon(\nabla\psi)^2}{8\pi} + \rho\psi + \sum_\alpha n_\alpha\phi_\alpha - \frac{1}{2}\sum_{\alpha\gamma} A_{\alpha\gamma}^{-1}\phi_\alpha\phi_\gamma + \sum_\alpha n_\alpha u_\alpha\right) + \int d\mathbf{r}\left(f(n_+, n_-) - \mu_+ n_+ - \mu_- n_-\right), \tag{4.39}$$

where f represents the free energy density, which depends on local ion concentrations $n_\pm$ and temperature T; $\mu_\pm$ are fixed chemical potentials of the ions, and $\rho = q_+n_+ + q_-n_-$ defines the local charge density of the ions. Then, the grand thermodynamic potential can be rewritten in the form

$$\Omega = \int d\mathbf{r}\left(-\frac{\varepsilon(\nabla\psi)^2}{8\pi} - \frac{1}{2}\sum_{\alpha\gamma} A_{\alpha\gamma}^{-1}\phi_\alpha\phi_\gamma\right) + \int d\mathbf{r}\left(f(n_+, n_-) - \bar{\mu}_+ n_+ - \bar{\mu}_- n_-\right), \tag{4.40}$$

and utilizing the local Legendre transformation

$$P(\bar{\mu}_+, \bar{\mu}_-) = \bar{\mu}_+ n_+ + \bar{\mu}_- n_- - f(n_+, n_-), \tag{4.41}$$

we arrive at the functional (4.38). By varying the latter with respect to ψ and $\phi_\pm$, we obtain the set of Eqs. (4.36)–(4.37).

To apply the framework mentioned above to nonuniform liquid-phase electrolytes, we need to determine the function $P(\bar{\mu}_+, \bar{\mu}_-)$, i.e., specify the reference system. For a mixture of ideal gases

$$P_R(\bar{\mu}_+, \bar{\mu}_-) = \frac{k_B T}{\Lambda_+^3} e^{\beta\bar{\mu}_+} + \frac{k_B T}{\Lambda_-^3} e^{\beta\bar{\mu}_-}, \tag{4.42}$$

where $\Lambda_\pm$ are the de Broglie thermal wavelengths, we arrive at previously obtained PB equation

$$\nabla^2\psi(\mathbf{r}) = -\frac{4\pi}{\varepsilon}\left(q_+ n_+^{(b)} e^{-\beta q_+\psi(\mathbf{r})} + q_- n_-^{(b)} e^{-\beta q_-\psi(\mathbf{r})}\right), \tag{4.43}$$

where we utilize the expression for the bulk chemical potentials of species, $\mu_\pm = k_B T \ln\left(\Lambda_\pm^3 n_\pm^{(b)}\right)$, with $n_\pm^{(b)}$ representing the bulk concentrations of the ions.

Another widely used reference system is the multi-component symmetric lattice gas [13, 82], for which

$$P_R(\bar{\mu}_+, \bar{\mu}_-) = \frac{k_B T}{v} \ln\left(1 + e^{\beta\bar{\mu}_+} + e^{\beta\bar{\mu}_-}\right), \tag{4.44}$$

where v is the lattice gas cell volume. In this case, when $q_\pm = \pm q$ and $\mu_\pm = k_B T \ln nv/(1 - 2nv)$, the self-consistent field equation takes the form [60, 86]

$$\Delta\psi(\mathbf{r}) = \frac{8\pi nq}{\varepsilon} \frac{\sinh\beta q\psi(\mathbf{r})}{1 + 2nv\left(\cosh\beta q\psi(\mathbf{r}) - 1\right)}, \tag{4.45}$$

where n represents the bulk concentration of ions. To take into account the asymmetry of the ions we can use the asymmetric lattice gas model, which will be discussed in details further, or the hard-sphere mixture model within the Percus-Yevick or Carnahan-Starling approximations [82]. The pressure and chemical potentials of species ($\alpha = \pm$) can be obtained within the Percus-Yevick approximation from the following relations

$$\beta P_R = n_0 \frac{1 + n_3 + n_3^2}{(1 - n_3)^3} + \frac{1/12\pi\, n_2^3 + n_1 n_2 (1 - n_3) - 3 n_0 n_3}{(1 - n_3)^3}, \tag{4.46}$$

$$\beta\bar{\mu}_\alpha = \ln\frac{\bar{n}_\alpha \Lambda_\alpha^3}{1 - n_3} + \frac{n_2 d_\alpha}{2(1 - n_3)} + \pi d_\alpha^2 \left(\frac{n_1}{1 - n_3} + \frac{n_2^2}{8\pi(1 - n_3)^2}\right) + \frac{\beta\pi d_\alpha^3 P_R}{6}, \tag{4.47}$$

where d_α represents the diameters of the ions and we introduce the following auxiliary variables: $n_0 = \bar{n}_+ + \bar{n}_-$, $n_1 = 1/2\, d_+ \bar{n}_+ + 1/2\, d_- \bar{n}_-$, $n_2 = \pi\, d_+^2 \bar{n}_+ + \pi\, d_-^2 \bar{n}_-$, $n_3 = \pi/6\, d_+^3 \bar{n}_+ + \pi/6\, d_-^3 \bar{n}_-$.

4.2 Equation of the Mechanical Equilibrium of the Ionic Fluid Within the Mean-Field Approximation. Stress Tensor

As demonstrated in the previous section, the grand thermodynamic potential of the electrolyte solution in the local approximation can be expressed as follows

$$\Omega = \int d\mathbf{r}\omega(\mathbf{r}), \tag{4.48}$$

where we introduce the density of the grand thermodynamic potential

$$\omega(\mathbf{r}) = \omega(\nabla\psi, \psi, \phi_\alpha) = -\frac{\varepsilon(\nabla\psi)^2}{8\pi} - \frac{1}{2}\sum_{\alpha\gamma} g_{\alpha\gamma}\phi_\alpha\phi_\gamma - P_R(\bar{\mu}_+, \bar{\mu}_-) \tag{4.49}$$

with $g_{\alpha\gamma} = A^{-1}_{\alpha\gamma}$. We assume that the dependence of the thermodynamic potential density on coordinates is realized through the potentials $\psi(\mathbf{r})$ and $\phi_\pm(\mathbf{r})$, so for simplicity, we consider $u_\pm(\mathbf{r}) = 0$. The case of $u_\pm(\mathbf{r}) \neq 0$ will be addressed later.

Variation of the functional (4.48) with respect to the potentials ψ and ϕ_α yields the set of self-consistent field equations (4.36) and (4.37), which can be expressed as:

$$\frac{\partial}{\partial x_i}\left(\frac{\partial\omega}{\partial\psi_{,i}}\right) - \frac{\partial\omega}{\partial\psi} = 0, \quad \frac{\partial\omega}{\partial\phi_\alpha} = 0, \tag{4.50}$$

where x_i are the Cartesian coordinates and $\psi_{,i} = \partial\psi/\partial x_i$. As stated in Chap. 1, we assume the summation over the repeating indexes, which enumerate the coordinates. Then, using the Eq. (4.50), we obtain

$$\frac{\partial\omega}{\partial x_i} = \frac{\partial}{\partial x_k}\left(\frac{\partial\omega}{\partial\psi_{,k}}\right)\psi_{,i} + \left(\frac{\partial\omega}{\partial\psi_{,k}}\right)\frac{\partial\psi_{,k}}{\partial x_i} = \frac{\partial}{\partial x_k}\left(\psi_{,i}\frac{\partial\omega}{\partial\psi_{,k}}\right) \tag{4.51}$$

or, rearranging all terms to one side of the equation, we get

$$\frac{\partial\sigma_{ik}}{\partial x_k} = 0, \tag{4.52}$$

where we introduce the following tensor

$$\sigma_{ik} = \omega\delta_{ik} - \psi_{,i}\frac{\partial\omega}{\partial\psi_{,k}}. \tag{4.53}$$

Equation (4.52) simply represents a mechanical equilibrium condition with a stress tensor σ_{ik}, which is determined by the following relation:

$$\sigma_{ik} = \frac{\varepsilon}{4\pi}\left(\mathcal{E}_i\mathcal{E}_k - \frac{1}{2}\mathcal{E}^2\delta_{ik}\right) - \delta_{ik}P, \tag{4.54}$$

where the first term defines the Maxwell stress tensor of the electrostatic field $\mathcal{E}_i = -\psi_{,i}$ in a continuous structureless dielectric medium, while the second term defines the stress tensor of the ionic fluid with the local osmotic pressure

$$P = P_0 + \frac{1}{2}\sum_{\alpha\gamma} g_{\alpha\gamma}\phi_\alpha\phi_\gamma = P_0 + \frac{1}{2}\sum_{\alpha\gamma} A_{\alpha\gamma}\bar{n}_\alpha\bar{n}_\gamma, \tag{4.55}$$

where $P_0 = P_R$ represents the contribution to the local osmotic pressure from the steric interactions of the ions, and δ_{ik} is the unit tensor (Kronecker delta).

Now, let us consider the scenario where ions are subjected to external potential fields with potential energies $u_\pm(\mathbf{r})$. These could be, for example, van der Waals forces acting on the ions from the side of an electrode or a membrane. In such case, the density of the thermodynamic potential explicitly depends on the coordinates through the potential energies $u_\pm(\mathbf{r})$. Similar evaluations lead to the following equation

$$\frac{\partial\sigma_{ik}}{\partial x_k} - \sum_\alpha \bar{n}_\alpha \frac{\partial u_\alpha}{\partial x_i} = 0, \tag{4.56}$$

which determines the mechanical equilibrium condition of the solution in the presence of the external volume potential forces.

Note that the force acting on a conducting or dielectric body placed in the ionic fluid can be calculated using the formula

$$F_i = \oint \sigma_{ik}e_k dS, \tag{4.57}$$

where e_k represents the surface normal vector at a given point, and dS is the surface element of the body. It is worth noting that in the absence of external volume forces acting on ions, integration can be performed over any closed surface containing a body immersed in the ionic fluid [87]. However, in the presence of external volume forces, the force acting on the body immersed in the ionic fluid must be calculated using the formula derived from the mechanical equilibrium condition of an arbitrary fluid volume

$$F_i = \oint \sigma_{ik}e_k dS + \int f_i d\mathbf{r}, \tag{4.58}$$

where $f_i = -\sum_\alpha \bar{n}_\alpha \partial u_\alpha/\partial x_i$ are the components of the non-electrostatic forces density, where the integration in the second term is carried out over the fluid volume, enclosed between the surface of the immersed body and the surface, over which the first term is integrated.

To calculate the force acting on the body immersed in the ionic fluid, we first need to solve the self-consistent field equations (4.36) and (4.37) with appropriate boundary conditions.[1] This allows us to determine the local pressure of the fluid and the electric field at and beyond the body's surface. And then we can evaluate the surface integral (4.57).

4.3 Disjoining Pressure

It is now constructive to discuss how one can obtain the expression for the disjoining pressure in the ionic fluid confined in the slit charged pore, using the expression derived in the previous section for the total stress tensor. Let us set the origin of the z-axis at one wall, so that the other wall is located at $z = H$. We can now evaluate the disjoining pressure using the following relation:

$$\Pi = -\frac{1}{\mathcal{A}}\frac{\partial \Omega}{\partial H} - P_b, \tag{4.59}$$

where $\mathcal{A}$ denotes the surface of the walls and P_b is the bulk osmotic pressure of the ionic fluid. Thus, we have

$$\frac{\partial \Omega}{\partial H} = \int_0^H dz \frac{\delta \Omega}{\delta \psi(z)} \frac{\partial \psi(z)}{\partial H} + \left(\omega - \psi' \frac{\partial \omega}{\partial \psi'}\right)\Big|_{z=H} - \int_0^H dz \sum_{\alpha} \frac{\partial P}{\partial u_\alpha} \frac{\partial u_\alpha}{\partial H}$$
$$= \int_0^H dz \sum_{\alpha} \bar{n}_\alpha(z) \frac{\partial u_\alpha(z)}{\partial H} + \sigma_{zz}(H), \tag{4.60}$$

where we take into account that $\partial P/\partial u_\alpha = -\partial P/\partial \bar{\mu}_\alpha = -\bar{n}_\alpha$ and $\delta \Omega/\delta \psi(z) = 0$. The latter can be derived as follows: We can write

$$\frac{d\Omega}{dH} = \frac{\partial \Omega}{\partial H} + \frac{\partial \Omega}{\partial \psi(H)} \psi'(H). \tag{4.61}$$

On the other hand, using theorem of differentiation of the integral with respect to a parameter, we get

$$\frac{d\Omega}{dH} = \frac{d}{dH} \int_0^H dz \omega(z) = \omega(H) + \int_0^H dz \frac{\partial \omega(z)}{\partial H} \tag{4.62}$$

[1] The boundary conditions are determined by the nature of the immersed body.

Moreover, we have

$$\frac{\partial \omega}{\partial H} = \frac{\delta \Omega}{\delta \psi(z)} \frac{\partial \psi(z)}{\partial H} + \sum_{\alpha} \frac{\partial \omega}{\partial u_{\alpha}} \frac{\partial u_{\alpha}}{\partial H} = \sum_{\alpha} \frac{\partial \omega}{\partial u_{\alpha}} \frac{\partial u_{\alpha}}{\partial H}, \tag{4.63}$$

since $\delta\Omega/\delta\psi(z) = 0$.

Thereby, we obtain

$$\frac{\partial \Omega}{\partial H} = \frac{d\Omega}{dH} - \frac{\partial \Omega}{\partial \psi(H)} \psi'(H) = \omega(H) + \int_{0}^{H} dz \sum_{\alpha} \frac{\partial \omega}{\partial u_{\alpha}} \frac{\partial u_{\alpha}}{\partial H} - \frac{\partial \Omega}{\partial \psi(H)} \psi'(H). \tag{4.64}$$

If we consider $\psi(H)$ as variable, then we can write after variation of Ω

$$\delta \Omega = \frac{\partial \omega(H)}{\partial \psi'(H)} \delta \psi(H) \tag{4.65}$$

that yields

$$\frac{\partial \Omega}{\partial \psi(H)} = \frac{\partial \omega(H)}{\partial \psi'(H)}. \tag{4.66}$$

Therefore, we eventually obtain

$$\frac{\partial \Omega}{\partial H} = \omega(H) - \frac{\partial \omega(H)}{\partial \psi'(H)} \psi'(H) + \int_{0}^{H} dz \sum_{\alpha} \frac{\partial \omega}{\partial u_{\alpha}} \frac{\partial u_{\alpha}}{\partial H} = \sigma_{zz}(H) + \int_{0}^{H} dz \sum_{\alpha} \bar{n}_{\alpha} \frac{\partial u_{\alpha}}{\partial H}. \tag{4.67}$$

Hence, we arrive at the expression

$$\Pi = -\sum_{\alpha} \int_{0}^{H} dz\, \bar{n}_{\alpha}(z) \frac{\partial u_{\alpha}(z)}{\partial H} - \sigma_{zz}(H) - P_b, \tag{4.68}$$

where $\sigma_{zz}(H) = \omega - \psi' \partial\omega/\partial\psi'|_{z=H} = \varepsilon_b \mathcal{E}^2(H)/2 - P_0(H)$ is a normal stress at $z = H$. Let us consider the case of two identical walls

$$u_{\alpha}(z) = \phi_{\alpha}(z) + \phi_{\alpha}(H - z), \tag{4.69}$$

where ϕ_{α} denotes the potential produced by one wall, and considering that $\bar{n}_{\alpha}(z) = \bar{n}_{\alpha}(H - z)$, we obtain

$$\Pi = -\sum_{\alpha} \int_{0}^{H} dz\, \bar{n}_{\alpha}(z) \phi'_{\alpha}(z) - \sigma_{zz}(H) - P_b. \tag{4.70}$$

We can rewrite Eq. (4.70) in a form more suitable for practical applications. Utilizing the mechanical equilibrium condition

$$\frac{d\sigma_{zz}}{dz} - \sum_{\alpha} \bar{n}_{\alpha}(z)u'_{\alpha}(z) = 0, \tag{4.71}$$

after the integration from $z = H/2$ to $z = H$, we obtain

$$\sigma_{zz}(H) = \sigma_{zz}\left(\frac{H}{2}\right) + \sum_{\alpha} \int_{H/2}^{H} dz\, \bar{n}_{\alpha}(z)\phi'_{\alpha}(z) - \sum_{\alpha} \int_{0}^{H/2} dz\, \bar{n}_{\alpha}(z)\phi'_{\alpha}(z), \tag{4.72}$$

where we take into account Eq. (4.69) and the condition $\bar{n}_{\alpha}(z) = \bar{n}_{\alpha}(H - z)$. Substituting (4.72) for $\sigma_{zz}(H)$ into the Eq. (4.70), after some algebra we obtain

$$\Pi = P_m - P_b - 2\sum_{\alpha} \int_{H/2}^{H} dz\, \bar{n}_{\alpha}(z)\phi'_{\alpha}(z), \tag{4.73}$$

where $P_m = P_0\,(H/2)$ denotes the osmotic pressure of the ions at the midpoint of the pore, and $\mathcal{E}(H/2) = 0$. In the scenario of negligible small interaction between ions and walls ($\phi_{\alpha}(z) = 0$), Eq. (4.73) reduces to the Langmuir equation, $\Pi = P_m - P_b$, for the disjoining pressure. This Eq. (4.73) presents a generalization of the previously derived expression (3.53) for the disjoining pressure of the Coulomb gas within the pore, accommodating arbitrary reference model of the fluid and taking into account interactions between the ions and pore walls.

4.4 Electric Double Layer Theory with Account of Steric Interactions of the Ions

In Chap. 3, we explored the simplest theory of the electric double layer at the metal-electrolyte solution interface—the Gouy-Chapman theory, which treats ions as point particles. In this section, we delve into a theory that accounts for the finite size of ions. Similar to the Gouy-Chapman theory, we consider a $1:1$ electrolyte solution in contact with an infinite flat charged metal electrode. Placing the origin of the z-axis at the electrode, with the solution located in the region $z > 0$, we express the grand thermodynamic potential per unit area of the electrode as follows:

$$\Omega/\mathcal{A} = -\int_{0}^{\infty} dz \left(\frac{\varepsilon\mathcal{E}^2}{8\pi} + P(\bar{\mu}_{+}, \bar{\mu}_{-})\right), \tag{4.74}$$

where $\mathcal{E}(z) = -\psi'(z)$ denotes the local electric field. For simplicity, we neglect the short-range specific interactions and non-electrostatic interactions between the ions and the electrode, i.e., we assume $A_{\alpha\gamma} = 0$ and $u_{\pm}(z) = 0$. As above, we introduce the local chemical potentials of the species $\bar{\mu}_{\pm}(z) = \mu_{\pm} \mp q\psi(z)$. For now, we don't specify the explicit dependence $P(\bar{\mu}_+, \bar{\mu}_-)$. The mechanical equilibrium condition (4.52) yields:

$$\frac{d}{dz}\left(\frac{\varepsilon\mathcal{E}^2}{8\pi} - P(\bar{\mu}_+, \bar{\mu}_-)\right) = 0 \tag{4.75}$$

or

$$\frac{\varepsilon\mathcal{E}^2}{8\pi} - P(\bar{\mu}_+, \bar{\mu}_-) = const. \tag{4.76}$$

The latter expression, which can be written as the condition of the constancy of the normal component of the stress tensor, i.e. $\sigma_{zz} = const$, is referred to as the *contact value theorem* [88].

The condition that far from the electrode, in the bulk phase, we have $\psi(\infty) = \mathcal{E}(\infty) = 0$ leads to the expression:

$$-\frac{\varepsilon\mathcal{E}^2}{8\pi} + P(\bar{\mu}_+, \bar{\mu}_-) = P(\mu_+, \mu_-) = P_b. \tag{4.77}$$

We should note that the condition (4.77) can be obtained as the first integral of the Euler-Lagrange equation. Indeed, since the integrand in (4.74) does not explicitly depend on the coordinate z, the Euler-Lagrange equation has the first integral, as is known from variational calculus

$$\psi'\frac{\partial\omega}{\partial\psi'} - \omega = const, \tag{4.78}$$

which, with the substitution of the expression $\omega = -\varepsilon\psi'^2/8\pi - P$ for the thermodynamic potential density, leads to the expression (4.77). Thus, expressing the local electric field from Eq. (4.77) as a function of the local potential, i.e., $\mathcal{E} = \mathcal{E}(\psi)$, we can evaluate the potential $\psi(z)$ by integrating the differential equation $\psi' = -\mathcal{E}(\psi)$ with the initial condition $\psi(0) = \psi_0$, where ψ_0 denotes the electrode surface potential.

The theory under discussion enables us to derive the expression for the electric differential capacitance as a function of the electrode surface potential. As known from the electrostatics [87], the surface charge density of the conducting electrode immersed into the dielectric can be expressed through the normal component of the electric displacement at the electrode, such that $\sigma = \varepsilon\mathcal{E}(0)/4\pi$. Hence, the condition for mechanical equilibrium at $z = 0$ is given by the expression:

$$\frac{2\pi\sigma^2}{\varepsilon} = P_s(\psi_0) - P_b, \tag{4.79}$$

where $P_s(\psi_0) = P(T, \mu_+ - q\psi_0, \mu_- + q\psi_0)$ represents the pressure exerted by the ions at the surface of the electrode. Therefore, differentiating σ with respect to ψ_0, we arrive at the expression, originally derived in Refs. [82, 89]:

$$C = \sqrt{\frac{\varepsilon}{8\pi}} \frac{q|n_{s-}(\psi_0) - n_{s+}(\psi_0)|}{\sqrt{P_s(\psi_0) - P_b}}, \tag{4.80}$$

where $n_{s\pm}(\psi_0) = \bar{n}_\pm(0)$ are the local concentrations of the ions at the electrode surface.

If we take the two-component symmetric lattice gas as the reference system, then pressure is expressed through the chemical potentials as follows:

$$P(\bar{\mu}_+, \bar{\mu}_-) = \frac{k_B T}{v} \ln\left(1 + e^{\beta\bar{\mu}_+} + e^{\beta\bar{\mu}_-}\right), \tag{4.81}$$

and the local concentrations are determined as

$$\bar{n}_\pm(z) = \frac{1}{v} \frac{e^{\beta\bar{\mu}_\pm}}{1 + e^{\beta\bar{\mu}_+} + e^{\beta\bar{\mu}_-}}, \tag{4.82}$$

where v denotes the lattice gas cell volume. Considering that in the bulk, the concentrations of both ionic types are equal, i.e., $\bar{n}_\pm(\infty) = n$, we arrive at the following expressions for the chemical potentials of the ions within the symmetric lattice gas model:

$$\mu_\pm = k_B T \ln\left(\frac{nv}{1 - 2nv}\right). \tag{4.83}$$

Applying the obtained relations, we arrive at the following analytical expression, originally obtained by Kornyshev [60]

$$C = C_{GC} \frac{\cosh \frac{q\psi_0}{2k_B T}}{1 + 4nv \sinh^2 \frac{q\psi_0}{2k_B T}} \sqrt{\frac{4nv \sinh^2 \frac{q\psi_0}{2k_B T}}{\ln\left(1 + 4nv \sinh^2 \frac{q\psi_0}{2k_B T}\right)}}. \tag{4.84}$$

This expression generalizes the Gouy-Chapman expression and incorporates the influence of excluded volume effects on the differential capacitance. It is worth noting that the Gouy-Chapman expression can be derived from Kornyshev's expression in the limit of the lattice gas cell volume approaching zero ($v \to 0$). As discussed earlier in the context of the Gouy-Chapman theory, neglecting excluded volume of the ions results in an exponential increase in the differential capacitance at high surface potentials, contradicting experimental observations. Kornyshev's expression, which considers excluded volume effects, predicts a decrease in capacitance according to the following power law in the region where $q|\psi_0|/(k_B T) \gg 1$:

$$C \simeq C_{GC}\sqrt{\frac{k_BT}{4nvq|\psi_0|}}. \quad (4.85)$$

The monotonic decrease in the differential capacitance arises because as the electrode potential increases to sufficiently large values, the number of ions in the electric double layer stops increasing due to their excluded volume (crowding regime). This phenomenon leads to the so-called saturation of the electric double layer, where further increases in potential barely affect the charge of the electric double layer. It is noteworthy that at zero electrode potential, the value of the differential capacitance predicted by Kornyshev's theory coincides with that predicted by the Gouy-Chapman theory, i.e., $C = C_{GC}$. However, this prediction significantly overestimates the values observed in experiments. To ensure that the differential capacitance at zero potential matches experimental observations, it is necessary to consider the effect of ion size asymmetry [82] or account for short-range specific interactions [85, 90]. These aspects will be revisited in Chap. 6. Figure 4.1 shows typical differential capacitance profiles obtained by Eq. (4.84). As can be seen, increase in the bulk volume fraction of ionic fluid leads to a crossover from 'camel'-shape (bimodal curve) to 'bell'-shape (unimodal curve) profiles which is observed in real ionic fluids [60].

Figure 4.2 shows typical electrostatic potential profiles obtained from the integration of the modified Poisson-Boltzmann equation within the symmetric lattice gas model in this case. As can be seen, for all packing fractions $\gamma = 2nv$ of the ionic fluid, there is a monotonic behavior of the electrostatic potential as a function of the distance from the electrode. The latter is not adequate to real ionic fluids with high packing fractions of the ions, for which oscillating behavior of local ionic densities and potential takes place [91]. To obtain, at least qualitatively, oscillatory behavior of the local ion concentrations and potential, it is necessary to take into account correlation effects beyond the pure mean-field approximation [83, 92, 93] (see also Chap. 6).

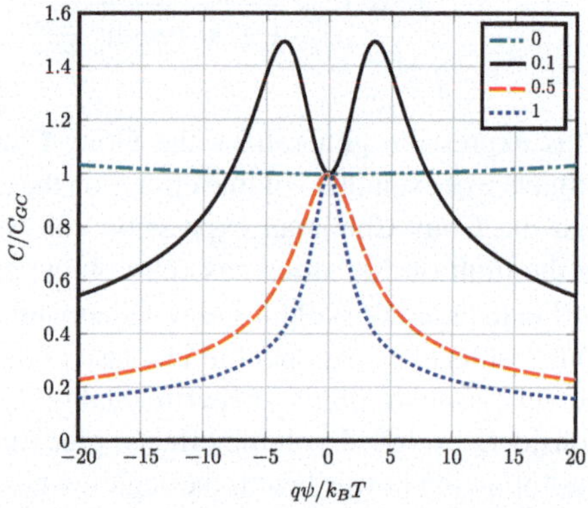

Fig. 4.1 Capacitance of the EDL as a function of electrode potential for different values of lattice-saturation parameter, $\gamma = 2nv$: transition from the bell-shaped profile to the camel-shape one. Case of $\gamma = 0$ corresponds to the Gouy-Chapman law

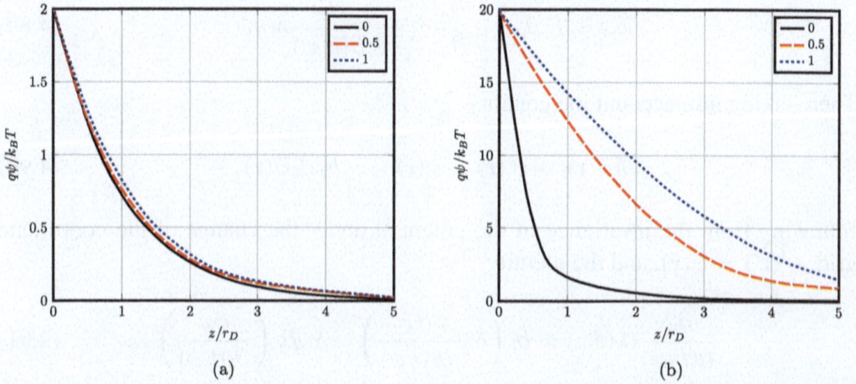

Fig. 4.2 Electrostatic potential profiles near a charged surface within the lattice-gas model, plotted for different values of the parameter $\gamma = 2nv$; profile with $\gamma = 0$ corresponds to the Gouy-Chapman theory. $\frac{q\psi_0}{k_BT} = 2$ (**a**); $\frac{q\psi_0}{k_BT} = 20$ (**b**)

4.5 Stress Tensor in Ionic Fluids. A Thermomechanical Approach

Earlier, we explored the mechanical stresses in inhomogeneous ionic fluids using mean-field theory. Now, Let us delve into this topic with a more rigorous approach. Specifically, we'll examine another method that enables us to derive the stress tensor of an inhomogeneous ionic fluid from the thermodynamic potential functional within the mean-field approximation—*a thermomechanical approach.*

Let us consider a general case of an inhomogeneous ionic fluid whose thermodynamic potential can be represented in the mean-field approximation in the form of the following functional

$$\Omega = \int d\mathbf{r}\,\omega(\mathbf{r}), \tag{4.86}$$

where the thermodynamic potential density depends on the coordinates through the electrostatic potential ψ and its first derivatives $\partial_i\psi$ (field components)

$$\omega(\mathbf{r}) = \omega(\psi(\mathbf{r}), \partial\psi(\mathbf{r})). \tag{4.87}$$

Now, let us subject the system to an infinitesimal constant shift by the vector **h**, i.e. let us substitute the variables

$$\mathbf{r}' = \mathbf{r} + \mathbf{h}. \tag{4.88}$$

The first-order variation of the thermodynamic potential density with respect to **h** has the following form

$$\delta\omega = h_i\partial_i\omega + \frac{\partial\omega}{\partial\psi}\delta\psi + \frac{\partial\omega}{\partial(\partial_i\psi)}\delta\partial_i\psi. \tag{4.89}$$

Then, taking into account the equality

$$\delta\psi(\mathbf{r}) = \psi'(\mathbf{r}) - \psi(\mathbf{r}) = -h_k\partial_k\psi(\mathbf{r}), \tag{4.90}$$

following from the invariance of the potential under the change of the coordinate grid, $\psi'(\mathbf{r}') = \psi(\mathbf{r})$, and the identity

$$\frac{\partial\omega}{\partial(\partial_i\psi)}\partial_i(\delta\psi) = \partial_i\left(\delta\psi\frac{\partial\omega}{\partial(\partial_i\psi)}\right) - \delta\psi\partial_i\left(\frac{\partial\omega}{\partial(\partial_i\psi)}\right), \tag{4.91}$$

we arrive at the following relation

$$\delta\Omega = h_k\int d\mathbf{r}\left(\partial_i\sigma_{ik} - \partial_k\psi\frac{\delta\Omega}{\delta\psi}\right), \tag{4.92}$$

where we introduce already known stress tensor

$$\sigma_{ik} = \omega\delta_{ik} - \frac{\partial\omega}{\partial(\partial_i\psi)}\partial_k\psi \tag{4.93}$$

and

$$\frac{\delta\Omega}{\delta\psi} = \frac{\partial\omega}{\partial\psi} - \partial_l\frac{\partial\omega}{\partial(\partial_l\psi)}. \tag{4.94}$$

Since the grand thermodynamic potential should be invariant under translation of the system as a whole by a constant vector, i.e., $\delta\Omega = 0$, and given that equality (4.92) holds for any arbitrary fluid volume, we arrive at the following equality:

$$\partial_i\sigma_{ik} = \partial_k\psi\frac{\delta\Omega}{\delta\psi}. \tag{4.95}$$

Thus, in the equilibrium configuration of the electrostatic field, when the electrostatic potential satisfies the self-consistent field equation (or the Euler-Lagrange equation), $\delta\Omega/\delta\psi = 0$, the condition of mechanical equilibrium of the fluid must be fulfilled

$$\partial_i\sigma_{ik} = 0. \tag{4.96}$$

Note that if the thermodynamic potential density depends on m scalar *order parameters* $\psi^{(s)}$, $s = 1, 2, ..., m$, and their first partial derivatives, then the expression for the stress tensor will have the form:

$$\sigma_{ik} = \omega\delta_{ik} - \sum_s\frac{\partial\omega}{\partial(\partial_i\psi^{(s)})}\partial_k\psi^{(s)}. \tag{4.97}$$

If the thermodynamic potential density depends explicitly on the coordinates through the potentials of external fields $u_\alpha(\mathbf{r})$ acting on the fluid particles, the relation (4.92) can be written a

$$\delta\Omega = h_k \int d\mathbf{r} \left(\partial_i \sigma_{ik} - \partial_k \psi \frac{\delta\Omega}{\delta\psi} - \sum_\alpha \partial_k u_\alpha \frac{\delta\Omega}{\delta u_\alpha} \right). \tag{4.98}$$

The last equality can be interpreted as the mechanical work done over the virtual displacement of the ionic fluid volume. It follows that the i-th component of the total force acting on the isolated volume of ionic fluid is equal to

$$F_i = -\frac{\partial\Omega}{\partial h_i} = -\int d\mathbf{r} \left(\partial_i \sigma_{ik} - \partial_k \psi \frac{\delta\Omega}{\delta\psi} - \sum_\alpha \partial_k u_\alpha \frac{\delta\Omega}{\delta u_\alpha} \right). \tag{4.99}$$

Since the fluid is in a state of mechanical equilibrium, $F_i = 0$, thus, due to the arbitrariness of the integration volume, we obtain

$$\partial_i \sigma_{ik} = \partial_k \psi \frac{\delta\Omega}{\delta\psi} + \sum_\alpha \partial_k u_\alpha \frac{\delta\Omega}{\delta u_\alpha}. \tag{4.100}$$

Taking into account that $\delta\Omega/\delta\psi = 0$ and $\delta\Omega/\delta u_\alpha = n_\alpha$, where n_α defines the equilibrium local ion concentrations, we obtain the already known condition of mechanical equilibrium of the ionic fluid in the presence of volume forces

$$\partial_i \sigma_{ik} - \sum_\alpha n_\alpha \partial_k u_\alpha = 0. \tag{4.101}$$

Now, we need to prove that the tensor σ_{ik} for the considered functionals is indeed a stress tensor. To do this, let us subject the system to an inhomogeneous infinitesimal shift according to the rule

$$\mathbf{r}' = \mathbf{r} + \mathbf{u}(\mathbf{r}), \tag{4.102}$$

where $\mathbf{u}$ is the vector field of deformation of the system. The only difference with the previous evaluations is that, unlike the constant shear, the deformation changes an infinitesimal volume according to the following rule [94]:

$$d\mathbf{r}' = d\mathbf{r}\,(1 + \partial_i u_i)\,. \tag{4.103}$$

Considering that the local electrostatic potential satisfies the Euler-Lagrange equation $\delta\Omega/\delta\psi = 0$, we obtain the change of the thermodynamic potential within the linear approximation over $\mathbf{u}$ in the form of

$$\delta\Omega = \int d\mathbf{r}\, \partial_i \left(u_i \sigma_{ik}\right) = \int d\mathbf{r}\, u_{ik} \sigma_{ik}, \tag{4.104}$$

which leads us to the well-known thermodynamic definition of the stress tensor [94]

$$\sigma_{ik} = \left.\frac{\delta\Omega}{\delta u_{ik}}\right|_{u_{ik}=0}. \tag{4.105}$$

When deriving the last expression, we take into account the equality (4.96), assume that the tensor σ_{ik} is symmetric, and also introduce the strain tensor

$$u_{ik} = \frac{1}{2}\left(\partial_i u_k + \partial_k u_i\right). \tag{4.106}$$

As noted, during the proof, we assumed the symmetry of the stress tensor σ_{ik}. However, in general, the above approach does not guarantee that the obtained tensor has to be symmetric. As we will show in the next section, in the case of ionic fluids described by functionals containing only the first derivatives of the potential, the stress tensor 4.93 is automatically symmetric.

4.6 Stress Tensor for Thermodynamic Potentials, Containing Second Derivatives of the Order Parameters

In practice, one often encounters functionals, containing order parameter with derivatives higher than the first order. However, in such cases, the approach outlined above typically results in an asymmetric tensor $\bar{\sigma}_{ik}$. Let us consider a scenario where the thermodynamic potential density depends on the second derivatives of the electrostatic potential

$$\omega(\mathbf{r}) = \omega(\psi(\mathbf{r}), \partial\psi(\mathbf{r}), \partial\partial\psi(\mathbf{r})). \tag{4.107}$$

Subjecting the system to a small deformation, and considering that the potential satisfies the Euler-Lagrange equation

$$\frac{\delta\Omega}{\delta\psi} = \frac{\partial\omega}{\partial\psi} - \partial_i\left(\frac{\partial\omega}{\partial\psi_{,i}}\right) + \partial_i\partial_j\left(\frac{\partial\omega}{\partial\psi_{,ij}}\right) = 0, \tag{4.108}$$

where $\psi_{,i} = \partial_i\psi$, $\psi_{,ij} = \partial_i\partial_j\psi$, after similar but more cumbersome calculations, we obtain

$$\delta\Omega = \int d\mathbf{r}\partial_i u_k \bar{\sigma}_{ik}, \tag{4.109}$$

where the tensor

$$\bar{\sigma}_{ik} = \omega\delta_{ik} - \frac{\partial\omega}{\partial\psi_{,i}}\psi^{,k} + \partial_j\left(\frac{\partial\omega}{\partial\psi_{,ij}}\right)\psi^{,k} - \frac{\partial\omega}{\partial\psi_{,ij}}\psi^{,jk} \tag{4.110}$$

is not symmetric in general case. However, it is always possible to carry out the symmetrization procedure according to the rule:

$$\sigma_{ik} = \bar{\sigma}_{ik} + \partial_l A_{ilk}, \tag{4.111}$$

where the third rank tensor A_{ilk}, antisymmetric with respect to the indices i and l, should be a function of the first and second partial derivatives of the potential. This substitution does not violate the mechanical equilibrium conditions (4.96) due to the antisymmetry of A_{ilk}. Taking this into account, we have:

$$\delta\Omega = \int d\mathbf{r}\, \partial_i u_k \bar{\sigma}_{ik} = \int d\mathbf{r}\, \partial_i u_k (\bar{\sigma}_{ik} + \partial_l A_{ilk}) = \int d\mathbf{r}\, u_{ik} \sigma_{ik}, \tag{4.112}$$

where we take into account that

$$\int d\mathbf{r}\, \partial_i u_k \partial_l A_{ilk} = -\int d\mathbf{r}\, u_k \partial_i \partial_l A_{ilk} = 0. \tag{4.113}$$

In the last equality we performed integration by parts and utilize the antisymmetry of the tensor A_{ilk} with respect to the indices i and l. Thus, we arrive at the same thermodynamic expression (4.105)

We can also symmetrize the tensor $\bar{\sigma}_{ik}$ differently. The method below will also allow us to find the explicit form of the tensor A_{ilk}. Let us take advantage of the fact that the thermodynamic potential must be invariant not only under constant shift, but also under rotation of the system as a whole. The transformation of infinitesimal rotation, as is known, has the form

$$x_i' = x_i + \omega_{ij} x_j, \quad \omega_{ij} = -\omega_{ji} = const. \tag{4.114}$$

The condition of invariance of the thermodynamic potential in this case leads to the following relation

$$\partial_j L_{jkl} = 0, \tag{4.115}$$

where we introduce the torque tensor

$$L_{jkl} = \bar{\sigma}_{jk} x_l - \bar{\sigma}_{jl} x_k + S_{kjl} - S_{ljk}, \tag{4.116}$$

where

$$S_{kjl} = \frac{\partial \omega}{\partial \psi_{,jk}} \psi_{,l}. \tag{4.117}$$

Utilizing (4.115), we obtain

$$\bar{\sigma}_{ik} - \bar{\sigma}_{ki} = 2\partial_l \varphi_{ikl}, \tag{4.118}$$

where

$$\varphi_{ikl} = \frac{1}{2}\left(S_{ilk} - S_{kli}\right). \tag{4.119}$$

Thereby,

$$\varphi_{ikl} = -\varphi_{kil}. \tag{4.120}$$

Thus, we can construct the following symmetric stress tensor

$$\sigma_{ik} = \frac{1}{2}\left(\bar{\sigma}_{ik} + \bar{\sigma}_{ki}\right) - \partial_l\left(\varphi_{ilk} + \varphi_{kli}\right), \tag{4.121}$$

which leads to the result:

$$\sigma_{ik} = \omega\delta_{ik} - \frac{1}{2}\left(\frac{\partial\omega}{\partial\psi_{,i}}\psi_{,k} + \frac{\partial\omega}{\partial\psi_{,k}}\psi_{,i}\right) - \partial_j\left(\frac{\partial\omega}{\partial\psi_{,ik}}\psi_{,j}\right) + \partial_j\left(\frac{\partial\omega}{\partial\psi_{,ij}}\right)\psi_{,k} + \partial_j\left(\frac{\partial\omega}{\partial\psi_{,kj}}\right)\psi_{,i}. \tag{4.122}$$

The condition (4.115) also leads us to the following tensor equality

$$\frac{1}{2}\left(\frac{\partial\omega}{\partial\psi_{,i}}\psi_{,k} - \frac{\partial\omega}{\partial\psi_{,k}}\psi_{,i}\right) + \frac{\partial\omega}{\partial\psi_{,ij}}\psi_{,jk} - \frac{\partial\omega}{\partial\psi_{,kj}}\psi_{,ji} = 0, \tag{4.123}$$

with the help of which the stress tensor (4.122) can be transformed into the form [95]

$$\begin{aligned}\sigma_{ik} = \omega\delta_{ik} &- \frac{\partial\omega}{\partial\psi_{,i}}\psi_{,k} + \partial_j\left(\frac{\partial\omega}{\partial\psi_{,ij}}\right)\psi_{,k} \\ &- \partial_j\left(\frac{\partial\omega}{\partial\psi_{,ik}}\psi_{,j}\right) + \partial_j\left(\frac{\partial\omega}{\partial\psi_{,jk}}\right)\psi_{,i} + \frac{\partial\omega}{\partial\psi_{,kj}}\psi_{,ji} - \frac{\partial\omega}{\partial\psi_{,ij}}\psi_{,jk}.\end{aligned} \tag{4.124}$$

This result straightforwardly transforms into the expression (4.93), when the thermodynamic potential density does not depend on the second derivatives. Moreover, formula (4.124) can be easily generalized to the case when the grand thermodynamic potential depends on m order parameters, $\psi^{(s)}$, $s = 1, 2, ..., m$.

Note that when the thermodynamic potential density does not depend on second derivatives, the condition of invariance of the thermodynamic potential with respect to rotations gives

$$\frac{\partial\omega}{\partial\psi_{,i}}\psi_{,k} = \frac{\partial\omega}{\partial\psi_{,k}}\psi_{,i}, \tag{4.125}$$

whence follows the symmetry of the tensor (4.93).

From the expression (4.124) it is straightforward to obtain the explicit form of the function A_{ilk}:

$$A_{ilk} = S_{ilk} - S_{lik}. \tag{4.126}$$

Note that the torque tensor can be used to calculate the macroscopic torque acting on a body immersed in a fluid using the following relations:

$$L_i = \varepsilon_{ikj} L_{kj}, \tag{4.127}$$

$$L_{kj} = \oint df e_l L_{lkj}, \tag{4.128}$$

where the integration is performed over the surface of the body, ε_{ikj} denotes the absolute antisymmetric Levi-Civita tensor (symbol), and e_i is the normal unit vector constructed at the chosen point on the body surface.

In conclusion of this section, we apply the formalism described above to obtain the stress tensor in the framework of the Bazant-Storey-Kornyshev (BSC) ionic fluid model, which accounts for short-range ionic correlations at the phenomenological level. The grand thermodynamic potential of the BSC model is given by:

$$\omega = -\frac{\varepsilon}{8\pi}\left((\nabla\psi)^2 + l_c^2\left(\nabla^2\psi\right)^2\right) + \rho\psi + f(\{n_\alpha\}) - \sum_\alpha \mu_\alpha n_\alpha, \tag{4.129}$$

where the first term, in contrast to the mean-field Poisson-Boltzmann model, contains an additional "nonlocal" term $\varepsilon l_c^2(\nabla^2\psi)^2/2$, which is associated with ionic correlations beyond the mean-field theory. Here, l_c represents the only phenomenological parameter of the model responsible for incorporating correlations of the ions. It is worth noting that l_c can be interpreted as a measure of the "quadrupolarizability" of the medium (see also Chap. 5). While the original BSC model [92] employs the symmetric lattice gas model as the reference fluid model, other models with different dependencies of the free energy density on the concentration $f(\{n_\alpha\})$ can be utilized for practical calculations. The Euler-Lagrange equations corresponding to this potential are as follows

$$\nabla^2\psi - l_c^2\Delta^2\psi = -\frac{4\pi}{\varepsilon}\sum_\alpha q_\alpha n_\alpha, \quad \frac{\partial f}{\partial n_\alpha} = \mu_\alpha - q_\alpha\psi. \tag{4.130}$$

Using (4.124), we arrive at the following stress tensor

$$\sigma_{ik} = \sigma_{ik}^{(h)} + \sigma_{ik}^{(el)}, \tag{4.131}$$

where $\sigma_{ik}^{(h)}$ is determined by previously obtained expression (6.13). The electrostatic stress tensor in this case consists of two terms:

$$\sigma_{ik}^{(el)} = \sigma_{ik}^{(M)} + \sigma_{ik}^{(cor)}, \tag{4.132}$$

where $\sigma_{ik}^{(M)}$ is the Maxwell stress tensor, as derived previously (see expression (6.12)). Finally, the tensor arising from the correlation of the ions is given by:

$$\sigma_{ik}^{(cor)} = \frac{\varepsilon l_c^2}{4\pi}\left(\left(\mathcal{E}\cdot\nabla^2\mathcal{E} + \frac{1}{2}(\nabla\cdot\mathcal{E})^2\right)\delta_{ik} - \mathcal{E}_i\nabla^2\mathcal{E}_k - \mathcal{E}_k\nabla^2\mathcal{E}_i\right). \quad (4.133)$$

We won't focus into concrete numerical results related to the BSC model here, as they are similar to those obtained within the Cahn-Hilliard-type model presented in Chap. 6. Interested readers can refer to the original papers for more details [92, 96].

4.7 Fluctuations of the Electrostatic Potential Within Variational Field Theory

Previously, we explored the self-consistent field theory of inhomogeneous ionic fluids, without considering the fluctuations of the electrostatic potential near the mean-field configuration induced by the thermal motion of ions. These fluctuations can be incorporated using the variational method outlined in Chap. 2 for simple fluids. In this section we discuss the details of this approach, known as variational field theory in the literature [39, 55, 57], as applied to ionic fluids.

Let us start by examining the variational field theory using the example of the Coulomb gas model within a grand canonical ensemble. The grand partition function can be expressed as the following functional integral:

$$\Xi = \int \frac{\mathcal{D}\varphi}{C_0}\exp\left[-\frac{\beta}{2}\left(\varphi G_0^{-1}\varphi\right) + i(\rho_{ext}\varphi) + \sum_\alpha \bar{z}_\alpha \int d\mathbf{r}\, e^{i\beta q_\alpha\varphi(\mathbf{r}) - \beta u_\alpha(\mathbf{r})}\right], \quad (4.134)$$

where as above, $\bar{z}_\alpha = z_\alpha e^{\beta q_\alpha^2 G_0(0)/2}$. Next, by shifting the field variable $\varphi \to \varphi + \varphi_0$ and transitioning to the new Green's function $G_0 \to G$, we arrive at the following expression:

$$\Xi = \frac{C}{C_0}\int \frac{\mathcal{D}\varphi}{C}\exp\Big[-\frac{\beta}{2}\left(\varphi G^{-1}\varphi\right) - \frac{\beta}{2}\left(\varphi[G_0^{-1} - G^{-1}]\varphi\right)$$
$$+ i(\rho_{ext}(\varphi+\varphi_0)) - \beta(\varphi G_0^{-1}\varphi_0) - \frac{\beta}{2}(\varphi_0 G_0^{-1}\varphi_0)$$
$$+ \sum_\alpha \bar{z}_\alpha \int d\mathbf{r}\, e^{i\beta q_\alpha\varphi_0(\mathbf{r}) + i\beta q_\alpha\varphi(\mathbf{r}) - \beta u_\alpha(\mathbf{r})}\Big], \quad (4.135)$$

where we introduce the notation for the normalization constant of new Gaussian measure

$$C = \int \mathcal{D}\varphi \exp\Big[-\frac{\beta}{2}\left(\varphi G^{-1}\varphi\right)\Big]. \quad (4.136)$$

Then, by applying the Bogolyubov's inequality, we obtain

$$\Xi \geq \exp W[G;\varphi_0], \quad (4.137)$$

where we introduce the following functional

$$W[G;\varphi_0] = \frac{1}{2}\ln\frac{DetG}{DetG_0} - \frac{1}{2}tr\left(G[G_0^{-1} - G^{-1}]\right) + i(\rho_{ext}\varphi_0) - \frac{\beta}{2}(\varphi_0 G_0^{-1}\varphi_0) + \sum_{\alpha} z_{\alpha}\int d\mathbf{r}\, e^{i\beta q_{\alpha}\varphi_0(\mathbf{r}) + \frac{\beta q_{\alpha}^2}{2}(G_0(0) - G(\mathbf{r},\mathbf{r})) - \beta u_{\alpha}(\mathbf{r})}. \quad (4.138)$$

The functions $\varphi_0 = i\psi$ and $G(\mathbf{r},\mathbf{r}')$ are determined from the equations

$$\left.\frac{\delta W}{\delta\varphi_0(\mathbf{r})}\right|_{\varphi_0 = i\psi} = 0, \quad \frac{\delta W}{\delta G(\mathbf{r},\mathbf{r}')} = 0. \quad (4.139)$$

The first one takes the following form

$$\nabla^2\psi(\mathbf{r}) = -\frac{4\pi}{\varepsilon}\sum_{\alpha} q_{\alpha} z_{\alpha} A_{\alpha}(\mathbf{r}) e^{-\beta q_{\alpha}\psi(\mathbf{r}) - \beta u_{\alpha}(\mathbf{r})}, \quad (4.140)$$

where

$$A_{\alpha}(\mathbf{r}) = \exp\left[\frac{\beta q_{\alpha}^2}{2}\left(G_0(0) - G(\mathbf{r},\mathbf{r})\right)\right]. \quad (4.141)$$

The second one is

$$G^{-1}(\mathbf{r},\mathbf{r}') - G_0^{-1}(\mathbf{r},\mathbf{r}') = \beta\sum_{\alpha} q_{\alpha}^2 z_{\alpha} A_{\alpha}(\mathbf{r}) e^{-\beta q_{\alpha}\psi(\mathbf{r}) - \beta u_{\alpha}(\mathbf{r})}\delta(\mathbf{r} - \mathbf{r}'). \quad (4.142)$$

Using the fact that

$$G_0^{-1}(\mathbf{r},\mathbf{r}') = -\frac{\varepsilon}{4\pi}\nabla^2\delta(\mathbf{r} - \mathbf{r}'), \quad (4.143)$$

and inverting the operator G^{-1}, the Eq. (4.142) can be rewritten as

$$\left(-\nabla^2 + \varkappa^2(\mathbf{r})\right) G(\mathbf{r},\mathbf{r}') = \frac{4\pi}{\varepsilon}\delta(\mathbf{r} - \mathbf{r}'). \quad (4.144)$$

Thus, the grand thermodynamic potential of the Coulomb gas can be calculated using the formula:

$$\Omega = -k_B T W[G; i\psi] = \int d\mathbf{r}\left(-\frac{\varepsilon(\nabla\psi)^2}{8\pi} + \rho_{ext}\psi\right) + k_B T\int d\mathbf{r}\,(n_+ + n_-) + \int d\mathbf{r}\,\mathcal{I}(\mathbf{r})\int_0^1 d\tau\left(G(\mathbf{r},\mathbf{r};\tau) - G(\mathbf{r},\mathbf{r})\right), \quad (4.145)$$

where

$$\mathcal{I}(\mathbf{r}) = \frac{1}{2}\sum_{\alpha} q_{\alpha}^{2} n_{\alpha}(\mathbf{r}) \tag{4.146}$$

and we introduce the notations for the local concentrations of the ions

$$n_{\alpha}(\mathbf{r}) = \frac{\delta \Omega}{\delta u_{\alpha}(\mathbf{r})} = z_{\alpha} e^{-\beta q_{\alpha}\psi(\mathbf{r}) + \frac{\beta q_{\alpha}^{2}}{2}(G_{0}(0) - G(\mathbf{r},\mathbf{r})) - \beta u_{\alpha}(\mathbf{r})}. \tag{4.147}$$

Equation (4.147) define the chemical equilibrium conditions for ions, which can be rewritten in terms of the equality of chemical potentials

$$\mu_{\alpha} = \bar{\mu}_{\alpha} + q_{\alpha}\psi + u_{\alpha}, \tag{4.148}$$

where the intrinsic chemical potentials take the form

$$\bar{\mu}_{\alpha} = k_{B}T \ln(n_{\alpha}\Lambda_{\alpha}^{3}) + \frac{q_{\alpha}^{2}}{2}(G_{0}(0) - G(\mathbf{r},\mathbf{r})). \tag{4.149}$$

The intermediate Green's function $G(\mathbf{r}, \mathbf{r}'; \tau)$ satisfies the equation

$$\left(-\nabla^{2} + \tau \varkappa^{2}(\mathbf{r})\right) G(\mathbf{r}, \mathbf{r}'; \tau) = \frac{4\pi}{\varepsilon}\delta(\mathbf{r} - \mathbf{r}'). \tag{4.150}$$

For the bulk phase case, where $u_{\alpha} = 0$ and $\rho_{ext} = 0$, $n_{\alpha} = const$,

$$\varkappa^{2}(\mathbf{r}) = \kappa^{2} = \frac{4\pi}{\varepsilon k_{B}T}\sum_{\alpha} q_{\alpha}^{2} n_{\alpha} \tag{4.151}$$

in the thermodynamic limit $\psi(\mathbf{r}) = 0$ and the Green's function becomes translationally invariant, i.e.

$$G(\mathbf{r}, \mathbf{r}') = G(\mathbf{r} - \mathbf{r}') = \int \frac{d\mathbf{k}}{(2\pi)^{3}} G(\mathbf{k}) e^{i\mathbf{k}(\mathbf{r}-\mathbf{r}')}. \tag{4.152}$$

As it follows from Eqs. (4.144) and (4.150)

$$G(\mathbf{k}) = \frac{4\pi}{\varepsilon(k^{2} + \kappa^{2})}, \tag{4.153}$$

$$G(\mathbf{k}; \tau) = \frac{4\pi}{\varepsilon(k^{2} + \tau\kappa^{2})}. \tag{4.154}$$

Thus, the relation (4.145) leads us to the expression (3.38) for the osmotic pressure of an Coulomb gas obtained earlier within the Gaussian approximation framework.

For a spatially inhomogeneous Coulomb gas, variational field theory generally yields results that differ from those obtained for the thermodynamic potential, local electrostatic potential, and ionic concentrations within the Gaussian approximation. However, for the bulk phase, both approximations yield the same results for the equation of state of the Coulomb gas, i.e., the Debye-Hückel approximation.

4.8 Variational Field Theory with Account of Nonlocality of Ionic Charge

In the previous section, we explored variational field theory formulated for point charges. However, in real electrolytes, It is more correct to consider the nonlocality of the charge distribution. To accommodate this, following Wang's work [39], we can introduce phenomenological charge densities (form-factors) of ions $\varrho_\alpha(\mathbf{r})$ that satisfy the normalization conditions:

$$\int d\mathbf{r}\, \varrho_\alpha(\mathbf{r}) = q_\alpha. \tag{4.155}$$

Thus, the total potential energy of the interactions between the ions has the form

$$U = \frac{1}{2}\left(\hat{\rho} G_0 \hat{\rho}\right) + \sum_\alpha (\hat{n}_\alpha u_\alpha), \tag{4.156}$$

where the following auxiliary notations are introduced as usual

$$\left(\hat{\rho} G_0 \hat{\rho}\right) = \int d\mathbf{r} \int d\mathbf{r}' \hat{\rho}(\mathbf{r}) G_0(\mathbf{r}, \mathbf{r}') \hat{\rho}(\mathbf{r}') \tag{4.157}$$

with the local charge density of the ions

$$\hat{\rho}(\mathbf{r}) = \sum_\alpha \sum_{j_\alpha=1}^{N_\alpha} \varrho_\alpha(\mathbf{r} - \mathbf{r}_{j_\alpha}). \tag{4.158}$$

Note that in the framework of this model, we account for the finite intrinsic electrostatic energy of ions, eliminating the need, as in the case of point charges, to subtract it from the total energy.

Then, applying the Hubbard-Stratonovich transformation, we obtain the following representation for the grand partition function:

$$\Xi = \int \frac{\mathcal{D}\varphi}{C_0} \exp\left[-\frac{\beta}{2}\left(\varphi G_0^{-1} \varphi\right) + \sum_\alpha z_\alpha \int d\mathbf{r}\, e^{i\beta \varrho_\alpha \varphi(\mathbf{r}) - \beta u_\alpha(\mathbf{r})}\right], \tag{4.159}$$

where

$$\varrho_\alpha \varphi(\mathbf{r}) = \int d\mathbf{r}' \varrho_\alpha(\mathbf{r}-\mathbf{r}')\varphi(\mathbf{r}'). \tag{4.160}$$

As in the previous section, by shifting $\varphi \to \varphi + \varphi_0$ and introducing the trial Green's function $G_0 \to G$, we obtain

$$\Xi = \frac{C}{C_0}\int \frac{\mathcal{D}\varphi}{C}\exp\Big[-\frac{\beta}{2}\left(\varphi G^{-1}\varphi\right) - \frac{\beta}{2}\left(\varphi[G_0^{-1} - G^{-1}]\varphi\right)$$
$$-\beta(\varphi_0 G_0^{-1}\varphi) - \frac{\beta}{2}(\varphi_0 G_0^{-1}\varphi_0) + \sum_\alpha z_\alpha \int d\mathbf{r}\, e^{i\beta\varrho_\alpha\varphi_0(\mathbf{r}) + i\beta\varrho_\alpha\varphi(\mathbf{r}) - \beta u_\alpha(\mathbf{r})}\Big]. \tag{4.161}$$

Utilizing the Gibbs-Bogolyubov inequality

$$\left\langle e^X\right\rangle \geq e^{\langle X\rangle}, \tag{4.162}$$

where

$$\langle(\ldots)\rangle = \int \frac{\mathcal{D}\varphi}{C}\exp\Big[-\frac{\beta}{2}\left(\varphi G^{-1}\varphi\right)\Big](\ldots), \tag{4.163}$$

we arrive at

$$\Xi \geq \exp W[G;\varphi_0], \tag{4.164}$$

where we introduce the following functional

$$W[G;\varphi_0] = \frac{1}{2}\ln\frac{DetG}{DetG_0} - \frac{1}{2}tr\left(G[G_0^{-1} - G^{-1}]\right) - \frac{\beta}{2}(\varphi_0 G_0^{-1}\varphi_0)$$
$$+\sum_\alpha z_\alpha \int d\mathbf{r}\, e^{i\beta\varrho_\alpha\varphi_0(\mathbf{r}) - \frac{\beta}{2}\varrho_\alpha G\varrho_\alpha - \beta u_\alpha(\mathbf{r})} \tag{4.165}$$

and take into account that

$$\left\langle e^{i\beta\varrho_\alpha\varphi(\mathbf{r})}\right\rangle = e^{-\frac{\beta}{2}\varrho_\alpha G\varrho_\alpha} = A_\alpha(\mathbf{r}), \tag{4.166}$$

$$\varrho_\alpha G\varrho_\alpha = \int d\mathbf{r}_1 \int d\mathbf{r}_2\, \varrho_\alpha(\mathbf{r}_1 - \mathbf{r})G(\mathbf{r}_1,\mathbf{r}_2)\varrho_\alpha(\mathbf{r}_2 - \mathbf{r}). \tag{4.167}$$

Functions $\varphi_0 = i\psi$ and $G(\mathbf{r},\mathbf{r}')$ can be determined from the self-consistent field equations

$$\left.\frac{\delta W}{\delta\varphi_0(\mathbf{r})}\right|_{\varphi_0 = i\psi} = 0, \quad \frac{\delta W}{\delta G(\mathbf{r},\mathbf{r}')} = 0. \tag{4.168}$$

First of which yields

$$\nabla^2\psi(\mathbf{r}) = -\frac{4\pi}{\varepsilon}\sum_\alpha z_\alpha \int d\mathbf{r}'\, \varrho_\alpha(\mathbf{r}-\mathbf{r}')A_\alpha(\mathbf{r}')e^{-\beta\varrho_\alpha\psi(\mathbf{r}')-\beta u_\alpha(\mathbf{r}')}. \tag{4.169}$$

Using similar calculations as those presented in the previous section, we arrive at the second equation in the following form:

$$G^{-1}(\mathbf{r},\mathbf{r}') - G_0^{-1}(\mathbf{r},\mathbf{r}') = \Sigma(\mathbf{r},\mathbf{r}'), \tag{4.170}$$

where

$$\Sigma(\mathbf{r},\mathbf{r}') = \beta\sum_\alpha z_\alpha \int d\mathbf{x}\, A_\alpha(\mathbf{x})e^{-\beta\varrho_\alpha\psi(\mathbf{x})-\beta u_\alpha(\mathbf{x})}\varrho_\alpha(\mathbf{r}-\mathbf{x})\varrho_\alpha(\mathbf{r}'-\mathbf{x}). \tag{4.171}$$

Equation 4.170 can be rewritten in the form

$$\frac{1}{4\pi}\left(-\varepsilon\nabla^2 + \Sigma\right)G(\mathbf{r},\mathbf{r}') = \delta(\mathbf{r}-\mathbf{r}'), \tag{4.172}$$

where Σ is the integral operator with the kernel (4.171). Thus, we obtain

$$\Omega = -k_BTW[G;i\psi] = -\int d\mathbf{r}\,\frac{\varepsilon(\nabla\psi)^2}{8\pi} - k_BT\int d\mathbf{r}\sum_\alpha \bar{n}_\alpha(\mathbf{r})$$
$$+k_BT\int d\mathbf{r}\int d\mathbf{r}'\,\Sigma(\mathbf{r},\mathbf{r}')\int_0^1 d\tau\,\left(G(\mathbf{r},\mathbf{r}';\tau) - G(\mathbf{r},\mathbf{r}')\right). \tag{4.173}$$

For the case of the point charge model, when $\varrho_\alpha(\mathbf{r}) = q_\alpha\delta(\mathbf{r})$, Eq. (4.173) simplifies to 4.145.

Intermediate Green's function $G(\mathbf{r},\mathbf{r}';\tau)$ is determined as

$$\frac{1}{4\pi}\left(-\varepsilon\nabla^2 + \tau\Sigma\right)G(\mathbf{r},\mathbf{r}';\tau) = \delta(\mathbf{r}-\mathbf{r}'). \tag{4.174}$$

In the bulk phase, where $u_\alpha = 0$ and n_α becomes constant, the electrostatic potential $\psi(\mathbf{r}) = 0$, resulting in a translationally invariant Green's function

$$G(\mathbf{r},\mathbf{r}') = G(\mathbf{r}-\mathbf{r}') = \int\frac{d\mathbf{k}}{(2\pi)^3}G(\mathbf{k})e^{i\mathbf{k}(\mathbf{r}-\mathbf{r}')}, \tag{4.175}$$

As follows from Eqs. (4.172) and (4.174), the Fourier transforms of Green's functions are given by:

$$G(\mathbf{k}) = \frac{4\pi}{\varepsilon(k^2 + \varkappa^2(\mathbf{k}))}, \tag{4.176}$$

$$G(\mathbf{k};\tau) = \frac{4\pi}{\varepsilon(k^2 + \tau\varkappa^2(\mathbf{k}))}, \tag{4.177}$$

where we introduce the screening function

$$\varkappa^2(\mathbf{k}) = \frac{4\pi}{k_B T\varepsilon}\sum_{\alpha} n_\alpha |\varrho_\alpha(\mathbf{k})|^2 \tag{4.178}$$

which we will address further below.

Thus, (4.173) yields the following expression for the osmotic pressure in the bulk phase $P_b = -\Omega/V$

$$P_b = k_B T\sum_{\alpha} n_\alpha + \frac{k_B T}{2}\int \frac{d\mathbf{k}}{(2\pi)^3}\left(\ln\left(1 + \frac{\varkappa^2(\mathbf{k})}{k^2}\right) - \frac{\varkappa^2(\mathbf{k})}{k^2 + \varkappa^2(\mathbf{k})}\right). \tag{4.179}$$

In the case of point charges when

$$\varkappa^2(\mathbf{k}) = \kappa^2 = \frac{4\pi}{k_B T\varepsilon}\sum_{\alpha} q_\alpha^2 n_\alpha \tag{4.180}$$

equation (4.179) reduces to the Debye-Hückel expression

$$P_b = k_B T\sum_{\alpha} n_\alpha - \frac{k_B T\kappa^3}{24\pi}. \tag{4.181}$$

4.9 Incorporation of Excluded Volume and Electrostatic Correlations of Ions

Above, we formulated a variational field theory for an Coulomb gas, where ions are modeled as point charges. The next step is to develop a theory that takes into account the excluded volume of ions and their electrostatic correlations. To achieve this, we can combine the formalism proposed earlier for deriving the modified Poisson-Boltzmann equations with the variational method. Let us assume that, in addition to Coulomb interactions, ions interact via repulsive potentials $U_{\alpha\gamma}(\mathbf{r})$. To maintain clarity in the narrative, we will exclude the consideration of the short-range specific interactions discussed earlier.

As before, applying the HS transformations, we arrive at the following representation of the grand partition function:

$$\Xi = \int \frac{\mathcal{D}\varphi}{C_0} \exp\left[-\frac{\beta}{2}\left(\varphi G_0^{-1}\varphi\right)\right] \Xi_R, \tag{4.182}$$

where the grand partition function of the reference system with repulsive potentials has the form

$$\Xi_R = \int \frac{\mathcal{D}\Phi}{\mathcal{N}_U} e^{-\frac{\beta}{2}\sum_{\alpha\gamma}(\Phi_\alpha U_{\alpha\gamma}^{-1}\Phi_\gamma) + W[\Phi;\chi]}, \tag{4.183}$$

with an auxiliary functional

$$W[\Phi;\chi] = \sum_\alpha \bar{z}_\alpha \int d\mathbf{r}\, e^{i\beta\Phi_\alpha + i\beta\chi_\alpha} \tag{4.184}$$

and following short-hand notation.

$$\chi_\alpha = q_\alpha\varphi + iu_\alpha \tag{4.185}$$

We also introduce "renormalazied" activities, $\bar{z}_\pm = e^{\frac{\beta U_{\pm\pm}(0)}{2} + \frac{\beta q_\pm^2 G_0(0)}{2}} z_\pm\theta_\pm$. As in previous section, Let us shift the field variable $\varphi \to \varphi + i\psi$ and consider a new Green's function, $G_0 \to G$, i.e.

$$\Xi = \int \frac{\mathcal{D}\Phi}{\mathcal{N}_U} e^{-\frac{\beta}{2}\sum_{\alpha\gamma}(\Phi_\alpha U_{\alpha\gamma}^{-1}\Phi_\gamma)} \frac{C}{C_0} \int \frac{\mathcal{D}\varphi}{C} e^{-\frac{\beta}{2}(\varphi G^{-1}\varphi)}$$
$$\times \exp\Bigg[-\frac{\beta}{2}\left(\varphi[G_0^{-1} - G^{-1}]\varphi\right) + i\beta(\rho_{ext}(\varphi + i\psi)) - i\beta(\varphi G_0^{-1}\psi)$$
$$+\frac{\beta}{2}(\psi G_0^{-1}\psi) + \sum_\alpha \bar{z}_\alpha \int d\mathbf{r}\, e^{i\beta\Phi_\alpha + i\beta q_\alpha\varphi - \beta q_\alpha\psi - \beta u_\alpha}\Bigg]. \tag{4.186}$$

Restricting to the first cumulant, we obtain

$$\Xi \approx \exp\left[\ln\frac{C}{C_0} - \frac{1}{2}tr\left(G[G_0^{-1} - G^{-1}]\right) - \beta(\rho_{ext}\psi) + \frac{\beta}{2}(\psi G_0^{-1}\psi)\right]$$
$$\times \int \frac{\mathcal{D}\Phi}{\mathcal{N}_U} e^{-\frac{\beta}{2}\sum_{\alpha\gamma}(\Phi_\alpha U_{\alpha\gamma}^{-1}\Phi_\gamma) + \sum_\alpha \tilde{z}_\alpha \int d\mathbf{r}\, A_\alpha(\mathbf{r}) e^{i\beta\Phi_\alpha - \beta q_\alpha\psi - \beta u_\alpha}}, \tag{4.187}$$

where, as before, $\tilde{z}_\alpha = z_\alpha e^{\frac{\beta}{2}U_{\alpha\alpha}(0)}$ and $A_\alpha(\mathbf{r}) = e^{\frac{\beta q_\alpha^2}{2}(G_0(0) - G(\mathbf{r},\mathbf{r}))}$. Now, using the same group expansion-based method we used to derive the modified PB equations above, we arrive at the following approximation for the grand partition function:

$$\Xi \approx \exp W[\psi; G], \tag{4.188}$$

where

$$W[\psi; G] = \ln \frac{C}{C_0} - \frac{1}{2} tr \left(G[G_0^{-1} - G^{-1}] \right) - \beta(\rho_{ext}\psi) + \frac{\beta}{2}(\psi G_0^{-1}\psi) - \beta \int d\mathbf{r}\, P_R\left(\{\bar{\mu}_\alpha\}\right), \tag{4.189}$$

where, as before, we introduce the local pressure of the reference fluid model, P_R, which depends on the intrinsic chemical potentials

$$\bar{\mu}_\alpha = \mu_\alpha - q_\alpha \psi + \frac{q_\alpha^2}{2}(G_0(0) - G(\mathbf{r}, \mathbf{r})) - u_\alpha. \tag{4.190}$$

For the theory to be self-consistent, it is necessary to define the "closure" relations for the functions ψ and G, which can be obtained from the extremum conditions of the functional $W[\psi; G]$, i.e.

$$\frac{\delta W}{\delta \psi(\mathbf{r})} = 0, \quad \frac{\delta W}{\delta G(\mathbf{r}, \mathbf{r}')} = 0, \tag{4.191}$$

which yield

$$\nabla^2 \psi(\mathbf{r}) = -\frac{4\pi}{\varepsilon} \left(\rho_{ext}(\mathbf{r}) + q_+ \bar{n}_+(\mathbf{r}) + q_- \bar{n}_-(\mathbf{r}) \right), \tag{4.192}$$

$$\left(-\nabla^2 + \varkappa^2(\mathbf{r}) \right) G(\mathbf{r}, \mathbf{r}') = \frac{4\pi}{\varepsilon} \delta(\mathbf{r} - \mathbf{r}'), \tag{4.193}$$

where $\varkappa^2(\mathbf{r}) = \beta \sum_\alpha q_\alpha^2 \bar{n}_\alpha(\mathbf{r})/\varepsilon$, and $\bar{n}_\alpha(\mathbf{r}) = \partial P_R / \partial \bar{\mu}_\alpha$ are the local concentrations of the ions.

The grand thermodynamic potentials has the following form

$$\Omega = \int d\mathbf{r} \left(-\frac{\varepsilon (\nabla \psi)^2}{8\pi} + \rho_{ext}\psi \right) + \int d\mathbf{r}\, P_0(\mathbf{r}) + \int d\mathbf{r}\, \mathcal{I}(\mathbf{r}) \int_0^1 d\tau \left(G(\mathbf{r}, \mathbf{r}; \tau) - G(\mathbf{r}, \mathbf{r}) \right), \tag{4.194}$$

where $\mathcal{I}(\mathbf{r}) = \sum_\alpha q_\alpha^2 \bar{n}_\alpha(\mathbf{r})/2$ and $P_0(\mathbf{r}) = P_R(\{\bar{\mu}_\alpha(\mathbf{r})\})$ is the local osmotic pressure.

Note that this approach, unlike the variational method applied in the last section to the Coulomb gas, does not guarantee the best variational estimate for a grand thermodynamic potential. Nevertheless, this approach can be seen as a natural extension of the mean-field theory that takes into account simultaneously electrostatic correlations and steric interactions of the ions. As in the case of the modified PB equations, this approach allows us to use different reference systems, including symmetric and asymmetric lattice gas, as well as a model of hard spheres mixture. For the case of a symmetric lattice gas, for which

$$P_R = \frac{k_B T}{v} \ln \left(1 + e^{\beta \mu_+} + e^{\beta \mu_-} \right), \tag{4.195}$$

the concentrations of the ions yield

$$\bar{n}_\alpha = \frac{1}{v} \frac{e^{\beta\bar{\mu}_\alpha}}{1 + e^{\beta\bar{\mu}_+} + e^{\beta\bar{\mu}_-}}. \tag{4.196}$$

The self-consistent field equations for the case of a symmetric lattice gas have also been derived in Ref. [97] within a different theoretical approach.

Repeating the same calculations from the previous section, it is easy to understand that for the case of the bulk phase, the theory with a symmetric lattice gas reference system leads to the following equation of state

$$P = -\frac{k_B T}{v} \ln\left(1 - n_+ v - n_- v\right) - \frac{k_B T \kappa^3}{24\pi}, \tag{4.197}$$

Where $n_\pm$ denotes the bulk concentration of the ions, satisfying the condition of electrical neutrality $q_+ n_+ + q_+ n_- n_- = 0$. Thus, for the bulk phase, this approximation yields the Debye-Hückel expression for the electrostatic contribution to the osmotic pressure.

4.10 Correlation Stress Tensor in Ionic Fluids

In this section, we would like to formulate a generalization of the method for obtaining the stress tensor in ionic fluids, considering fluctuations of the electrostatic potential. Let us start from expressing the previously known grand partition function of the Coulomb gas:

$$\Xi = \int \frac{\mathcal{D}\varphi}{C_0} \exp\left[- \beta \mathcal{H}[\varphi] \right], \tag{4.198}$$

where the functional $\mathcal{H}$ is determined by the following relation

$$\mathcal{H} = \frac{\varepsilon}{8\pi} \int d\mathbf{r} (\nabla\varphi(\mathbf{r}))^2 - \int d\mathbf{r}\, \hat{P}(\mathbf{r}). \tag{4.199}$$

with fluctuating osmotic pressure of the ions

$$\hat{P}(\mathbf{r}) = k_B T \sum_\alpha \bar{z}_\alpha e^{i\beta q_\alpha \varphi(\mathbf{r}) - \beta u_\alpha(\mathbf{r})}. \tag{4.200}$$

First, we consider the case when $u_\alpha = 0$. Additionally, we assume that the external charges are surface charges, and their influence is incorporated through the boundary conditions of the self-consistent field equations. Similar to previous derivations, we perform a virtual displacement of the system by an infinitesimal vector **h**, i.e.

$$x_k' = x_k + h_k, \tag{4.201}$$

and require the invariance of the grand partition function, i.e.

$$\delta \ln \Xi = -\beta \Big\langle \delta \mathcal{H}[\varphi] \Big\rangle = 0. \tag{4.202}$$

Thus, after straightforward transformations, we obtain

$$\partial_i \Big\langle \hat{\sigma}_{ik} h_k \Big\rangle = \Big\langle \frac{\delta \mathcal{H}[\varphi]}{\delta \varphi} h_k \partial_k \varphi \Big\rangle, \tag{4.203}$$

where we introduce a stress tensor, defined on the fluctuating electrostatic field

$$\hat{\sigma}_{ik} = \frac{\varepsilon}{8\pi} \delta_{ik} \partial_j \varphi \partial_j \varphi - \frac{\varepsilon}{4\pi} \partial_i \varphi \partial_k \varphi - \hat{P} \delta_{ik}. \tag{4.204}$$

Due to the arbitrariness of h_k we obtain

$$\partial_i \langle \hat{\sigma}_{ik} \rangle = \Big\langle \frac{\delta \mathcal{H}[\varphi]}{\delta \varphi} \partial_k \varphi \Big\rangle, \tag{4.205}$$

where averaging is understood in the usual sense:

$$\langle (\ldots) \rangle = \frac{1}{\Xi} \int \frac{\mathcal{D}\varphi}{C_0} \exp\Big(-\beta \mathcal{H}[\varphi] \Big) (\ldots). \tag{4.206}$$

Further, taking into account the identity (1.203)

$$\Big\langle \frac{\delta \mathcal{H}[\varphi]}{\delta \varphi(\mathbf{r})} \partial_k \varphi(\mathbf{r}) \Big\rangle = \partial_k \frac{\delta \varphi(\mathbf{r})}{\delta \varphi(\mathbf{r})} = \partial_k \delta(\mathbf{0}) = 0, \tag{4.207}$$

we arrive at the condition of mechanical equilibrium of ionic fluid taking into account thermal fluctuations of electrostatic potential

$$\partial_i \langle \hat{\sigma}_{ik} \rangle = 0. \tag{4.208}$$

In the presence of external potential fields, $u_\alpha(\mathbf{r}) \neq 0$, similar calculations lead to the following condition of mechanical equilibrium

$$\partial_i \langle \hat{\sigma}_{ik} \rangle - \sum_\alpha \langle \hat{n}_\alpha \rangle \partial_k u_\alpha = 0, \tag{4.209}$$

where

$$\hat{n}_\alpha(\mathbf{r}) = -\partial \hat{P}/\partial u_\alpha = \bar{z}_\alpha e^{i\beta q_\alpha \varphi(\mathbf{r}) - \beta u_\alpha(\mathbf{r})} \tag{4.210}$$

are fluctuating concentrations of the ions.

Let us calculate the stress tensor within the framework of the variational field theory formulated above. For this purpose, let us represent the field variable as $\varphi(\mathbf{r}) = i\psi(\mathbf{r}) + \eta(\mathbf{r})$, where $\psi(\mathbf{r})$ is the solution of the Eq. (4.140), and η is the fluctuation over which we need to integrate. Thus, we obtain

$$\sigma_{ik} = -P\delta_{ik} + \sigma_{ik}^{(M)} + \sigma_{ik}^{(cor)}, \tag{4.211}$$

where $\sigma_{ik}^{(M)}$ is already introduced the Maxwell stress tensor, $P = \left\langle \hat{P} \right\rangle$ is the average osmotic pressure of the ions, and

$$\sigma_{ik}^{(cor)} = \frac{\varepsilon}{8\pi}\delta_{ik}\left\langle \partial_j \eta \partial_j \eta \right\rangle - \frac{\varepsilon}{4\pi}\left\langle \partial_i \eta \partial_k \eta \right\rangle \tag{4.212}$$

is the correlation stress tensor, determined by correlations of fluctuations of electrostatic field components. Using the averaging approximation, dictated by the variational field theory

$$\frac{1}{\Xi}\int \frac{\mathcal{D}\varphi}{C_0} \exp\left[-\beta\mathcal{H}[\varphi]\right](\ldots) \approx \int \frac{\mathcal{D}\eta}{C} \exp\left[-\frac{\beta}{2}(\eta G^{-1}\eta)\right](\ldots), \tag{4.213}$$

performing functional integration, we arrive at the following stress tensor [3]

$$\sigma_{ik}^{(corr)} = \frac{\varepsilon}{4\pi}\left(\frac{1}{2}\mathcal{D}_{ll}(\mathbf{r})\delta_{ik} - \mathcal{D}_{ik}(\mathbf{r})\right), \tag{4.214}$$

where we introduce tensor

$$\mathcal{D}_{ik}(\mathbf{r}) = \lim_{\mathbf{r}' \to \mathbf{r}} \partial_i \partial_k' G(\mathbf{r}, \mathbf{r}'). \tag{4.215}$$

For practical purposes, it is necessary to utilize the regularized tensor

$$\mathcal{D}_{ik}(\mathbf{r}) = \lim_{\mathbf{r}' \to \mathbf{r}} \partial_i \partial_k' \bar{G}(\mathbf{r}, \mathbf{r}'), \tag{4.216}$$

where

$$\bar{G}(\mathbf{r}, \mathbf{r}') = G(\mathbf{r}, \mathbf{r}') - G_0(\mathbf{r} - \mathbf{r}'), \tag{4.217}$$

where we introduce the Green's function of the Poisson equation for the infinite dielectric medium

$$G_0(\mathbf{r} - \mathbf{r}') = \frac{1}{\varepsilon|\mathbf{r} - \mathbf{r}'|}. \tag{4.218}$$

The osmotic pressure of ions, according to the averaging rule (4.213), is as follows:

$$P = k_B T \sum_{\alpha} \bar{z}_{\alpha} e^{-\frac{\beta q_{\alpha}^2}{2} G(\mathbf{r},\mathbf{r}) - \beta q_{\alpha}\psi(\mathbf{r}) - \beta u_{\alpha}(\mathbf{r})}. \quad (4.219)$$

For the ionic fluid model incorporating excluded volume effects, we can construct the following stress tensor within the formalism presented above, ensuring consistency with Eqs. (4.192) and (4.193):

$$\sigma_{ik} = -P\left(\{\bar{\mu}_{\alpha}\}\right)\delta_{ik} + \frac{\varepsilon}{4\pi}\left(\partial_i\psi\partial_k\psi - \frac{1}{2}\delta_{ik}\partial_l\psi\partial_l\psi\right) + \frac{\varepsilon}{4\pi}\left(\frac{1}{2}\mathcal{D}_{ll}(\mathbf{r})\delta_{ik} - \mathcal{D}_{ik}(\mathbf{r})\right), \quad (4.220)$$

where the osmotic pressure of the ions, expressed as a function of the intrinsic chemical potentials, $\bar{\mu}_{\alpha}$, as given by the relation (4.190), is determined by a reference fluid model.

4.11 Interaction of Charged Conductive and Dielectric Plates in Electrolyte Solutions

Let us examine the Coulomb gas confined within a slit-like pore with flat charged infinite walls. Let us assume that these walls generate external potentials

$$u_{\alpha}(z) = \phi_{\alpha}(z) + \phi_{\alpha}(H - z), \quad (4.221)$$

where $\phi_{\alpha}(z)$ represents the potential of the external field acting on the α-th ion at the point $z \in [0, H]$. The identical walls are situated at $z = 0$ and $z = H$. The disjoining pressure is given by the expression

$$\Pi = -\frac{\partial(\Omega/\mathcal{A})}{\partial H} - P_b, \quad (4.222)$$

where $\mathcal{A}$ represents the total area of the walls' surface. By considering that solutions of Eqs. (4.192) and (4.193) minimize the functional $\Omega = -k_B T W[G; \psi]$, we obtain

$$\Pi = -\sum_{\alpha} \int_{-\infty}^{\infty} dz\, \bar{n}_{\alpha}(z)\phi_{\alpha}'(z) - P_b, \quad (4.223)$$

where we utilized the relation $\bar{n}_{\alpha}(H - z) = \bar{n}_{\alpha}(z)$, which follows from the symmetry of the problem, i.e., the identity of the pore walls. Integrating the mechanical equilibrium condition,

$$\frac{d\sigma_{zz}}{dz} - \sum_{\alpha} \bar{n}_{\alpha}(z)u_{\alpha}'(z) = 0, \quad (4.224)$$

from $z = H/2$ to $z = \infty$ and considering the condition $\sigma_{zz}(\infty) = 0$, we arrive at

$$-\sigma_{zz}(H/2) = \sum_\alpha \int_{H/2}^{\infty} dz\, \bar{n}_\alpha(z) u'_\alpha(z) = \sum_\alpha \int_{H/2}^{\infty} dz\, \bar{n}_\alpha(z)\phi'_\alpha(z) - \sum_\alpha \int_{-\infty}^{H/2} dz\, \bar{n}_\alpha(z)\phi'_\alpha(z). \tag{4.225}$$

Therefore,

$$\Pi = -\sum_\alpha \int_{-\infty}^{\infty} dz\, \bar{n}_\alpha(z)\phi'_\alpha(z) - P_b = -\sum_\alpha \int_{-\infty}^{H/2} dz\, \bar{n}_\alpha(z)\phi'_\alpha(z) - \sum_\alpha \int_{H/2}^{\infty} dz\, \bar{n}_\alpha(z)\phi'_\alpha(z) - P_b$$
$$= -\sigma_{zz}(H/2) - P_b - 2\sum_\alpha \int_{H/2}^{\infty} dz\, \bar{n}_\alpha(z)\phi'_\alpha(z). \tag{4.226}$$

Thus, given that $\bar{n}\alpha(z) = 0$ at $z \geq H$ (impermeable wall), we finally obtain the already known expression (see (6.23))

$$\Pi = P_n - P_b - 2\sum_\alpha \int_{H/2}^{H} dz\, \bar{n}_\alpha(z)\phi'_\alpha(z), \tag{4.227}$$

where $P_n = -\sigma_{zz}(H/2)$ is the normal pressure in the middle of the pore. If the pore is wide enough and the range of the potential ϕ_α is small enough, we can neglect the integral on the right side of the expression (4.227).

Now Let us specify the normal pressure in the middle of the pore. For a slit-shaped pore, the Green's function can be expressed as

$$G(\mathbf{r}, \mathbf{r}') = G(\boldsymbol{\rho} - \boldsymbol{\rho}'|z, z'), \tag{4.228}$$

where $\boldsymbol{\rho}$ is a two-dimensional vector lying in the plane parallel to the pore walls, and the z-axis is perpendicular to the pore wall. Let us examine the two-dimensional Fourier transform:

$$G(\boldsymbol{\rho} - \boldsymbol{\rho}'|z, z') = \int \frac{d^2\mathbf{p}}{(2\pi)^2} e^{-i\mathbf{p}(\boldsymbol{\rho}-\boldsymbol{\rho}')} G(p|z, z'), \tag{4.229}$$

where $G(p|z, z')$ is an even function of $\mathbf{p}$, depending only on the absolute value $p = |\mathbf{p}|$. Nondiagonal elements of the tensor $\mathcal{D}_{ik}(z)$ are

$$\mathcal{D}_{xy}(z) = \mathcal{D}_{xz}(z) = \mathcal{D}_{yz}(z) = 0, \tag{4.230}$$

whereas the diagonal ones have the form

$$\mathcal{D}_{xx}(z) = \int \frac{d^2\mathbf{p}}{(2\pi)^2} p_x^2 Q(q,z), \quad \mathcal{D}_{yy}(z) = \int \frac{d^2\mathbf{p}}{(2\pi)^2} p_y^2 Q(q,z), \quad \mathcal{D}_{zz}(z) = \int \frac{d^2\mathbf{p}}{(2\pi)^2} \mathcal{D}_{zz}(q,z), \tag{4.231}$$

where

$$\mathcal{D}_{zz}(p,z) = k_B T \lim_{z' \to z} \partial_z \partial_{z'} G(p|z,z'), \tag{4.232}$$

$$Q(p,z) = k_B T \lim_{z' \to z} G(p|z,z'), \tag{4.233}$$

so

$$\mathcal{D}_{//}(z) = \int \frac{d^2\mathbf{p}}{(2\pi)^2} \left(p^2 Q(p,z) + \mathcal{D}_{zz}(p,z) \right). \tag{4.234}$$

The Fourier transform of the Green's function, $G(p|z,z')$, can be found from the following equation

$$\frac{\varepsilon}{4\pi} \left(-\partial_z^2 + p^2 + \varkappa^2(z) \right) G(p|z,z') = \delta(z-z'). \tag{4.235}$$

Hence, the normal pressure in (4.227) has the following form

$$P_n = P_m + \frac{\varepsilon}{4\pi} \left(\mathcal{D}_{zz}\left(\frac{H}{2}\right) - \frac{1}{2} \mathcal{D}_{//}\left(\frac{H}{2}\right) \right), \tag{4.236}$$

where $P_m = P_0(H/2)$ denotes the osmotic pressure of the ions at the midpoint of the pore, and we also assume that $\psi'(H/2) = 0$.

Finally, let us estimate the asymptotics of the second term on the right-hand side of Eq. (4.236) corresponding to the electrostatic field fluctuations as $H \to \infty$. In this asymptotic regime, we have $\varkappa(z) \simeq \kappa$, and since

$$G(p|z,z') \simeq \frac{2\pi e^{-\sqrt{p^2+\kappa^2}|z-z'|}}{\varepsilon \sqrt{p^2+\kappa^2}}, \quad G_0(p|z,z') = \frac{2\pi e^{-p|z-z'|}}{\varepsilon p}, \tag{4.237}$$

we obtain

$$\frac{\varepsilon}{4\pi} \left(\mathcal{D}_{zz}\left(\frac{H}{2}\right) - \frac{1}{2} \mathcal{D}_{//}\left(\frac{H}{2}\right) \right) \simeq \frac{k_B T}{8\pi} \int_0^\infty dp p \left(2p - \sqrt{p^2+\kappa^2} - \frac{p^2}{\sqrt{p^2+\kappa^2}} \right) = -\frac{k_B T \kappa^3}{24\pi}, \tag{4.238}$$

that is, the Debye-Hückel expression, as expected.

Let us now calculate the disjoining pressure between charged identical walls located at a sufficiently large distance from each other, such that we can neglect the contribution of specific interactions of ions with the walls. We can express it as

$$\Pi = P_n - P_b = P_m + \frac{\varepsilon}{4\pi} \left(\mathcal{D}_{zz}\left(\frac{H}{2}\right) - \frac{1}{2} \mathcal{D}_{//}\left(\frac{H}{2}\right) \right) - P_b. \tag{4.239}$$

Let us estimate the asymptotic behavior of the disjoining pressure as $H \to \infty$, where $\varkappa(z) \simeq \kappa$. In this case, the Green's function is defined by the following equation:

$$\frac{\varepsilon}{4\pi}\left(- \partial_z^2 + p^2 + \kappa^2 \right) G(p|z, z') = \delta(z - z'). \tag{4.240}$$

Metal Walls

First, Let us consider the case of metal walls (the case of dielectric walls will be considered below), for which the boundary conditions are defined as follows

$$\begin{gathered} G(p|0, z') = G(p|H, z') = 0, \\ \frac{\varepsilon}{4\pi}\left(\frac{\partial G(p|z'-0, z')}{\partial z} - \frac{\partial G(p|z'+0, z')}{\partial z} \right) = 1, \\ G(p|z'-0, z') = G(p|z'+0, z'). \end{gathered} \tag{4.241}$$

The solution is as follows

$$G(p|z, z') = \begin{cases} \frac{4\pi \sinh \kappa_p z (\cosh \kappa_p z' - \coth \kappa_p H \sinh \kappa_p z')}{\varepsilon \kappa_p}, & z < z', \\ \frac{4\pi \sinh \kappa_p z' (\cosh \kappa_p z - \coth \kappa_p H \sinh \kappa_p z)}{\varepsilon \kappa_p}, & z > z', \end{cases} \tag{4.242}$$

where $\kappa_p = \sqrt{p^2 + \kappa^2}$. Normal pressure in the midpoint of the pore can be determined by the expression

$$\begin{aligned} \frac{\varepsilon}{4\pi}\left(\mathcal{D}_{zz}\left(\frac{H}{2}\right) - \frac{1}{2}\mathcal{D}_{ll}\left(\frac{H}{2}\right)\right) &= \frac{k_B T}{4}\int \frac{d^2\mathbf{p}}{(2\pi)^2}\left(2p - \frac{\kappa_p}{\tanh \frac{\kappa_p H}{2}} - \frac{p^2 \tanh \frac{\kappa_p H}{2}}{\kappa_p} \right) \\ &= \frac{k_B T}{4}\int \frac{d^2\mathbf{p}}{(2\pi)^2}\left(2p - \sqrt{p^2+\kappa^2} - \frac{p^2}{\sqrt{p^2+\kappa^2}} \right) \\ &- \frac{k_B T}{4}\int \frac{d^2\mathbf{p}}{(2\pi)^2}\left(\kappa_p\left(\coth \frac{\kappa_p H}{2} - 1 \right) + \frac{p^2}{\kappa_p}\left(\tanh \frac{\kappa_p H}{2} - 1 \right) \right). \end{aligned} \tag{4.243}$$

Furthermore,

$$P_n \simeq P_0 + \frac{\varepsilon \kappa^2 \psi_m^2}{8\pi} + \delta P_n, \tag{4.244}$$

where $P_0 = k_B T \sum_\alpha n_\alpha$ is the bulk pressure without taking into account the electrostatic interactions of the ions, δP_n is its perturbation, arising from the electrostatic field fluctuations, and $\psi_m = \psi(H/2)$

$$\delta P_n = -\frac{1}{2}\sum_\alpha q_\alpha^2 n_\alpha \int \frac{d^2\mathbf{p}}{(2\pi)^2}\delta G\left(p \middle| \frac{H}{2}, \frac{H}{2}\right). \tag{4.245}$$

We also introduce the notation

$$\delta G\left(p\left|\frac{H}{2},\frac{H}{2}\right.\right)=G\left(q\left|\frac{H}{2},\frac{H}{2}\right.\right)-\lim_{H\to\infty}G\left(p\left|\frac{H}{2},\frac{H}{2}\right.\right)=\frac{2\pi(\tanh\frac{\kappa_p H}{2}-1)}{\varepsilon\kappa_p} \tag{4.246}$$

and take into account that $n_\alpha = \partial P/\partial\mu_\alpha$. Consequently, considering (4.243) and

$$P_b = P_0 - \frac{k_B T\kappa^3}{24\pi}, \tag{4.247}$$

and substituting (4.244) into the Eq. (4.239), we obtain

$$\Pi = \Pi_{MF} + \Pi_{corr}, \tag{4.248}$$

where the mean-field contribution to the disjointing pressure has already been obtained earlier (see expression (3.57))

$$\Pi_{MF} = \frac{2\pi\sigma^2}{\varepsilon}\frac{1}{\sinh^2\left(\frac{\kappa H}{2}\right)} \simeq \frac{8\pi\sigma^2}{\varepsilon}e^{-\kappa H}, \tag{4.249}$$

whereas the correlation contribution to the disjoining pressure takes the form:

$$\Pi_{corr} = -\frac{k_B T}{2}\int\frac{d^2\mathbf{p}}{(2\pi)^2}\kappa_p\left(\coth\kappa_p H - 1\right) = -\frac{k_B T}{2\pi}\int\limits_0^\infty dp\, p\frac{\kappa_p}{e^{2\kappa_p H}-1}. \tag{4.250}$$

The latter expression is coincides with Eq. (3.74) obtained in Chap. 3. As already mentioned in Chap. 3, this contribution take into account the classical van der Waals interactions between metals.

Dielectric Walls

Now let us consider the case of two identical dielectric walls with dielectric constant ε_c. In this case, the boundary conditions (4.241) should be replaced by the following ones

$$\begin{aligned}\varepsilon_s\frac{\partial G(p|+0,z')}{\partial z} &= \varepsilon_c\frac{\partial G(p|-0,z')}{\partial z},\\ \varepsilon_s\frac{\partial G(p|H-0,z')}{\partial z} &= \varepsilon_c\frac{\partial G(p|H+0,z')}{\partial z}.\end{aligned} \tag{4.251}$$

Taking into account the latter boundary conditions, the Green's function has the form

$$G(p|z,z') = \begin{cases}\dfrac{4\pi(\kappa_p\varepsilon\cosh\kappa_p(H-z')+(\varepsilon_c p+\kappa_p\varepsilon)\sinh\kappa_p(H-z'))(\kappa_p\varepsilon\cosh\kappa_p z+(\varepsilon_c p+\kappa_p\varepsilon)\sinh\kappa_p z)}{\varepsilon\kappa_p((\kappa_p^2\varepsilon^2+(\varepsilon_c p+\kappa_p\varepsilon)^2)\sinh\kappa_p H+2\kappa_p\varepsilon(\varepsilon_c p+\kappa_p\varepsilon)\cosh\kappa_p H)}, & z<z',\\[2ex] \dfrac{4\pi(\kappa_p\varepsilon\cosh\kappa_p(H-z)+(\varepsilon_c p+\kappa_p\varepsilon)\sinh\kappa_p(H-z))(\kappa_p\varepsilon\cosh\kappa_p z'+(\varepsilon_c p+\kappa_p\varepsilon)\sinh\kappa_p z')}{\varepsilon\kappa_p((\kappa_p^2\varepsilon^2+(\varepsilon_c p+\kappa_p\varepsilon)^2)\sinh\kappa_p H+2\kappa_p\varepsilon(\varepsilon_c p+\kappa_p\varepsilon)\cosh\kappa_p H)}, & z>z'.\end{cases} \tag{4.252}$$

After similar but more cumbersome calculations, the correlation contribution to the disjoining pressure takes the following form:

$$\Pi_{corr} = -\frac{k_B T}{2\pi}\int_0^\infty dp\, p\Delta(p; H)\frac{\kappa_p}{e^{2\kappa_p H} - 1}, \tag{4.253}$$

where

$$\Delta(q; H) = \frac{(\kappa_p \varepsilon - p\varepsilon_c)^2}{\kappa_p^2 \varepsilon^2 + (\varepsilon_c p + \kappa_p \varepsilon)^2 + 2\kappa_p \varepsilon(\varepsilon_c p + \kappa_p \varepsilon)\coth \kappa_p H}. \tag{4.254}$$

It is evident that the case of metal walls is encompassed within the formula (4.253). To derive the formula (4.250), we need to take the formal limit $\varepsilon_c \to \infty$ in (4.253)

4.12 Surface Tension of Electrolyte Solutions: Beyond Onsager-Samaras Theory

Let us apply the formalism described above to characterize the interface between an electrolyte solution and a nonpolar medium, calculating the contribution of ions' electrostatic interactions to the surface tension. We consider two phases, modeled as continuous dielectric media, with constant permittivities: ε_s for the polar medium ($z > 0$) and ε_m for the nonpolar medium ($z < 0$). We model the interface as a plane at $z = 0$, with the z-axis perpendicular to it. The highly polar medium ($\varepsilon_s \gg 1$) contains ions with charges q_α, where α denotes the ion species. The nonpolar medium, with permeability $\varepsilon_m \sim 1$, contains no ions. We'll calculate the surface tension using the stress tensor (4.220).

More specifically, let us use a mechanical interpretation of surface tension defined as the difference between the normal P_n and tangential P_τ pressures of the fluid in the layer [25, 98]

$$\Delta\alpha = \int_0^\infty dz\,(P_n - P_\tau) = \int_0^\infty dz\,(\sigma_{xx} - \sigma_{zz}), \tag{4.255}$$

where $P_n = -\sigma_{zz}$ and $P_\tau = -\sigma_{xx} = -\sigma_{yy}$. Thus, we obtain (see (4.232) and (4.233))

$$\sigma_{xx} - \sigma_{zz} = -\frac{\varepsilon_s \psi'^2}{4\pi} + \frac{\varepsilon_s}{4\pi}\int \frac{d^2\mathbf{p}}{(2\pi)^2}\left(\mathcal{D}_{zz}(p, z) - \frac{1}{2}q^2 \mathcal{Q}(p, z)\right). \tag{4.256}$$

To utilize the formula for surface tension (4.255), we first need to derive and solve the self-consistent field equations. As demonstrated earlier, the grand thermodynamic potential, Ω, of an electrolyte solution within the Coulomb gas model can be approximated as:

$$-\frac{\Omega}{k_BT} = W[G;\psi] = \frac{1}{2}\ln\frac{DetG}{DetG_0} - \frac{1}{2}tr\left(G[G_0^{-1} - G^{-1}]\right) + \frac{\beta}{2}(\psi G_0^{-1}\psi)$$
$$+\sum_\alpha z_\alpha \int d\mathbf{r}\,\theta(z)A_\alpha(z)e^{-\beta q_\alpha\psi(\mathbf{r})-\beta u_\alpha(z)} = \frac{1}{2}\ln\frac{DetG}{DetG_0} - \frac{1}{2}tr\left(G[G_0^{-1} - G^{-1}]\right) + \frac{\beta}{2}(\psi G_0^{-1}\psi)$$
$$+\sum_\alpha z_\alpha \int d\mathbf{r}\,\theta(z)A_\alpha(z)e^{-\beta q_\alpha\psi(\mathbf{r})} + \sum_\alpha z_\alpha \int d\mathbf{r}\,\theta(z)A_\alpha(z)e^{-\beta q_\alpha\psi(\mathbf{r})}(e^{-\beta u_\alpha(z)} - 1). \quad (4.257)$$

The operator for the Green's function of the Poisson equation in this case takes the following form:

$$G_0^{-1}(\mathbf{r},\mathbf{r}') = -\frac{1}{4\pi}\nabla\varepsilon(\mathbf{r})\nabla\delta(\mathbf{r}-\mathbf{r}') \quad (4.258)$$

where the local dielectric permittivity is

$$\varepsilon(\mathbf{r}) = \varepsilon(z) = \begin{cases} \varepsilon_m, & z < 0 \\ \varepsilon_s, & z > 0, \end{cases} \quad (4.259)$$

We also introduce the effective potentials, $u_\alpha(z)$, of ions-interface interaction, arising from van der Waals and hydration forces. We consider these effects implicitly at the phenomenological level. Assuming that the potentials u_α are non-zero only within a narrow layer near the interface, we make the following approximation:

$$e^{-\beta u_\alpha(z)} - 1 \approx b_\alpha\delta(z), \quad (4.260)$$

so we obtain

$$W[G;\psi] = \frac{1}{2}\ln\frac{DetG}{DetG_0} - \frac{1}{2}tr\left(G[G_0^{-1} - G^{-1}]\right) + \frac{\beta}{2}(\psi G_0^{-1}\psi)$$
$$+\sum_\alpha z_\alpha \int d\mathbf{r}\,\theta(z)A_\alpha(z)e^{-\beta q_\alpha\psi(\mathbf{r})} + \sum_\alpha z_\alpha b_\alpha \int d\mathbf{r}\,A_\alpha(0)e^{-\beta q_\alpha\psi(\mathbf{r})}\delta(z), \quad (4.261)$$

where b_α have a dimension of length and are defined by integrals

$$b_\alpha = \int_0^\infty dz\left(e^{-\beta u_\alpha(z)} - 1\right). \quad (4.262)$$

The phenomenological parameters b_α, which are generally functions of temperature, determine the measure of adhesion of the ion of the α-th sort to the interface.

Thus, varying the functional W with respect to ψ and G, i.e., considering the Euler-Lagrange equations

$$\frac{\delta W}{\delta \psi(\mathbf{r})} = 0, \quad \frac{\delta W}{\delta G(\mathbf{r}, \mathbf{r}')} = 0, \tag{4.263}$$

applying the 2D Fourier transformation

$$G(\mathbf{r}, \mathbf{r}') = G(\rho - \rho'|z, z') = \int \frac{d^2\mathbf{p}}{(2\pi)^2} e^{-i\mathbf{p}(\rho-\rho')} G(p|z, z'), \tag{4.264}$$

and taking into account that $\psi(\mathbf{r}) = \psi(z)$, we arrive at the following self-consistent field equations

$$\frac{1}{4\pi}\partial_z \varepsilon(z) \partial_z \psi(z) = -\sum_\alpha q_\alpha \bar{n}_\alpha(z) - \sigma_s \delta(z), \tag{4.265}$$

$$\frac{1}{4\pi}\left(-\partial_z \varepsilon(z)\partial_z + \varepsilon(z)(p^2 + \varkappa^2(z)) + \varepsilon_s \kappa_s \delta(z)\right) G(p|z, z') = \delta(z - z'), \tag{4.266}$$

where we introduce average concentrations of the ions

$$\bar{n}_\alpha(z) = z_\alpha A_\alpha(z) e^{-\beta q_\alpha \psi(z)} \theta(z), \tag{4.267}$$

with auxiliary functions

$$A_\alpha(z) = \exp\left[-\frac{\beta q_\alpha^2 G(0|z, z)}{2}\right], \tag{4.268}$$

and following notations

$$\varkappa^2(z) = \frac{4\pi}{\varepsilon_s}\sum_\alpha q_\alpha^2 z_\alpha A_\alpha(z) e^{-\beta q_\alpha \psi(z)} \theta(z), \tag{4.269}$$

$$\kappa_s = \frac{4\pi}{\varepsilon_s}\sum_\alpha q_\alpha^2 z_\alpha b_\alpha A_\alpha(0) e^{-\beta q_\alpha \psi(0)}. \tag{4.270}$$

The effective surface charge density at the interface, arising due to the adsorption of ions, is determined by the relation:

$$\sigma_s = \sum_\alpha q_\alpha z_\alpha b_\alpha A_\alpha(0) e^{-\beta q_\alpha \psi(0)}. \tag{4.271}$$

Within the mean-field approximation, where $A_\alpha(z) \approx 1$, and in the regime of weak electrostatics, Eqs. (4.265) and (4.266) simplify to

$$\psi''(z) - \varkappa^2(z)\psi(z) = 0, \tag{4.272}$$

$$\left(-\partial_z \varepsilon(z)\partial_z + \varepsilon(z)(p^2 + \varkappa^2(z)) + \varepsilon_s \kappa_s \delta(z)\right) G(p|z, z') = 4\pi\delta(z - z'), \tag{4.273}$$

where

$$\varkappa^2(z) \approx \kappa^2\theta(z), \quad \kappa^2 = \frac{4\pi}{\varepsilon_s k_B T}\sum_{\alpha} q_{\alpha}^2 n_{\alpha}, \quad \kappa_s = \frac{4\pi}{\varepsilon_s}\sum_{\alpha} q_{\alpha}^2 n_{\alpha} b_{\alpha}(1 - \beta q_{\alpha}\psi_0), \tag{4.274}$$

and we take into account that in the mean-field approximation $z_{\alpha} = n_{\alpha}$ (n_{α} are the bulk concentrations). Equation (4.272) represents the linearized Poisson-Boltzmann equation for $z > 0$ and the Laplace equation for $z < 0$. In the region $z > 0$ we have the following expression

$$\psi'' - \kappa^2\psi = 0 \tag{4.275}$$

with the following boundary condition

$$\varepsilon_s \psi'(0) = -4\pi\sigma_s, \quad \psi'(\infty) = 0, \tag{4.276}$$

where

$$\sigma_s = \sum_{\alpha} q_{\alpha} n_{\alpha} b_{\alpha} e^{-\beta q_{\alpha}\psi_0} \tag{4.277}$$

and

$$\kappa = \left(\frac{4\pi}{\varepsilon_s k_B T}\sum_{\alpha} q_{\alpha}^2 n_{\alpha}\right)^{1/2} \tag{4.278}$$

is nothing more than the inverse Debye radius. The solution of the equation (4.275) is

$$\psi(z) = \psi_0 e^{-\kappa z}, \tag{4.279}$$

where the surface potential ψ_0 is determined by the first boundary condition from (4.276), resulting in the following nonlinear equation:

$$\psi_0 = \frac{1}{\varepsilon_s \kappa}\sum_{\alpha} q_{\alpha} n_{\alpha} b_{\alpha} e^{-\beta q_{\alpha}\psi_0}. \tag{4.280}$$

Within the linear approximation, where $\beta q_{\alpha}\psi_0 \ll 1$, the latter expression yields

$$\psi_0 = \frac{\frac{4\pi}{\varepsilon_s \kappa}\sum_{\alpha} q_{\alpha} n_{\alpha} b_{\alpha}}{1 + \frac{4\pi}{\varepsilon_s \kappa k_B T}\sum_{\alpha} q_{\alpha}^2 n_{\alpha} b_{\alpha}}. \tag{4.281}$$

Let us proceed to the evaluation of the Green's function. For $z < 0$, the Green's function takes the form:

$$G(p|z, z') = Ae^{pz}. \tag{4.282}$$

For $z > 0$ Green's function satisfies the following equation (see also (4.273))

$$\varepsilon_s(-\partial_z^2 + \kappa_p^2 + \kappa_s\delta(z))G(p|z, z') = 4\pi\delta(z - z'), \tag{4.283}$$

where $\varkappa_p = \sqrt{\kappa^2 + p^2}$ and

$$\kappa_s = \frac{4\pi}{\varepsilon_s}\sum_\alpha q_\alpha^2 n_\alpha b_\alpha (1 - \beta q_\alpha \psi_0). \tag{4.284}$$

The solution of Eq. (4.283) yields

$$G(p|z, z') = \begin{cases} B\sinh\kappa_p z + C\cosh\kappa_p z, & 0 < z < z', \\ D\,e^{-\kappa_p z}, & z > z'. \end{cases} \tag{4.285}$$

Coefficients A, B, C, D can be found from the following boundary conditions

$$\begin{gathered} G(p|-0, z') - G(p|+0, z') = 0, \\ \varepsilon_m\frac{\partial G(p|-0, z')}{\partial z} - \varepsilon_s\frac{\partial G(p|+0, z')}{\partial z} + \varepsilon_s\kappa_s G(p|0, z') = 0, \\ G(p|z' - 0, z') - G(p|z' + 0, z') = 0, \\ \frac{\varepsilon_s}{4\pi}\left(\frac{\partial G(p|z' - 0, z')}{\partial z} - \frac{\partial G(p|z' + 0, z')}{\partial z}\right) = 1, \end{gathered} \tag{4.286}$$

which lead to the following solution

$$G(p|z, z') = \begin{cases} \frac{\varepsilon_m e^{-\kappa_p z'}}{\varepsilon_m p + \varepsilon_s(\kappa_p + \kappa_s))} e^{pz}, & z < 0, \\ \frac{4\pi e^{-\kappa_p z'}\left((\varepsilon_m p + \varepsilon_s\kappa_s)\sinh\kappa_p z + \varepsilon_s\kappa_p\cosh\kappa_p z\right)}{\varepsilon_s\kappa_p\left(\varepsilon_m p + \varepsilon_s(\kappa_s + \kappa_p)\right)}, & 0 < z < z', \\ \frac{4\pi e^{-\kappa_p z}\left((\varepsilon_m p + \varepsilon_s\kappa_s)\sinh\kappa_p z' + \varepsilon_s\kappa_p\cosh\kappa_p z'\right)}{\varepsilon_s\kappa_p\left(\varepsilon_m p + \varepsilon_s(\kappa_s + \kappa_p)\right)}, & z > z' > 0. \end{cases} \tag{4.287}$$

For further evaluations, we need to subtract the Green's function of the infinite medium, $G(p|z, z') = 2\pi e^{-\kappa_p|z - z'|}/\varepsilon_s\kappa_p$, from the obtained Green's function in the region $z > 0$, which yields the result:

$$G(p|z, z') = \frac{2\pi e^{-\kappa_p(z+z')}\left(\varepsilon_s(\kappa_p - \kappa_s) - \varepsilon_m p\right)}{\varepsilon_s\kappa_p\left(\varepsilon_m p + \varepsilon_s(\kappa_s + \kappa_p)\right)}. \tag{4.288}$$

Hence, we obtain

$$Q(p,z) = k_B T \frac{2\pi\left(\varepsilon_s(\kappa_p-\kappa_s)-\varepsilon_m q\right)}{\varepsilon_s\kappa_p\left(\varepsilon_m p+\varepsilon_s(\kappa_s+\kappa_p)\right)} e^{-2\kappa_p z}, \tag{4.289}$$

$$\mathcal{D}_{zz}(p,z) = k_B T \frac{2\pi\kappa_p\left(\varepsilon_s(\kappa_p-\kappa_s)-\varepsilon_m p\right)}{\varepsilon_s\left(\varepsilon_m p+\varepsilon_s(\kappa_s+\kappa_p)\right)} e^{-2\kappa_p z}. \tag{4.290}$$

Thus, substituting (4.289), (4.290) and (4.279) into (4.256) and integrating over z in (4.255) we obtain

$$\Delta\alpha = -\frac{\varepsilon_s\kappa\psi_0^2}{8\pi} + \frac{k_B T}{8}\int \frac{d^2\mathbf{p}}{(2\pi)^2} \frac{\left(\varepsilon_s(\sqrt{\kappa^2+p^2}-\kappa_s)-\varepsilon_m p\right)(2\kappa^2+p^2)}{(\kappa^2+p^2)\left(\varepsilon_m p+\varepsilon_s(\kappa_s+\sqrt{\kappa^2+p^2})\right)}. \tag{4.291}$$

In the absence of ions ($\psi_0 = 0$, $\kappa_s = \kappa = 0$), the relation (4.291) transfers to

$$\Delta\alpha^{(0)} = \frac{k_B T}{8}\int \frac{d^2\mathbf{p}}{(2\pi)^2} \frac{\varepsilon_s-\varepsilon_m}{\varepsilon_s+\varepsilon_m}, \tag{4.292}$$

which determines the classical contribution of van der Waals interactions (zero Matsubara frequency) [63, 64, 99] to the surface tension and must be subtracted from the final result. Thus, we obtain [100]:

$$\Delta\alpha = -\frac{\varepsilon_s\kappa\psi_0^2}{8\pi} + \frac{k_B T}{16\pi}\frac{\varepsilon_s-\varepsilon_m}{\varepsilon_s+\varepsilon_m}\int_0^{\Lambda} dp\, p \left(\frac{(\varepsilon_s+\varepsilon_m)\left(\varepsilon_s(\sqrt{\kappa^2+p^2}-\kappa_s)-\varepsilon_m p\right)(2\kappa^2+p^2)}{(\varepsilon_s-\varepsilon_m)(\kappa^2+p^2)\left(\varepsilon_m p+\varepsilon_s(\kappa_s+\sqrt{\kappa^2+p^2})\right)} - 1\right), \tag{4.293}$$

where we introduce an ultraviolet cutoff parameter, Λ, which is related to the length scale of the order of the ion size. Turning to the dimensionless integration variable $x = p/\kappa$ in (4.293), we obtain:

$$\Delta\alpha = -\frac{\varepsilon_s\kappa\psi_0^2}{8\pi} + \frac{k_B T\kappa^2}{16\pi}\frac{\varepsilon_s-\varepsilon_m}{\varepsilon_s+\varepsilon_m} f(\tilde{\Lambda},\delta,\varepsilon_{sm}), \tag{4.294}$$

where we introduce the auxiliary function

$$f(\tilde{\Lambda},\delta,\varepsilon_{sm}) = \int_0^{\tilde{\Lambda}} dx\, x \left(\frac{(\varepsilon_{sm}+1)\left(\varepsilon_{sm}(\sqrt{1+x^2}-\delta)-x\right)(2+x^2)}{(\varepsilon_{sm}-1)(1+x^2)\left(x+\varepsilon_{sm}(\delta+\sqrt{1+x^2})\right)} - 1\right), \tag{4.295}$$

with the arguments $\tilde{\Lambda} = \Lambda/\kappa$, $\delta = \kappa_s/\kappa$ and $\varepsilon_{sm} = \varepsilon_s/\varepsilon_m$.

If we ignore the adsorption of the ions ($\kappa_s = 0$, $\psi_0 = 0$), we obtain

$$\Delta\alpha = \frac{k_B T \kappa^2}{16\pi} \frac{\varepsilon_s - \varepsilon_m}{\varepsilon_s + \varepsilon_m} g(\tilde{\Lambda}, \varepsilon_{sm}), \tag{4.296}$$

where $g(\tilde{\Lambda}, \varepsilon_{sm}) = f(\tilde{\Lambda}, 0, \varepsilon_{sm})$. At sufficiently low concentrations, when $\tilde{\Lambda} \gg 1$, the function $g(\tilde{\Lambda}, \varepsilon_{sm})$ has the following asymptotics

$$g(\tilde{\Lambda}, \varepsilon_{sm}) \simeq (1 + O(\varepsilon_m/\varepsilon_s)) \ln \tilde{\Lambda}, \tag{4.297}$$

which leads to the classical result of Onsager and Samaras [101]

$$\Delta\alpha_{OS} \simeq \frac{k_B T \kappa^2}{16\pi} \frac{\varepsilon_s - \varepsilon_m}{\varepsilon_s + \varepsilon_m} \ln \tilde{\Lambda}. \tag{4.298}$$

Let us now clarify the physical meaning of the Green's function $G(p|z, z')$ for this problem. As can be seen from the relation (4.267), the local ion concentrations can be approximated as follows:

$$\bar{n}_\alpha(z) = n_\alpha e^{-\beta q_\alpha \psi(z) - \frac{\beta q_\alpha^2 G(0|z,z)}{2}}, \tag{4.299}$$

where we define the function

$$G(0|z, z) = 2\pi \int \frac{d^2\mathbf{p}}{(2\pi)^2} \frac{e^{-2\kappa_p z} \left(\varepsilon_s(\kappa_p - \kappa_s) - \varepsilon_m p\right)}{\varepsilon_s \kappa_p \left(\varepsilon_m p + \varepsilon_s(\kappa_s + \kappa_p)\right)}, \tag{4.300}$$

taking into account the subtracted Green's function of the infinite medium, and we also consider that

$$n_\alpha = z_\alpha \exp\left[-\frac{\beta\pi q_\alpha^2}{\varepsilon_s} \int \frac{d^2\mathbf{p}}{(2\pi)^2} \left(\frac{1}{\kappa_p} - \frac{1}{p}\right)\right] = z_\alpha \exp\left[\frac{\beta q_\alpha^2 \kappa}{2\varepsilon_s}\right], \tag{4.301}$$

and

$$G_0(0) = \frac{2\pi}{\varepsilon_s} \int \frac{d^2\mathbf{p}}{(2\pi)^2} \frac{1}{p}. \tag{4.302}$$

From (4.301) we obtain an expression for the bulk chemical potentials of ions within the Debye-Hückel approximation

$$\mu_\alpha = k_B T \ln(n_\alpha \Lambda_\alpha^3) - \frac{q_\alpha^2 \kappa}{2\varepsilon_s}. \tag{4.303}$$

In the absence of adsorbed ions ($\kappa_s = 0$) and at $\kappa z \ll 1$ we obtain

$$G(0|z, z) = \frac{1}{2\varepsilon_s}\frac{\varepsilon_s - \varepsilon_m}{\varepsilon_s + \varepsilon_m}\frac{1}{z}, \tag{4.304}$$

which determines nothing more than the potential energy of the image forces [87]

$$u_\alpha^{(im)}(z) = \frac{q_\alpha^2}{2}G(0|z, z) = \frac{\varepsilon_s - \varepsilon_m}{\varepsilon_s + \varepsilon_m}\frac{q_\alpha^2}{4\varepsilon_s z}. \tag{4.305}$$

When $\varepsilon_s \gg \varepsilon_c$, evaluation of the integral in 4.300 yields

$$G(0|z, z) = \frac{1}{2\varepsilon_s}\frac{e^{-2\kappa z}}{z}, \tag{4.306}$$

which gives the following expression for the potential of image forces

$$u_\alpha^{(im)}(z) = \frac{q_\alpha^2}{2}G(0|z, z) = \frac{q_\alpha^2}{4\varepsilon_s z}e^{-2\kappa z}, \tag{4.307}$$

for the first time established heuristically by Onsager and Samaras [101].

Thus, the quantity

$$u_\alpha^{(im)}(z) = \frac{q_\alpha^2}{2}G(0|z, z) = \pi q_\alpha^2 \int \frac{d^2\mathbf{p}}{(2\pi)^2}\frac{e^{-2\kappa_p z}\left(\varepsilon_s(\kappa_p - \kappa_s) - \varepsilon_m q\right)}{\varepsilon_s \kappa_p\left(\varepsilon_m q + \varepsilon_s(\kappa_s + \kappa_p)\right)} \tag{4.308}$$

can be considered as an expression for the potential energy of the ions image force, considering the influence of ions adsorption at the interface within the framework of variational field theory.

Numerical results. Let us now consider a case of the one-component aqueous 1:1 electrolyte solution coexisting with a nonpolar dielectric media. In this case, $q_\pm = \pm e$. From the literature data it is known that only anions adsorb onto the interface. In this case we assume $b_+ = 0$ and $b_- = -b$. Thus, Eq. (4.281) transforms into

$$\psi_0 = \frac{enb}{\varepsilon_s \kappa\left(1 - \frac{\kappa b}{2}\right)}, \tag{4.309}$$

where $\kappa = (2e^2 n/k_B T\varepsilon_s)^{1/2}$ and $n = n_\pm$ is the bulk concentration of the ions. As it follows from (4.281) and (4.284), the parameter $\delta = \kappa_s/\kappa$ is determined by

$$\delta = -\frac{\kappa b}{4\left(1 - \frac{\kappa b}{2}\right)}. \tag{4.310}$$

We adopt the following relation for the ultraviolet cut-off parameter [102]: $\Lambda = 2\sqrt{\pi}/a$. Here, a represents the average minimal distance of approach between ions, determined by the summation of the hydration radii, $r_{\pm}$. The values for $r_{\pm}$ are obtained from [103]. Then, using b as an adjustable parameter, we can fit a number of available in literature experimental data with the help of Eq. (4.294).

In Figs. 4.3 and 4.4 we present a comparison of calculated excess surface tension values with experimental results for a set of electrolyte solutions with Na^+ and K^+ cations at water-air interface, and also plot a well-known result of the OS theory (see Fig. 4.3a). Overall, we can conclude that the model successfully describes available data, with the exception of the region of rather small concentration. This, as we assume, can be related to the difficulties of experimental measurements in this range. Moreover, the experimental data from different sources are rather scattered even

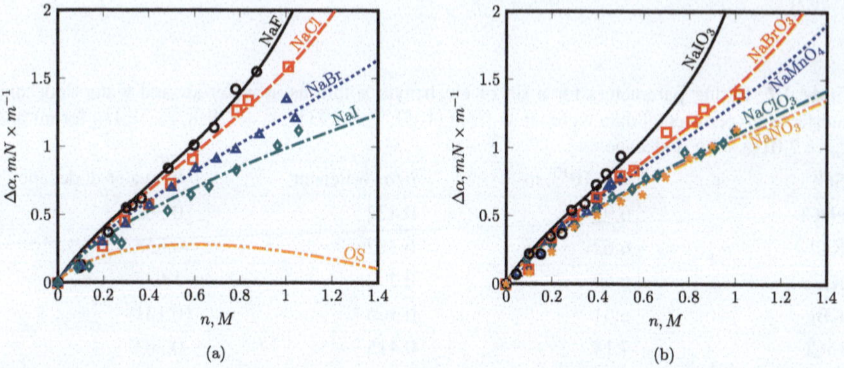

Fig. 4.3 Fit of experimental [104] excess surface tension of electrolyte solutions containing sodium cation, T = 298 K, $\varepsilon_s = 78\varepsilon_0$, $\varepsilon_m = 1\varepsilon_0$ (ε_0 is the vacuum permittivity)

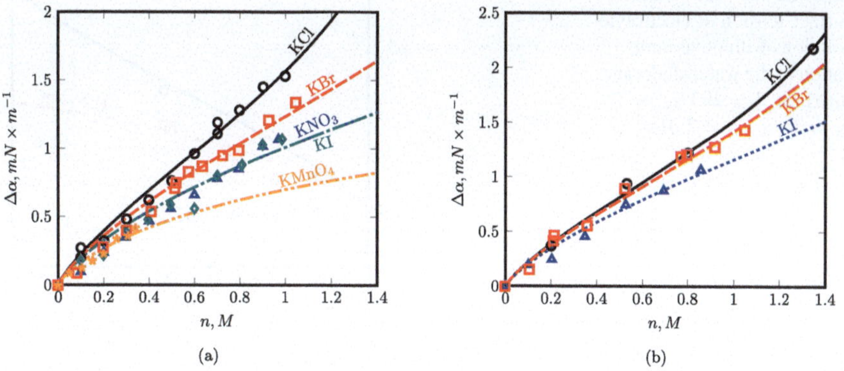

Fig. 4.4 Fit of experimental [104, 105] excess surface tension of electrolyte solutions containing potassium cation, T = 298 K (4.3a) and 293 K (4.3b), $\varepsilon_s = 78\varepsilon_0$, $\varepsilon_m = 1\varepsilon_0$

Table 4.1 Parameters of fitting for sodium and potassium containing aqueous solutions; experimental data were taken from [104]; T = 298 K, $\varepsilon_s = 78\varepsilon_0$, $\varepsilon_m = 1\varepsilon_0$

Salt	$a \times 10^{10}$, m	b/a	Salt	$a \times 10^{10}$, m	b/a
NaCl	6.9	0.397	KCl	6.62	0.397
NaF	7.1	0.438			
NaBr	6.88	0.319	KBr	6.61	0.315
NaI	6.89	0.236	KI	6.62	0.232
$NaNO_3$	6.93	0.267	KNO_3	6.66	0.234
$NaMnO_4$	7.03	0.352	$KMnO_4$	6.76	0.081
$NaClO_3$	6.99	0.283			
$NaBrO_3$	7.09	0.378			
$NaIO_3$	7.32	0.454			

Table 4.2 Fitting parameters for a set of electrolyte solutions at water-air and water-dodecane interfaces; experimental data were taken from [105]; T = 293 K, $\varepsilon_s = 78\varepsilon_0$, $\varepsilon_m = 1\varepsilon_0$ for air and $\varepsilon_m = 2.01\varepsilon_0$—for dodecane

Salt	$a \times 10^{10}$, m	b/a, water-air	b/a, water-dodecane
NaCl	6.9	0.422	0.361
KCl	6.62	0.387	0.358
KI	6.62	0.292	−1.037
KBr	6.61	0.366	0.131
LiCl	7.14	0.415	0.363

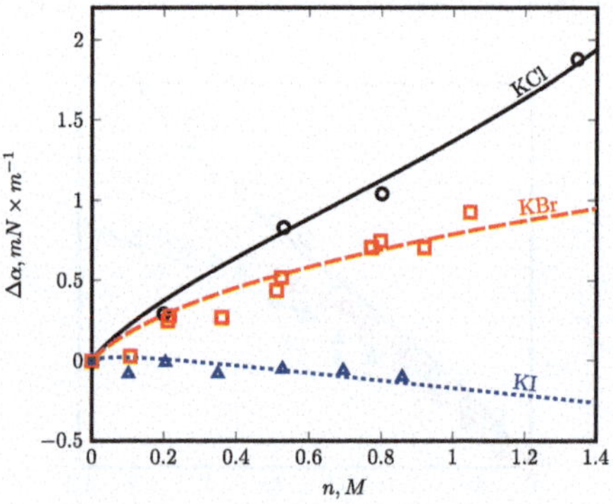

Fig. 4.5 Fit of the experimental [105] excess surface tension of electrolyte solutions with potassium cation at the water-dodecane interface, T = 293 K, $\varepsilon_s = 78\varepsilon_0$, $\varepsilon_m = 2.01\varepsilon_0$

across the entire range of concentrations (refer, for instance, to the visual comparison at Figs. 4.4, and check the fitting parameter values in Tables 4.1 and 4.2, where we fit the experiments from Refs. [104] and [105], correspondingly).

Table 4.3 Fitting parameters for a set of different systems, experimental data for which were taken from [104]; T = 298 K, $\varepsilon_s = 78\varepsilon_0$, $\varepsilon_m = 1\varepsilon_0$

Salt	$a \times 10^{10}$, m	b/a
LiCl	7.14	0.373
$LiNO_3$	7.17	0.279
NH_4NO	6.66	0.166
NH_4Cl	6.63	0.292
NH_4Br	6.61	0.273
NH_4I	6.62	−0.083
CsCl	6.61	0.408
$CuSO_4$	7.98	0.429
$MgSO_4$	8.07	0.413
$NiSO_4$	7.83	0.425

Within described model we can also accurately describe the behavior of the surface tension of 1:1 electrolyte solutions at the oil-water interface (see Fig. 4.5 and Table 4.2). Moreover, we successfully fitted the surface tension of a number of 1:1 electrolytes, data for which were taken from Ref. [104]. The parameters of calculations can be found in the Table 4.3.

Chapter 5
Nonlocal Statistical Theory of Solutions of Multipolar Molecules

In this chapter, we formulate field-theoretic models of fluids that consider the details of the internal charge distribution of their particles, and utilize these models to evaluate the thermodynamic functions of the fluid. Throughout the chapter, we will use the canonical ensemble framework as in Chap. 3.

5.1 Nonlocal Statistical Theory of Dilute Solutions of Polar Molecules

As in Chap. 3, let's consider an ionic fluid with N_+ point cations with charges $q_+ > 0$, N_- point anions with charges $q_- < 0$. Additionally, we introduce an impurity of N_d polar particles, modeled as dimers composed of point ionic groups with charges $\pm q$ (zwitterions). The number of ions $N_\pm$ satisfies the condition of electrical neutrality, $q_+N_+ + q_-N_- = 0$. As before, we assume these particles are immersed in a solvent, represented as a continuous dielectric medium with a constant dielectric permittivity ε. We postulate that each dipole particle is associated with a probability distribution function $g(\mathbf{r})$ describing the distance between the charge centres.

Adopting these modelling assumptions, we can write the configuration integral of the mixture in the form of[1]

$$Q = \int d\Gamma_d \int d\Gamma_s \exp\left[-\beta U\right], \tag{5.1}$$

where

[1] Here, we begin with the configuration integral defining the excess thermodynamic properties of the system compared to the ideal gas, omitting the partition function of the mixture of ideal gases, which does not affect the following discussions.

Y. A. Budkov and N. N. Kalikin, *Statistical Field Theory of Ion-Molecular Fluids*,
https://doi.org/10.1007/978-3-031-68364-0_5

$$\int d\Gamma_d(\ldots) = \int \ldots \int \prod_{j=1}^{N_d} \frac{d\mathbf{r}_j^{(+)} d\mathbf{r}_j^{(-)}}{V} g\left(\mathbf{r}_j^{(+)} - \mathbf{r}_j^{(-)}\right)(\ldots) \tag{5.2}$$

is the integration measure over the coordinates $\mathbf{r}_j^{(\pm)}$ of the ionic groups with the above-mentioned probability distribution function $g(\mathbf{r})$, which, by definition, must satisfy the normalization condition:

$$\int g(\mathbf{r}) d\mathbf{r} = 1 \tag{5.3}$$

and

$$\int d\Gamma_s(\ldots) = \int \prod_{l=1}^{N_+} \frac{d\mathbf{r}_l^{(+)}}{V} \int \prod_{k=1}^{N_-} \frac{d\mathbf{r}_k^{(-)}}{V} (\ldots) \tag{5.4}$$

is the integration measure over the ions coordinates. The Hamiltonian of the electrostatic interactions can be written as follows:

$$U = \frac{1}{2} \int d\mathbf{r} \int d\mathbf{r}' \, \hat{\rho}(\mathbf{r}) G_0(\mathbf{r} - \mathbf{r}') \hat{\rho}(\mathbf{r}') = \frac{1}{2} \left(\hat{\rho} G_0 \hat{\rho}\right), \tag{5.5}$$

where

$$\hat{\rho}(\mathbf{r}) = \hat{\rho}_d(\mathbf{r}) + \hat{\rho}_i(\mathbf{r}) + \rho_{ext}(\mathbf{r}), \tag{5.6}$$

and we introduce the local bound charge density of the polar particles

$$\hat{\rho}_d(\mathbf{r}) = q \sum_{j=1}^{N_d} \left(\delta\left(\mathbf{r} - \mathbf{r}_j^{(+)}\right) - \delta\left(\mathbf{r} - \mathbf{r}_j^{(-)}\right)\right), \tag{5.7}$$

and the local charge density of the ions

$$\hat{\rho}_i(\mathbf{r}) = q_+ \sum_{k=1}^{N_+} \delta\left(\mathbf{r} - \mathbf{r}_k^{(+)}\right) + q_- \sum_{l=1}^{N_-} \delta\left(\mathbf{r} - \mathbf{r}_l^{(-)}\right). \tag{5.8}$$

As usual, $\rho_{ext}(\mathbf{r})$ defines the density of external charges. Additionally, we note that in the Hamiltonian, we do not take into account non-Coulombian contributions. The validity of this assumption can be justified by sufficiently low concentrations of the ions and polar particles. The possibility to incorporate the steric interactions will be discussed later.

Thus, we can rewrite the configuration integral in the following form:

$$Q = \int d\Gamma_d \int d\Gamma_s \exp\left[-\frac{\beta}{2}(\hat{\rho} G_0 \hat{\rho})\right]. \tag{5.9}$$

Utilizing the Hubbard–Stratonovich transformation

$$\exp\left[-\frac{\beta}{2}(\hat{\rho}G_0\hat{\rho})\right] = \int \frac{\mathcal{D}\varphi}{C_0}\exp\left[-\frac{\beta}{2}(\varphi G_0^{-1}\varphi) + i\beta(\hat{\rho}\varphi)\right], \tag{5.10}$$

we arrive at the following representation of the configuration integral in the form of functional integral:

$$Q = \int \frac{\mathcal{D}\varphi}{C_0}\exp\left[-\frac{\beta}{2}(\varphi G_0^{-1}\varphi) + i\beta(\rho_{ext}\varphi)\right] Q_d^{N_d}[\varphi]Q_+^{N_+}[\varphi]Q_-^{N_-}[\varphi] \tag{5.11}$$

with the one-particle configuration integrals

$$Q_d[\varphi] = \frac{1}{V}\int d\mathbf{r}\int d\mathbf{r}'\, g(\mathbf{r}-\mathbf{r}')\exp(i\beta q(\varphi(\mathbf{r}) - \varphi(\mathbf{r}'))), \tag{5.12}$$

$$Q_\pm[\varphi] = \int \frac{d\mathbf{r}}{V}\exp(i\beta q_\pm\varphi(\mathbf{r})). \tag{5.13}$$

In thermodynamic limit, $V \to \infty$, $N_{d,\pm} \to \infty$, $N_{d,\pm}/V \to n_{d,\pm}$, we obtain:

$$Q_d^{N_d}[\varphi] \simeq \exp\left[n_d\int d\mathbf{r}\int d\mathbf{r}'\, g(\mathbf{r}-\mathbf{r}')(\exp\left[i\beta q(\varphi(\mathbf{r}) - \varphi(\mathbf{r}'))\right] - 1)\right] \tag{5.14}$$

and

$$Q_\pm^{N_\pm}[\varphi] \simeq \exp\left[n_\pm\int d\mathbf{r}\,(\exp(i\beta q_\pm\varphi(\mathbf{r})) - 1)\right]. \tag{5.15}$$

Thus, we can rewrite (5.11) as

$$Q = \int \frac{\mathcal{D}\varphi}{C_0}\exp\left[-S[\varphi]\right], \tag{5.16}$$

where we introduce the following functional

$$S[\varphi] = \frac{\beta}{2}(\varphi G_0^{-1}\varphi) - i\beta(\rho_{ext}\varphi) - n_d\int d\mathbf{r}\int d\mathbf{r}'\, g(\mathbf{r}-\mathbf{r}')(\exp\left[i\beta q(\varphi(\mathbf{r}) - \varphi(\mathbf{r}'))\right] - 1)$$
$$- n_+\int d\mathbf{r}\,(\exp\left[i\beta q_+\varphi(\mathbf{r})\right] - 1) - n_-\int d\mathbf{r}\,(\exp\left[i\beta q_-\varphi(\mathbf{r})\right] - 1). \tag{5.17}$$

Let's examine the results of the formulated theory in the case of sufficiently small dipole lengths. Defining the probability distribution function for the polar particles with a fixed distance between charged centers $|\mathbf{r}-\mathbf{r}'| = l$ as follows:

$$g(\mathbf{r}-\mathbf{r}') = \int \frac{d\mathbf{n}}{4\pi}\delta(\mathbf{r}-\mathbf{r}'-\mathbf{l}), \tag{5.18}$$

where the integration $\int d\mathbf{n}/4\pi(\ldots)$ defines averaging over the orientations of the unit vector $\mathbf{n} = \mathbf{l}/|\mathbf{l}|$ (or averaging over the solid angle), we obtain the following functional

$$S[\varphi] = \frac{\beta}{2}(\varphi G_0^{-1}\varphi) - n_d \int \frac{d\mathbf{n}}{4\pi} \int d\mathbf{r}\left(\exp\left[i\beta q(\varphi(\mathbf{r}) - \varphi(\mathbf{r}-\mathbf{l}))\right] - 1\right)$$
$$- n_+ \int d\mathbf{r}\left(\exp\left[i\beta q_+\varphi(\mathbf{r})\right] - 1\right) - n_- \int d\mathbf{r}\left(\exp\left[i\beta q_-\varphi(\mathbf{r})\right] - 1\right) - i\beta(\rho_{ext}\varphi). \tag{5.19}$$

As $\mathbf{l} \to 0$, expanding the exponential function into a Taylor series up to the first order with respect to $\mathbf{l}$ and averaging over the orientations of the vector $\mathbf{l}$ leads to a functional of the so-called Poisson–Boltzmann–Langevin theory [56]:

$$S_{PBL}[\varphi] = \frac{\beta}{2}(\varphi G_0^{-1}\varphi) - n_d \int d\mathbf{r}\left(\frac{\sin\beta p|\nabla\varphi(\mathbf{r})|}{\beta p|\nabla\varphi(\mathbf{r})|} - 1\right) - i\beta(\rho_{ext}\varphi)$$
$$- n_+ \int d\mathbf{r}\left(\exp\left[i\beta q_+\varphi(\mathbf{r})\right] - 1\right) - n_- \int d\mathbf{r}\left(\exp\left[i\beta q_-\varphi(\mathbf{r})\right] - 1\right), \tag{5.20}$$

where $p = ql$ is the dipole moment.

Mean-field approximation. Now, let's examine the mean-field approximation for the configuration integral. At first we set the variational derivative of the functional (5.17) to zero:

$$\frac{\delta S[\varphi]}{\delta\varphi(\mathbf{r})} = 0, \tag{5.21}$$

which yields the following expression

$$\int d\mathbf{r}'\, G_0^{-1}(\mathbf{r}-\mathbf{r}')\varphi(\mathbf{r}') - i\rho_{ext}(\mathbf{r}) + 2n_d q \int d\mathbf{r}'\, g(\mathbf{r}-\mathbf{r}') \sin\frac{q(\varphi(\mathbf{r}) - \varphi(\mathbf{r}'))}{k_B T}$$
$$- i\left(q_+ n_+ \exp\left[i\beta q_+\varphi(\mathbf{r})\right] + q_- n_- \exp\left[i\beta q_-\varphi(\mathbf{r})\right]\right) = 0. \tag{5.22}$$

Next, introducing the mean-field electrostatic potential $\varphi^{(MF)}(\mathbf{r}) = i\psi(\mathbf{r})$ and utilizing the expression (3.12) for the inverse Green's function of the Poisson equation, we derive the following nonlinear integro-differential self-consistent field equation:

$$-\frac{\varepsilon}{4\pi}\Delta\psi(\mathbf{r}) = 2n_d q \int d\mathbf{r}'\, g(\mathbf{r}-\mathbf{r}') \sinh\frac{q(\psi(\mathbf{r}') - \psi(\mathbf{r}))}{k_B T}$$
$$+ q_+ n_+ \exp\left[-\beta q_+\psi(\mathbf{r})\right] + q_- n_- \exp\left[-\beta q_-\psi(\mathbf{r})\right] + \rho_{ext}(\mathbf{r}). \tag{5.23}$$

The electrostatic free energy within the mean-field approximation is given by

$$F_{el}^{(MF)} = k_B T\, S[i\psi]. \tag{5.24}$$

Point charge in an ion-dipole environment. Let's now utilize the obtained Eq. (5.23) to evaluate the electrostatic potential of the electric field generated by a point ion with the charge q_0, surrounded by dipole particles in the absence of other ions (i.e., $n_\pm = 0$). Placing the point charge at the origin and considering that $\rho_{ext}(\mathbf{r}) = q_0\delta(\mathbf{r})$, we can express the self-consistent field equation as follows:

$$-\frac{\varepsilon}{4\pi}\Delta\psi(\mathbf{r}) = 2n_d q \int d\mathbf{r}'\, g(\mathbf{r}-\mathbf{r}') \sinh\frac{q(\psi(\mathbf{r}')-\psi(\mathbf{r}))}{k_B T} + q_0\delta(\mathbf{r}). \quad (5.25)$$

To understand the qualitative changes in the Coulomb electrostatic potential of an isolated point charge induced by the presence of polar molecules, we analyze the regime of weak electrostatic interactions, when $q\psi(\mathbf{r})/k_B T \ll 1$. Consequently, we obtain

$$-\frac{\varepsilon}{4\pi}\Delta\psi(\mathbf{r}) = -\frac{2n_d q^2}{k_B T}\psi(\mathbf{r}) + \frac{2n_d q^2}{k_B T}\int d\mathbf{r}'\, g(\mathbf{r}-\mathbf{r}')\psi(\mathbf{r}') + q_0\delta(\mathbf{r}). \quad (5.26)$$

Turning to the Fourier transform of the electrostatic potential

$$\tilde{\psi}(\mathbf{k}) = \int d\mathbf{r}\, \psi(\mathbf{r})e^{-i\mathbf{k}\mathbf{r}}, \quad (5.27)$$

we can rewrite the expression (5.26) in the following algebraic form

$$\frac{\varepsilon}{4\pi}k^2\tilde{\psi}(\mathbf{k}) = -\frac{2n_d q^2}{k_B T}(1-g(\mathbf{k}))\tilde{\psi}(\mathbf{k}) + q_0, \quad (5.28)$$

where $k = |\mathbf{k}|$.

After straightforward algebra, we arrive at

$$\tilde{\psi}(\mathbf{k}) = \frac{4\pi q_0}{\varepsilon(k^2 + \varkappa^2(\mathbf{k}))}, \quad (5.29)$$

where

$$\varkappa^2(\mathbf{k}) = \frac{8\pi n_d q^2}{\varepsilon k_B T}(1-g(\mathbf{k})) \quad (5.30)$$

is the so-called screening function, and

$$g(\mathbf{k}) = \int d\mathbf{r}\, e^{-i\mathbf{k}\mathbf{r}} g(\mathbf{r}) \quad (5.31)$$

is the Fourier transform of the distribution function over the distance between the ionic groups. Thus, the electrostatic potential of a point charge takes the form:

$$\psi(\mathbf{r}) = \frac{4\pi q_0}{\varepsilon} \int \frac{d\mathbf{k}}{(2\pi)^3} \frac{e^{i\mathbf{kr}}}{k^2 + \varkappa^2(\mathbf{k})}. \tag{5.32}$$

To compute the electrostatic potential, we need to define the probability distribution function $g(\mathbf{r})$. For polar particles with constant dipole moments and a probability distribution function defined by expression (5.18), the characteristic function takes the form:

$$g(\mathbf{k}) = \frac{\sin kl}{kl}, \tag{5.33}$$

where $k = |\mathbf{k}|$. However, the relation (5.33) can't provide us an analytical expression for the electrostatic potential. Nonetheless, we can employ a different model characteristic function, which allows us to obtain analytical results without significantly altering the properties of the potential of a test point charge in an ion-dipole environment

$$g(\mathbf{k}) = \frac{1}{1 + \frac{k^2 l^2}{6}}, \tag{5.34}$$

defining the following probability distribution function

$$g(\mathbf{r}) = \frac{3}{2\pi l^2 r} \exp\left[-\frac{\sqrt{6}r}{l}\right], \tag{5.35}$$

where $r = |\mathbf{r}|$. It is worth emphasising that the distributions defined by the characteristic functions (5.33) and (5.34) have the same root-mean-square distance between the charged centres of polar particles, l.

Thus, substituting the expression (5.34) into (5.30) and integrating (5.32), we arrive at the result

$$\psi(\mathbf{r}) = \frac{q_0}{\varepsilon r} \frac{1 + y_d \exp\left(-\frac{r}{l_s}\right)}{1 + y_d}, \tag{5.36}$$

where $y_d = 4\pi n_d q^2 l^2/(3\varepsilon k_B T) = l^2/6r_D^2$ is a dimensionless parameter defining the "strength" of the dipole-dipole electrostatic interaction and $l_s = l/\sqrt{6(1 + y_d)}$ is a length whose physical meaning will be revealed below; $r_D = \left(8\pi n_d q^2/\varepsilon k_B T\right)^{-1/2}$ is the Debye radius referring to the charge centres of the polar particles. It is interesting to examine the behavior of the electrostatic potential for the case of dipoles with very long length, i.e. when $y_d \gg 1$ (or $l \gg r_D$). In this case, the relation (5.36) takes the form of the Debye–Hückel potential discussed above:

$$\psi(\mathbf{r}) \simeq \frac{q_0}{\varepsilon r} \exp\left(-\frac{r}{r_D}\right). \tag{5.37}$$

Therefore, when the effective dipole length is much larger than the Debye screening radius, the charge centers of polar particles can be considered as unbound ions, which screen the charge of the test ion.

It is worth noting that the relation (5.36) allows us to formally introduce the local dielectric permittivity as [106, 107]

$$\varepsilon(\mathbf{r}) = \frac{\varepsilon(1 + y_d)}{1 + y_d \exp\left(-\frac{r}{l_s}\right)}. \tag{5.38}$$

Hence, the length l_s defines the radius of the sphere around the point charge, within which the dielectric constant is lower than its value in the bulk solution

$$\varepsilon_b = \varepsilon(1 + y_d) = \varepsilon + \frac{4\pi n_d q^2 l^2}{3k_B T}. \tag{5.39}$$

In other words, the length l_s can be interpreted as the effective solvation radius of a point charge surrounded by dipoles within the linear theory.

The electrostatic free energy within the linear approximation can be expressed as:

$$\begin{aligned} F_{el} = k_B T S[i\psi] = & -\frac{1}{2}\int d\mathbf{r}\int d\mathbf{r}'\,\psi(\mathbf{r}) G_0^{-1}(\mathbf{r}, \mathbf{r}')\psi(\mathbf{r}') \\ & + \frac{n_d q^2}{k_B T}\int d\mathbf{r}\int d\mathbf{r}'\, g(\mathbf{r} - \mathbf{r}')(\psi(\mathbf{r}) - \psi(\mathbf{r}'))^2 + \int d\mathbf{r}\,\rho_{ext}(\mathbf{r})\psi(\mathbf{r}) \\ = & -\frac{1}{2}\int d\mathbf{r}\int d\mathbf{r}'\,\psi(\mathbf{r}) G^{-1}(\mathbf{r}, \mathbf{r}')\psi(\mathbf{r}') + \int d\mathbf{r}\,\rho_{ext}(\mathbf{r})\psi(\mathbf{r}), \end{aligned} \tag{5.40}$$

where we introduce the inverse Green's function, renormalized by the presence of polar particles as

$$G^{-1}(\mathbf{r}, \mathbf{r}') = G_0^{-1}(\mathbf{r}, \mathbf{r}') + \frac{2n_d q^2}{k_B T}(\delta(\mathbf{r} - \mathbf{r}') - g(\mathbf{r} - \mathbf{r}')). \tag{5.41}$$

Then, taking into account that

$$\rho_{ext}(\mathbf{r}) = \int d\mathbf{r}'\, G^{-1}(\mathbf{r}, \mathbf{r}')\psi(\mathbf{r}'), \tag{5.42}$$

we arrive at the expression

$$F_{el} = \frac{1}{2}\int d\mathbf{r}\,\rho_{ext}(\mathbf{r})\psi(\mathbf{r}). \tag{5.43}$$

The obtained expression allows us to calculate the solvation free energy of the test point charge. For this purpose, we need to subtract the electrostatic self-energy of the point charge from (5.43), i.e.

$$\Delta F_{solv} = \frac{1}{2}\int d\mathbf{r}\rho_{ext}(\mathbf{r})(\psi(\mathbf{r}) - \psi_{ext}(\mathbf{r})). \tag{5.44}$$

Considering that

$$\psi_{ext}(\mathbf{r}) = \frac{q_0}{\varepsilon r}, \tag{5.45}$$

we arrive at

$$\Delta F_{solv} = -\left(\frac{3}{2}\right)^{1/2}\frac{q_0^2}{\varepsilon l}\frac{y_d}{\sqrt{1+y_d}}. \tag{5.46}$$

Now, let's calculate the electrostatic potential of the point charge at non-zero salt concentration in the bulk solution ($n_{\pm} \neq 0$) within the linear approximation, i.e., under the condition $q\psi/k_BT \ll 1$. In this case, the self-consistent field equation takes the following form:

$$-\frac{\varepsilon}{4\pi}\Delta\psi(\mathbf{r}) = -\frac{2q^2 I}{k_BT}\psi(\mathbf{r}) - \frac{2n_d q^2}{k_BT}\psi(\mathbf{r}) + \frac{2n_d q^2}{k_BT}\int d\mathbf{r}'\, g(\mathbf{r}-\mathbf{r}')\psi(\mathbf{r}') + q_0\delta(\mathbf{r}), \tag{5.47}$$

where $I = \left(q_+^2 n_+ + q_-^2 n_-\right)/2q^2$ is the ionic strength of the solution (e denotes the elementary charge).

Turning to the Fourier representation, after some algebraic manipulations, we obtain an expression for the Fourier transform of the potential:

$$\tilde{\psi}(\mathbf{k}) = \frac{4\pi q_0}{\varepsilon\left(k^2 + \varkappa^2(\mathbf{k})\right)} \tag{5.48}$$

with the following screening function

$$\varkappa^2(\mathbf{k}) = \kappa_s^2 + \frac{8\pi n_d q^2}{\varepsilon k_BT}(1 - g(\mathbf{k})), \tag{5.49}$$

where $\kappa_s = \sqrt{8\pi q^2 I/(\varepsilon k_BT)}$ is the inverse Debye length, related to the ionic component.

Using the same probability distribution function and the same characteristic function $g(\mathbf{k}) = 1/\left(1 + k^2 l^2/6\right)$ and performing the inverse Fourier transform, we obtain the following analytical relation for the electrostatic potential [106, 107]:

$$\psi(\mathbf{r}) = \frac{q_0}{\varepsilon r}\left(u(y_d, y_s)e^{-\kappa_1(y_d,y_s)r} + (1 - u(y_d, y_s))\, e^{-\kappa_2(y_d,y_s)r}\right), \tag{5.50}$$

where

$$\kappa_{1,2}(y_d, y_s) = \frac{\sqrt{3}}{l}\left(1 + y_s + y_d \pm \sqrt{(1 + y_s + y_d)^2 - 4y_s}\right)^{1/2},$$
$$u(y_d, y_s) = \frac{y_s + y_d + \sqrt{(1 + y_s + y_d)^2 - 4y_s} - 1}{2\sqrt{(1 + y_s + y_d)^2 - 4y_s}}, \tag{5.51}$$

and we also introduce the parameter $y_s = \kappa_s^2 l^2/6$. In the absence of ions ($y_s = 0$), expression (5.50) reduces to the previously obtained expression (5.36). In the absence of polar particles ($y_d = 0$), we obtain the standard Debye–Hückel potential:

$$\psi(\mathbf{r}) = \frac{q_0}{\varepsilon r} e^{-\kappa_s r}. \tag{5.52}$$

To derive an expression for the solvation free energy of a test charge in an ion-dipole environment, we need to perform mathematical transformations similar to those in the previous section. In this case, we arrive at the same expression for the electrostatic free energy in the mean-field approximation:

$$F_{el}^{(MF)} = -\frac{1}{2}\int d\mathbf{r} \int d\mathbf{r}'\, \psi(\mathbf{r}) G^{-1}(\mathbf{r}, \mathbf{r}')\psi(\mathbf{r}') + \int d\mathbf{r}\, \rho_{ext}(\mathbf{r})\psi(\mathbf{r}) = \frac{1}{2}\int d\mathbf{r}\, \rho_{ext}(\mathbf{r})\psi, \tag{5.53}$$

where

$$G^{-1}(\mathbf{r}, \mathbf{r}') = G_0^{-1}(\mathbf{r}, \mathbf{r}') + \frac{2Iq^2}{k_B T}\delta(\mathbf{r} - \mathbf{r}') + \frac{2n_d q^2}{k_B T}(\delta(\mathbf{r} - \mathbf{r}') - g(\mathbf{r} - \mathbf{r}')) \tag{5.54}$$

is the inverse Green's function, renormalized due to the presence of ions and polar particles. Subtracting the self-energy of the ion from (5.53), we obtain the following analytical expression:

$$\Delta F_{solv} = -\frac{q_0^2}{2\varepsilon}\left(u(y_d, y_s)(\kappa_1(y_d, y_s) - \kappa_2(y_d, y_s)) + \kappa_2(y_d, y_s)\right), \tag{5.55}$$

which reduces to the previously obtained Eq. (5.46) when there are no ions ($y_s = 0$).

Now let's run to the analysis of numerical results. We introduce the following dimensionless parameters: $\tilde{n}_{d,\pm} = n_{d,\pm} l^3$. For simplicity, we consider the case of $1:1$ electrolyte, i.e., $q_+ = |q_-| = e$. Hence, due to the global electroneutrality condition, the concentrations of ions in the bulk solution are equal, i.e., $n_+ = n_- = n_s$. We also introduce the dimensionless salt concentration $\tilde{n}_s = n_s l^3$.

Figure 5.1 illustrates the dependence of the local dielectric permittivity, as determined by expression (5.38), normalized by the its bulk value, on the distance from the test point ion across various concentrations of polar particles. Notably, the local dielectric permittivity near the point ion registers lower values compared to those in the bulk solution. Moreover, increasing the concentration of polar particles intensifies the increase in local dielectric permittivity near the test ion as one moves away from

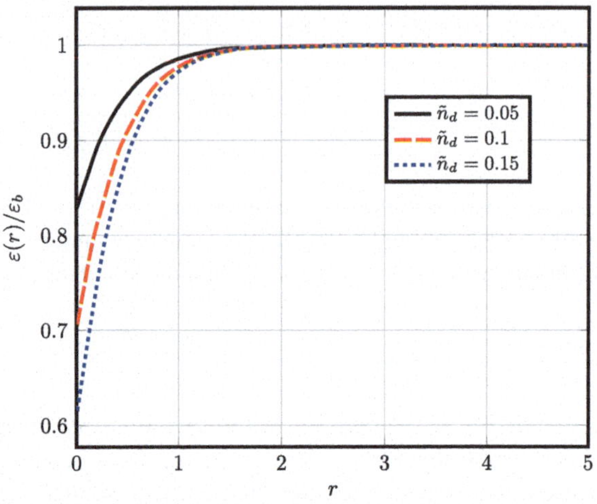

Fig. 5.1 Dependencies of the local dielectric permittivity (normalized to its bulk value) on the distance from a test point ion at various reduced concentrations $\tilde{n}_d = n_d l^3$ of polar particles

it. The critical distance at which the local dielectric permittivity practically equals its bulk value corresponds to the length l_s, as previously discussed.

Now, let's discuss how addition of salt affects the electrostatic potential profile of the test charge in the ion-dipole environment. In Fig. 5.2, we observe the dependency of the point charge potential (5.50) within the dipole and salt ion environment, with a fixed concentration of $\tilde{n}_s = 0.2$ at various concentrations of polar particles $\tilde{n}_d$.

As noted above, in the absence of polar particles ($n_d = 0$) the potential takes the form of the standard Debye–Hückel potential. When polar particles are added, the value of the potential decreases in the region of small distances. This reduction stems from the additional screening effect on the test ion's charge, attributed to the nearby charge centres of the polar particles.

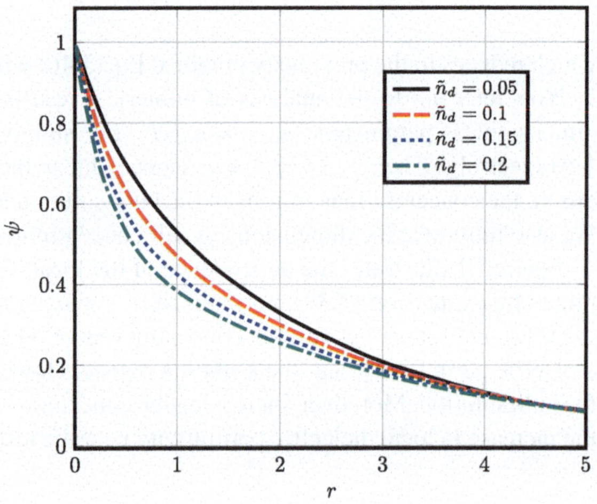

Fig. 5.2 Dependencies of the point charge potential at non-zero salt ion concentration ($\tilde{n}_s = 0.01$) on distance at different concentrations of polar particles

Gaussian approximation for the electrostatic free energy of the ion-dipole mixture. Here, we derive the expression for the electrostatic free energy of an ion-dipole mixture within the Gaussian approximation. By expanding the functional $S[\varphi]$ from (5.16) into a power series around the mean field $\varphi(\mathbf{r}) = i\psi(\mathbf{r})$ and neglecting terms higher than the second order, we obtain:

$$Q \approx \exp\left[-S[i\psi]\right] \int \frac{\mathcal{D}\Lambda}{C_0} \exp\left[-\frac{\beta}{2}\left(\Lambda \mathcal{G}^{-1}\Lambda\right)\right], \tag{5.56}$$

where

$$\mathcal{G}^{-1}(\mathbf{r},\mathbf{r}'|\psi) = k_B T \frac{\delta^2 S[i\psi]}{\delta\varphi(\mathbf{r})\delta\varphi(\mathbf{r}')} = G_0^{-1}(\mathbf{r},\mathbf{r}')$$
$$+ \frac{1}{k_B T}\left(q_+^2 n_+ \exp\left[-\beta q_+ \psi(\mathbf{r})\right] + q_-^2 n_- \exp\left[-\beta q_- \psi(\mathbf{r})\right]\right)\delta(\mathbf{r}-\mathbf{r}')$$
$$+ \frac{2n_d q^2}{k_B T}\int d\mathbf{r}''\, g(\mathbf{r}-\mathbf{r}'')\cosh\left(\frac{q\left(\psi(\mathbf{r})-\psi(\mathbf{r}'')\right)}{k_B T}\right)\left(\delta(\mathbf{r}-\mathbf{r}')-\delta(\mathbf{r}''-\mathbf{r}')\right) \tag{5.57}$$

is the renormalized inverse Green's function with the mean-field electrostatic potential $\psi(\mathbf{r})$ satisfying the Eq. (5.23). Therefore, upon formally evaluating the Gaussian functional integral, we arrive at the previously established relation for the configuration integral within the Gaussian approximation

$$Q \approx \exp\left\{-S[i\psi] + \frac{1}{2} tr\left(\ln \mathcal{G} - \ln G_0\right)\right\}. \tag{5.58}$$

When no external charges are present in the system (i.e., $\rho_{ext}(\mathbf{r}) = 0$), the electrostatic potential $\psi(\mathbf{r}) = 0$, so the mean field contribution to the electrostatic free energy is $F_{el}^{(MF)} = k_B T S[0] = 0$. Therefore, in this case, the electrostatic free energy is solely determined by thermal fluctuations of the electrostatic potential around its zero value. Consequently, the renormalized inverse Green's function adopts a simple form:

$$G^{-1}(\mathbf{r},\mathbf{r}') = G_0^{-1}(\mathbf{r},\mathbf{r}') + \frac{8\pi I q^2}{k_B T}\delta(\mathbf{r}-\mathbf{r}') + \frac{8\pi n_d q^2}{k_B T}\left(\delta(\mathbf{r}-\mathbf{r}') - g(\mathbf{r}-\mathbf{r}')\right), \tag{5.59}$$

whereas the electrostatic free energy of the ion-dipole mixture within the GA can be calculated as follows:

$$F_{el} \approx \frac{k_B T}{2} tr\left(\ln G_0 - \ln G\right) = \frac{V k_B T}{2}\int \frac{d\mathbf{k}}{(2\pi)^3} \ln \frac{G_0(\mathbf{k})}{G(\mathbf{k})}, \tag{5.60}$$

where $G_0(\mathbf{k}) = 4\pi/(\varepsilon k^2)$ and $G(\mathbf{k}) = 4\pi/(\varepsilon(k^2 + \varkappa^2(\mathbf{k})))$ are the Fourier transforms of the Green's functions. Note that the expression (5.60) can be easily obtained using the relation (1.198) from the Chap. 1. Therefore, the electrostatic free energy of an ion-dipole mixture within the Gaussian approximation can be written in the following form:

$$F_{el} = \frac{Vk_BT}{2}\int\frac{d\mathbf{k}}{(2\pi)^3}\left(\ln\left(1+\frac{\varkappa^2(\mathbf{k})}{k^2}\right)-\frac{\varkappa^2(\mathbf{k})}{k^2}\right), \tag{5.61}$$

where the screening function $\varkappa^2(\mathbf{k})$ is defined by the expression (5.49). Note that we have subtracted the intrinsic electrostatic energy of the particles of the mixture from the electrostatic free energy.

For the distribution function defined by the expression (5.34), the integral from (5.61) can be evaluated analytically only in the absence of ions in the system ($n_{\pm} = 0$):

$$F_{el} = -\frac{Vk_BT}{l^3}\sigma(y_d), \tag{5.62}$$

where we introduce the auxiliary dimensionless function:

$$\sigma(y_d) = \frac{\sqrt{6}}{4\pi}(2(1+y_d)^{3/2} - 2 - 3y_d). \tag{5.63}$$

It is important to analyze the electrostatic free energy in two limiting regimes following from the relations (5.62)–(5.63), namely:

$$\frac{F_{el}}{Vk_BT} = \begin{cases} -\frac{\sqrt{6}\pi q^4 l n_d^2}{3(\varepsilon k_BT)^2}, & y_d \ll 1 \\ -\frac{1}{12\pi r_D^3}, & y_d \gg 1. \end{cases} \tag{5.64}$$

The first regime describes the case where the dipole length l is much smaller than the Debye screening radius $r_D = \left(8\pi n_d q^2/\varepsilon k_BT\right)^{-1/2}$, which is associated with the ionic groups of polar particles. In this scenario, electrostatic correlations are reduced to the effective pairwise Keesom interaction of dipole particles [30, 108]. Conversely, in the opposite scenario where the dipole length is much larger than the Debye radius, the electrostatic free energy can be described by the standard Debye–Hückel limit law. Therefore, charged groups of sufficiently long polar particles can be modeled as unbound ions.

Now, we can analyze the limiting behavior of the electrostatic free energy in the presence of ions for $y_d \ll 1$ and $y_d \gg 1$. For $y_d \ll 1$, we have

$$\frac{F_{el}}{Vk_BT} = -\frac{\kappa_s^3}{12\pi} - \frac{3\sqrt{6}}{2\pi l^3}\frac{y_s}{1+\sqrt{y_s}}y_d - \frac{3\sqrt{6}}{16\pi l^3}\frac{1+3\sqrt{y_s}+y_s}{(1+\sqrt{y_s})^3}y_d^2 + O(y_d^3), \tag{5.65}$$

where the first term on the right-hand side describes the contribution from the ionic components of the mixture to the electrostatic free energy within the Debye–Hückel approximation. The second and third terms describe, respectively, the contributions of ion-dipole and dipole-dipole pair correlations. In the opposite regime, when $y_d \gg 1$, we obtain the standard Debye–Hückel approximation:

$$\frac{F_{el}}{Vk_BT} \simeq -\frac{1}{12\pi}\left(\frac{8\pi(n_d q^2 + Iq^2)}{\varepsilon k_B T}\right)^{3/2}. \tag{5.66}$$

As mentioned earlier, in this case, the charged groups of polar particles act as unbound ions, contributing to the charge screening alongside free ions.

Modified Gaussian approximation for the electrostatic free energy of the solution of polar particles. Expression (5.62), representing the electrostatic free energy of polar particles within a continuous dielectric medium resembles the well-established Debye–Hückel limit law applicable to ionic systems like electrolyte solutions and equilibrium high-temperature plasma. Similar to the Debye–Hückel limit law, Eq. (5.62) is expected to hold under conditions of high temperatures or low concentrations. To estimate the free energy of a polar fluid under strong electrostatic interactions, we can employ a modified GA, as discussed in Chap. 3 in relation to the OCP and HCOCP models.

Let's evaluate the electrostatic free energy of a solution of polar particles within the modified Gaussian approximation:

$$F_{el} \simeq \frac{Vk_BT}{2}\int\limits_{|\mathbf{k}|<\Lambda} \frac{d\mathbf{k}}{(2\pi)^3}\left(\ln\left(1+\frac{\varkappa^2(\mathbf{k})}{k^2}\right)-\frac{\varkappa^2(\mathbf{k})}{k^2}\right), \tag{5.67}$$

where the screening function is determined by the expression (5.30).

To exclude the nonphysical modes from the summation (integration) over the wave vectors $\mathbf{k}$, arising from small distances between the dipole particles, we introduce an ultraviolet cutoff by setting a maximum wave vector $\Lambda = 2\pi/r_s$, where $r_s = (3/4\pi n_d)^{1/3}$ denotes the radius of a sphere containing exactly one polar particle (the Wigner–Seitz radius). Alternatively, another cutoff criterion, similar to that used for the OCP and HCOCP models, could be employed, leading to the same cutoff parameter up to a prefactor of the order of unity. However, as it shown in Ref. [79], the criterion associated with the Wigner–Seitz radius provides a better estimate when compared with the results of molecular dynamics simulation.

Then, utilizing the characteristic function defined previously as

$$g(\mathbf{k}) = \frac{1}{1+\frac{k^2l^2}{6}}, \tag{5.68}$$

taking into account that $\Lambda = (6\pi^2 n_d)^{1/3}$ and calculating the integral (5.67), we arrive at the following analytical relation for the electrostatic free energy of the solution of dipole particles within the modified Gaussian approximation

$$F_{el} = \frac{k_BTV}{l^3}\sigma(y_d, \theta), \tag{5.69}$$

where we introduce following auxiliary function

$$\sigma(y_d,\theta) = \frac{\sqrt{6}}{2\pi^2}\theta^3 \ln\left(1+\frac{y_d}{1+\theta^2}\right) + \frac{\sqrt{6}}{2\pi^2}\left((2+3y_d)\arctan\theta - 2(1+y_d)^{3/2}\arctan\left(\frac{\theta}{\sqrt{1+y_d}}\right) - \theta y_d\right), \tag{5.70}$$

with the arguments $y_d = \kappa^2 l^2/6$ and $\theta = \Lambda l/\sqrt{6}$. Here it would be constructive to discuss the limiting regimes following from the interpolation formula (5.69):

$$\frac{F_{el}}{Vk_BT} \simeq \begin{cases} -\left(\frac{48}{\pi}\right)^{1/3} \frac{q^2 n_d^{4/3}}{\varepsilon k_B T}, & 1 \ll \theta^2 \ll y_d \\ -\frac{1}{12\pi r_D^3}, & 1 \ll y_d \ll \theta^2 \\ -\frac{\sqrt{6}\pi e^4 l n_d^2}{3(\varepsilon k_B T)^2}, & y_d \ll 1. \end{cases} \tag{5.71}$$

The first regime, which is new compared to the standard Gaussian approximation, occurs when the dipole length l is much larger than the Wigner–Seitz radius r_s, and r_s is much larger than the Debye radius r_D ($r_D \ll r_s \ll l$). In this scenario, we observe an amorphous structure of densely packed charged groups behaving as unbound ions. The second regime arises when $r_s \ll r_D \ll l$. Here, similar to the first regime, the charged groups can be regarded as free ions. However, in this case, they must be organized akin to a dilute electrolyte solution, leading to the electrostatic free energy being described by the limiting law of Debye–Hückel. Lastly, the third regime emerges under the condition $r_D \gg l$, where charge screening is entirely absent, and the electrostatic correlations of the polar particles reduce to the effective pair-additive Keesom attraction of dipole particles.

5.2 Nonlocal Theory of the Solution of the Multipolar Colloidal Particles

Theory of dilute solutions of multipolar particles. In the preceding discussion, we introduced a general nonlocal statistical theory for salt solutions comprising polar molecules, which can be used to describe the thermodynamic properties of solutions of dipole macromolecules such as proteins and betaines. However, there is a broad class of macromolecules, which internal electrical structure cannot be reduced to just two charge centers. A natural extension of the dipole model emerges in the form of a "star-like" configuration, wherein a central charge q is surrounded by peripheral charges q_α ($\alpha = 1, 2, \ldots, m$), with either fixed or fluctuating positions. Such configurations find relevance in systems like polymer stars, colloidal particles in low-polarity solvents, and solutions of metal-organic complexes, among others.

Let's consider an electrolyte solution containing N_+ point cations with charges $q_+ > 0$, N_- point anions with charges $q_- < 0$, and N_c electrically neutral colloidal particles confined in a volume V at temperature T. We model the solvent as a continuous dielectric medium with a dielectric constant ε. Each colloidal particle is assumed to consist of a central charge q and m peripheral charges q_α, $\alpha = 1, 2, \ldots, m$, positioned at fluctuating distances $\boldsymbol{\xi}\alpha$ from the central charge q, described by the distribution functions $g_\alpha(\boldsymbol{\xi}_\alpha)$. Since the particles are electrically neutral, the relation $q + \sum_{\alpha=1}^{m} q_\alpha = 0$ holds. We assume that the colloidal solution is sufficiently dilute so that all interactions except electrostatic interactions can be neglected. The method of accounting for excluded volume effects is outlined below.

Taking into account all modeling assumptions, we write the configuration integral of the system in the following form

$$Q = \int d\Gamma_s \int d\Gamma_c \exp[-\beta U], \tag{5.72}$$

where

$$\int d\Gamma_c (\ldots) = \int \ldots \int \prod_{j=1}^{N_c} d\Gamma_j (\ldots) \tag{5.73}$$

is the integration measure over the internal configurations Γ_j of the colloidal particles, i.e.

$$\int d\Gamma_j (\ldots) = \int \frac{d\mathbf{R}_j}{V} \int \ldots \int \prod_{\alpha=1}^{m} d\boldsymbol{\xi}_j^{(\alpha)}\, g_\alpha(\boldsymbol{\xi}_j^{(\alpha)})(\ldots), \tag{5.74}$$

where $\boldsymbol{\xi}_j^{(\alpha)}$ are the vectors of displacements of peripheral charges from positions $\mathbf{R}_j$ of central charges. The integration measure over the coordinate of salt ions $\mathbf{r}_k^{(\pm)}$ is defined as

$$\int d\Gamma_s (\ldots) = \int \cdots \int \prod_{l=1}^{N_+} \frac{d\mathbf{r}_l^{(+)}}{V} \prod_{k=1}^{N_-} \frac{d\mathbf{r}_k^{(-)}}{V} (\ldots). \tag{5.75}$$

The Hamiltonian of electrostatic interactions can be written as follows

$$U = \frac{1}{2} \int d\mathbf{r} \int d\mathbf{r}'\, \hat{\rho}(\mathbf{r}) G_0(\mathbf{r} - \mathbf{r}') \hat{\rho}(\mathbf{r}') = \frac{1}{2} \left(\hat{\rho} G_0 \hat{\rho}\right), \tag{5.76}$$

where $G_0(\mathbf{r} - \mathbf{r}') = 1/(\varepsilon|\mathbf{r} - \mathbf{r}'|)$ is the Green's function of the Poisson equation and

$$\hat{\rho}(\mathbf{r}) = \hat{\rho}_c(\mathbf{r}) + \hat{\rho}_i(\mathbf{r}) + \rho_{ext}(\mathbf{r}) \tag{5.77}$$

determines the local charge density of the system, where

$$\hat{\rho}_c(\mathbf{r}) = \sum_{\alpha=1}^{m} q_\alpha \sum_{j=1}^{N_c} \left(\delta\left(\mathbf{r} - \mathbf{R}_j - \boldsymbol{\xi}_j^{(\alpha)}\right) - \delta\left(\mathbf{r} - \mathbf{R}_j\right)\right) \tag{5.78}$$

is the bound charge density of colloidal particles;

$$\hat{\rho}_i(\mathbf{r}) = q_+ \sum_{k=1}^{N_+} \delta\left(\mathbf{r} - \mathbf{r}_k^{(+)}\right) + q_- \sum_{l=1}^{N_-} \delta\left(\mathbf{r} - \mathbf{r}_l^{(-)}\right) \tag{5.79}$$

is the local charge density of the ions; $\rho_{ext}(\mathbf{r})$ denotes the density of the external charges.

Utilizing the Hubbard–Stratonovich transformation:

$$\exp\left[-\frac{\beta}{2}(\hat{\rho} G_0 \hat{\rho})\right] = \int \frac{\mathcal{D}\varphi}{C} \exp\left[-\frac{\beta}{2}(\varphi G_0^{-1} \varphi) + i\beta(\hat{\rho}\varphi)\right], \tag{5.80}$$

we arrive at the following representation of the configuration integral in the form of the functional integral

$$Q = \int \frac{\mathcal{D}\varphi}{C} \exp\left[-\frac{\beta}{2}(\varphi G_0^{-1} \varphi) + i\beta(\rho_{ext}\varphi)\right] Q_c^{N_c}[\varphi] Q_+^{N_+}[\varphi] Q_-^{N_-}[\varphi] \tag{5.81}$$

with the one-particle configuration integrals

$$Q_c[\varphi] = \int \frac{d\mathbf{R}}{V} \int d\sigma(\boldsymbol{\xi}_1, \ldots, \boldsymbol{\xi}_m)\, e^{i\beta \sum\limits_{\alpha=1}^{m} q_\alpha \left(\varphi(\mathbf{R}+\boldsymbol{\xi}_\alpha) - \varphi(\mathbf{R})\right)} \tag{5.82}$$

and

$$Q_\pm[\varphi] = \int \frac{d\mathbf{r}}{V} \exp(i\beta q_\pm \varphi(\mathbf{r})), \tag{5.83}$$

with normalized integration measure over the peripheral charges displacements

$$\int d\sigma(\boldsymbol{\xi}_1, \ldots, \boldsymbol{\xi}_m)(\ldots) = \int d\boldsymbol{\xi}_1 \ldots \int d\boldsymbol{\xi}_m \prod_{\alpha=1}^{m} g_\alpha(\boldsymbol{\xi}_\alpha)(\ldots), \tag{5.84}$$

i.e. $\int d\sigma\,(\ldots) = 1$. In thermodynamic limit $V \to \infty,\ N_{c,\pm} \to \infty,\ N_{c,\pm}/V \to n_{c,\pm}$ we obtain

$$Q = \int \frac{\mathcal{D}\varphi}{C} \exp\left[-S[\varphi]\right], \tag{5.85}$$

where we introduce the following functional

$$S[\varphi] = \frac{\beta}{2}(\varphi G_0^{-1} \varphi) - i\beta(\rho_{ext}\varphi) - W[\varphi] \tag{5.86}$$

with an auxiliary functional

$$W[\varphi] = n_c \int d\mathbf{R} \int d\sigma(\boldsymbol{\xi}_1, \ldots, \boldsymbol{\xi}_m) \left(e^{i\beta \sum_{\alpha=1}^{m} q_\alpha (\varphi(\mathbf{R}+\boldsymbol{\xi}_\alpha) - \varphi(\mathbf{R}))} - 1 \right)$$
$$+ n_+ \int d\mathbf{r}\, (e^{i\beta q_+ \varphi(\mathbf{r})} - 1) + n_- \int d\mathbf{r}\, (e^{i\beta q_- \varphi(\mathbf{r})} - 1). \quad (5.87)$$

Next, let's derive an expression for the electrostatic free energy of the solution within the Gaussian approximation. Expanding the functional $S[\varphi]$ from (5.86) into a power series around the mean field potential $\varphi^{(MF)}(\mathbf{r}) = i\psi(\mathbf{r})$, which satisfies the Euler–Lagrange equation:

$$\left. \frac{\delta S[\varphi]}{\delta \varphi(\mathbf{r})} \right|_{\varphi = i\psi} = 0 \quad (5.88)$$

and neglecting the terms above the second order of φ, we obtain

$$Q \approx \exp\left[-S[i\psi]\right] \int \frac{\mathcal{D}\Lambda}{C} \exp\left[-\frac{\beta}{2} \left(\Lambda \mathcal{G}^{-1} \Lambda \right) \right]. \quad (5.89)$$

Here, we introduce the following renormalized inverse Green's function

$$\mathcal{G}^{-1}(\mathbf{r}, \mathbf{r}'|\psi) = k_B T \frac{\delta^2 S[i\psi]}{\delta \varphi(\mathbf{r}) \delta \varphi(\mathbf{r}')} = G_0^{-1}(\mathbf{r}, \mathbf{r}') + \beta S_i(\mathbf{r}, \mathbf{r}') + \beta S_c(\mathbf{r}, \mathbf{r}'), \quad (5.90)$$

with

$$S_i(\mathbf{r}, \mathbf{r}') = \left(q_+^2 n_+ e^{-\beta q_+ \psi(\mathbf{r})} + q_-^2 n_- e^{-\beta q_- \psi(\mathbf{r})} \right) \delta(\mathbf{r} - \mathbf{r}'), \quad (5.91)$$

$$S_c(\mathbf{r}, \mathbf{r}') = n_c \sum_{\delta\gamma} q_\delta q_\gamma \left\langle e^{-\beta \sum_{\alpha=1}^{m} q_\alpha (\psi(\mathbf{r}+\boldsymbol{\xi}_\alpha - \boldsymbol{\xi}_\gamma) - \psi(\mathbf{r} - \boldsymbol{\xi}_\gamma))} (\delta(\mathbf{r} - \mathbf{r}' + \boldsymbol{\xi}_\delta - \boldsymbol{\xi}_\gamma) - \delta(\mathbf{r} - \mathbf{r}' - \boldsymbol{\xi}_\gamma)) \right\rangle_\xi$$
$$- n_c \sum_{\delta\gamma} q_\delta q_\gamma \left\langle e^{-\beta \sum_{\alpha=1}^{m} q_\alpha (\psi(\mathbf{r}+\boldsymbol{\xi}_\alpha) - \psi(\mathbf{r}))} (\delta(\mathbf{r} - \mathbf{r}' + \boldsymbol{\xi}_\delta) - \delta(\mathbf{r} - \mathbf{r}')) \right\rangle_\xi \quad (5.92)$$

and the mean-field electrostatic potential $\psi(\mathbf{r})$ satisfying the following self-consistent field equation, derived from (5.88)

$$\Delta \psi(\mathbf{r}) = -\frac{4\pi}{\varepsilon} \left(\rho_{ext}(\mathbf{r}) + \bar{\rho}_i(\mathbf{r}) + \bar{\rho}_c(\mathbf{r}) \right), \quad (5.93)$$

where

$$\bar{\rho}_i(\mathbf{r}) = q_+ n_+ e^{-\beta q_+ \psi(\mathbf{r})} + q_- n_- e^{-\beta q_- \psi(\mathbf{r})} \quad (5.94)$$

is the average charge density of the ions and

$$\bar{\rho}_c(\mathbf{r}) = n_c \sum_{\gamma=1}^{m} q_\gamma \left\langle e^{-\beta \sum_{\alpha=1}^{m} q_\alpha (\psi(\mathbf{r}+\boldsymbol{\xi}_\alpha-\boldsymbol{\xi}_\gamma)-\psi(\mathbf{r}-\boldsymbol{\xi}_\gamma))} - e^{-\beta \sum_{\alpha=1}^{m} q_\alpha (\psi(\mathbf{r}+\boldsymbol{\xi}_\alpha)-\psi(\mathbf{r}))} \right\rangle_\xi \tag{5.95}$$

is the average bound charge density of the colloidal particle. Note that in expressions (5.92) and (5.95) we introduce the averaging over the configurations of peripheral charges as:

$$\langle (\ldots) \rangle_\xi = \int d\sigma(\boldsymbol{\xi}_1, \ldots, \boldsymbol{\xi}_m)\, (\ldots). \tag{5.96}$$

Thus, formally calculating the Gaussian functional integral, we arrive at the following general expression for the configuration integral within the GA:

$$Q \approx \exp\left\{ -S[i\psi] + \frac{1}{2} tr\, (\ln \mathcal{G} - \ln G_0) \right\}, \tag{5.97}$$

where, as previously mentioned, the symbol $tr(\ldots)$ denotes the trace of the integral operator. When there are no external charges in the solution (i.e., $\rho_{ext}(\mathbf{r}) = 0$), the electrostatic potential of the self-consistent field $\psi(\mathbf{r}) = 0$, resulting in the mean-field contribution to the electrostatic free energy being $F_{el}^{(MF)} = k_B T S[0] = 0$. Thus, the electrostatic free energy is entirely determined by the thermal fluctuations of the electrostatic potential around its zero value. In this case, the renormalized inverse Green's function takes a simple form:

$$\begin{aligned} \mathcal{G}^{-1}(\mathbf{r}, \mathbf{r}'|0) &= G^{-1}(\mathbf{r}, \mathbf{r}') = G_0^{-1}(\mathbf{r}, \mathbf{r}') + 8\pi\beta I q^2 \delta(\mathbf{r}-\mathbf{r}') \\ &+ \beta n_c \sum_{\delta\gamma} q_\delta q_\gamma \left\langle \delta(\mathbf{r}-\mathbf{r}'+\boldsymbol{\xi}_\delta-\boldsymbol{\xi}_\gamma) - 2\delta(\mathbf{r}-\mathbf{r}'-\boldsymbol{\xi}_\gamma) + \delta(\mathbf{r}-\mathbf{r}') \right\rangle_\xi, \end{aligned} \tag{5.98}$$

while the electrostatic free energy of the ion-dipole mixture in the Gaussian approximation can be calculated as follows:

$$F_{el} \approx \frac{k_B T}{2} tr\, (\ln G_0 - \ln G) = \frac{V k_B T}{2} \int \frac{d\mathbf{k}}{(2\pi)^3} \ln \frac{G_0(\mathbf{k})}{G(\mathbf{k})}, \tag{5.99}$$

where $G_0(\mathbf{k}) = 4\pi/(\varepsilon k^2)$ and $G(\mathbf{k}) = 4\pi/(\varepsilon(k^2 + \varkappa^2(\mathbf{k})))$ are the Fourier transforms of the Green's functions. Hence, the electrostatic free energy of the solution within the Gaussian approximation has the form

$$F_{el} = \frac{V k_B T}{2} \int \frac{d\mathbf{k}}{(2\pi)^3} \left(\ln\left(1 + \frac{\varkappa^2(\mathbf{k})}{k^2}\right) - \frac{\varkappa^2(\mathbf{k})}{k^2} \right) \tag{5.100}$$

with the screening function

$$\varkappa^2(\mathbf{k}) = \kappa_s^2 + \frac{4\pi n_c}{\varepsilon k_B T}\left(|q(\mathbf{k})|^2 + \sum_{\alpha=1}^{m} q_\alpha^2(1 - |g_\alpha(\mathbf{k})|^2)\right), \tag{5.101}$$

where we introduce the following auxiliary function

$$q(\mathbf{k}) = \sum_{\alpha=1}^{m} q_\alpha(1 - g_\alpha(\mathbf{k})) \tag{5.102}$$

and the functions

$$g_\alpha(\mathbf{k}) = \int d\mathbf{r}\, g_\alpha(\mathbf{r}) e^{-i\mathbf{kr}}. \tag{5.103}$$

Now, let's derive an expression for the electrostatic potential of external charges in the colloidal solution medium for the case of weak electrostatic interactions. Within this approximation, as follows from the Eq. (5.93), the potential is determined by the relation:

$$\psi(\mathbf{r}) = \frac{4\pi}{\varepsilon}\int \frac{d\mathbf{k}}{(2\pi)^3}\frac{\tilde{\rho}_{ext}(\mathbf{k})e^{i\mathbf{kr}}}{k^2 + \varkappa^2(\mathbf{k})}, \tag{5.104}$$

where

$$\tilde{\rho}_{ext}(\mathbf{k}) = \int d\mathbf{r} \rho_{ext}(\mathbf{r}) e^{-i\mathbf{kr}} \tag{5.105}$$

is the Fourier transform of the external charges density.

The case of identical peripheral charges. Excess free energy and oscillations of the electrostatic potential. In this section we consider the case of identical peripheral charges $q_\alpha = -q/m$ with identical spherically symmetric distribution functions $g_\alpha(\mathbf{r}) = g(|\mathbf{r}|)$. In this case the screening function has the form

$$\varkappa^2(\mathbf{k}) = \kappa_s^2 + \frac{4\pi q^2 n_c(m+1)}{\varepsilon k_B T m}(1 - g(\mathbf{k}))\left(1 - \frac{m-1}{m+1}g(\mathbf{k})\right), \tag{5.106}$$

where $\kappa_s^2 = 8\pi I q^2/(\varepsilon k_B T)$ is the square of the inverse Debye radius of the ions, and $I = (q_+^2 n_+ + q_-^2 n_-^2 n_-)/2q^2$ is the ionic strength of the solution. For the case of polar particles ($m = 1$), the expression (5.106) reduces to previously obtained one (see (5.49))

$$\varkappa^2(\mathbf{k}) = \kappa_s^2 + \frac{8\pi q^2 n_c}{\varepsilon k_B T}(1 - g(\mathbf{k})). \tag{5.107}$$

Let's also consider another limiting case $m \gg 1$, where peripheral charges form a continuous cloud surrounding the central charges. In this case, the screening function takes the form:

$$\varkappa^2(\mathbf{k}) \simeq \kappa_s^2 + \frac{4\pi q^2 n_c}{\varepsilon k_B T}(1 - g(\mathbf{k}))^2. \tag{5.108}$$

Note that in the regime of weak electrostatic interactions in the presence of "quadrupolar" particles in solution, the self-consistent field equation has the following form [106, 109, 110]

$$\Delta\psi(\mathbf{r}) - L_Q^2\Delta^2\psi(\mathbf{r}) - \kappa_s^2\psi(\mathbf{r}) = -\frac{4\pi}{\varepsilon}\rho_{ext}(\mathbf{r}), \tag{5.109}$$

where

$$L_Q^2 = \frac{4\pi q^2 n_c R^4}{\varepsilon k_B T} \tag{5.110}$$

is the so-called quadrupole length, introduced for the first time by Slavchov and Ivanov [111, 112] as a phenomenological parameter; R denotes the characteristic size of the charge cloud of colloidal particles.

Using the Eq. (5.109), we can calculate the potential of the test charge q_0 in a salt solution of quadrupole particles. Placing the test charge at the origin, taking into account that in this case $\rho_{ext}(\mathbf{r}) = q_0\delta(\mathbf{r})$ and using the Eq. (5.109) in the Fourier representation, we obtain

$$\psi(\mathbf{r}) \simeq \frac{4\pi q_0}{\varepsilon}\int\frac{d\mathbf{k}}{(2\pi)^3}\frac{e^{i\mathbf{k}\mathbf{r}}}{k^2+\kappa_s^2+L_Q^2k^4}, \tag{5.111}$$

which after integration leads to the following cases

$$\psi(\mathbf{r}) = \frac{q_0}{\varepsilon r}\times\begin{cases}\frac{\exp(-\kappa_- r)-\exp(-\kappa_+ r)}{\sqrt{1-4\kappa_s^2L_Q^2}}, & \kappa_s L_Q < \frac{1}{2}\\ \frac{r}{\sqrt{2}L_Q}\exp\left(-\frac{r}{\sqrt{2}L_Q}\right), & \kappa_s L_Q = \frac{1}{2}\\ \frac{\exp(-\kappa_0 r)}{\sqrt{4\kappa_s^2L_Q^2-1}}\sin\left(\frac{r}{\lambda}\right), & \kappa_s L_Q > \frac{1}{2},\end{cases} \tag{5.112}$$

where $r = |\mathbf{r}|$, and we introduce the following notations

$$\kappa_\pm = \frac{\left(1\pm\sqrt{1-4\kappa_s^2L_Q^2}\right)^{1/2}}{\sqrt{2}L_Q}, \quad \kappa_0 = \kappa_s\sqrt{1+1/(2\kappa_s L_Q)}, \quad \lambda^{-1} = \kappa_s\sqrt{1-1/(2\kappa_s L_Q)}. \tag{5.113}$$

Therefore, we can conclude that when the quadrupole length is less than half of the Debye radius of the salt ions, the potential decreases monotonically at large distances. However, when the quadrupole length is greater than half the Debye radius, the potential decreases, oscillating with the characteristic wavelength λ. It's worth noting that this behavior of the potential was first obtained in [111] within the framework of phenomenological theory. As will be shown in Chap. 6, similar oscillations of the potential can be observed in polyelectrolyte fluids.

For the model distribution function we are using, $g(r) = 1/(4\pi R^2 r)\exp(-r/R)$, which is defined by the following characteristic function

$$g(\mathbf{k}) = \frac{1}{1 + k^2 R^2}, \tag{5.114}$$

using Eq. (5.100), we can derive an analytical expression for the electrostatic free energy for a salt-free solution ($\kappa_s = 0$), i.e.

$$\frac{F_{el}}{V k_B T} = -\frac{1}{6\pi R^3}\left((1 + y_c)\sqrt{1 + \frac{y_c}{4}} - \frac{9y_c}{8} - 1\right), \tag{5.115}$$

where $y_c = L_Q^2/R^2 = 2\pi q^2 n_c l^2/(3\varepsilon k_B T) = \kappa_c^2 l^2/6$, $R = l/\sqrt{6}$, and $\kappa_c = (4\pi q^2 n_c/\varepsilon k_B T)^{1/2}$ is the inverse Debye length associated with the central charges of the quadrupole particles. It's informative to analyze the limiting cases derived from (5.115). We obtain

$$\frac{F_{el}}{V k_B T} = \begin{cases} -\frac{5\pi q^4 l}{16\sqrt{6}(\varepsilon k_B T)^2} n_c^2, & y_c \ll 1 \\ -\frac{\kappa_c^3}{12\pi}, & y_c \gg 1. \end{cases} \tag{5.116}$$

The first regime occurs when the size R of the charge cloud is much smaller than the Debye length κ_c^{-1}. In this scenario, the electrostatic correlations of colloidal particles lead to an effective van der Waals interaction, specifically, the Kirkwood–Schumaker one [110, 113]. This interaction arises from the spatial fluctuations of charged clouds of colloidal particles situated at significant distances from each other. Conversely, in the opposite case, the electrostatic free energy can be described by the Debye–Hückel limit law. Here, the colloidal solution can be described by the OCP model, where the system consists of central charges q moving within a compensating homogeneous charge background formed by overlapping charge clouds.

Now, let's determine the electrostatic free energy of the salt solution when the concentration of colloidal particles is much less than the salt concentration, i.e., when $y_c \ll 1$. The result of this calculation is:

$$\frac{F_{el}}{V k_B T} = -\frac{\kappa_s^3}{12\pi} - \frac{\sqrt{6}\pi q^2 n_c}{\varepsilon k_B T l} h_1(y_s) - \frac{\pi q^4 l n_c^2}{16\sqrt{6}(\varepsilon k_B T)^2} h_2(y_s) + O(n_c^3), \tag{5.117}$$

where $y_s = \kappa_s^2 R^2$ and we introduce the following auxiliary functions

$$h_1(y_s) = \frac{y_s(1 + 2\sqrt{y_s})}{(1 + \sqrt{y_s})^2}, \tag{5.118}$$

$$h_2(y_s) = \frac{40 y_s^{3/2} + 8 y_s^2 + 48 y_s + 25\sqrt{y_s} + 5}{(1 + \sqrt{y_s})^5}. \tag{5.119}$$

In Eq. (5.117), the first term describes the well-known Debye–Hückel limiting law for the excess free energy of a dilute electrolyte solution. The second term accounts for the electrostatic interactions between ions and colloidal particles. The third term represents the contribution of electrostatic interactions of colloidal particles, considering the screening effect of the salt ions. In the region where $y_s \gg 1$ and $y_c \gg 1$, we obtain the following limiting regime:

$$\frac{F_{el}}{Vk_BT} = -\frac{\kappa^3}{12\pi}, \tag{5.120}$$

where $\kappa = (\kappa_s^2 + \kappa_c^2)^{1/2}$; the excess free energy (5.120) defines the Debye–Hückel limiting law for the free energy of OCP in the presence of salt ions.

Incorporation of the excluded volume in multipolar colloidal particle solutions. In this section, we derive an expression for the electrostatic free energy of a salt-free solution of multipolar colloidal particles, incorporating excluded volume interactions (steric factor), within the Gaussian approximation. We assume that the central charge q is situated at the center of a solid sphere with diameter d, while the peripheral charges q_α ($\alpha = 1, ..., m$) surrounding it can penetrate the hard core of the molecules. The configuration integral of the system takes the form:

$$Q = \int d\Gamma_c \exp\left[-\beta U_{hs} - \beta U_{el}\right], \tag{5.121}$$

where we integrate over the configurations of multipolar colloidal particles according to the definitions (5.73) and (5.74). The first term in the exponent of the integrand accounts for the excluded volume interactions between multipolar colloidal particles, represented by

$$U_{hs} = \frac{1}{2} \sum_{i,j \neq j} U_{hc}(\mathbf{R}_i - \mathbf{R}_j), \tag{5.122}$$

where

$$U_{hc}(\mathbf{r}) = \begin{cases} 0, |\mathbf{r}| > 0 \\ \infty, |\mathbf{r}| \leq 0 \end{cases} \tag{5.123}$$

represents the hard-core potential. The second term determines the potential energy of the electrostatic interactions between the multipolar particles, given by

$$U_{el} = \frac{1}{2}\left(\hat{\rho}_c G_0 \hat{\rho}_c\right) = \frac{1}{2} \int d\mathbf{r} \int d\mathbf{r}'\, \hat{\rho}_c(\mathbf{r}) G_0(\mathbf{r} - \mathbf{r}') \hat{\rho}_c(\mathbf{r}'), \tag{5.124}$$

where $\hat{\rho}_c(\mathbf{r})$ is the local charge density of colloidal particles defined by the relation (5.78). Thus, using the system of hard spheres as a reference system, we rewrite the configuration integral in a form suitable for constructing the thermodynamic perturbation theory, i.e.

$$Q = Q_R \left\langle \exp\left[-\frac{1}{2}\left(\hat{\rho}_c G_0 \hat{\rho}_c\right)\right]\right\rangle_R, \tag{5.125}$$

where

$$Q_R = \int d\Gamma_c \exp[-\beta U_{hs}] \tag{5.126}$$

is the configuration integral of the reference system, and

$$\langle(...)\rangle_R = \frac{1}{Q_R}\int d\Gamma_c \exp[-\beta U_{hs}]\,(...) \tag{5.127}$$

defines the averaging over the microstates of the latter. Now, utilizing the Hubbard–Stratonovich transformation, we obtain the following representation of the configuration integral

$$Q = Q_R \int \frac{\mathcal{D}\varphi}{C} \exp\left[-\frac{\beta}{2}(\varphi G_0^{-1}\varphi)\right]\left\langle \exp\left[i\beta(\hat{\rho}_c\varphi)\right]\right\rangle_R \tag{5.128}$$

where C is the normalization constant of the Gaussian measure, introduced previously. The cumulant expansion (see Chap. 1) yields

$$\left\langle \exp\left[i\beta(\hat{\rho}_c\varphi)\right]\right\rangle_R = \exp\left[i\beta(\langle\hat{\rho}_c\rangle_R\,\varphi) - \frac{\beta^2}{2}(\varphi\mathcal{K}\varphi) + ...\right], \tag{5.129}$$

where

$$\mathcal{K}(\mathbf{r}-\mathbf{r}') = \langle\hat{\rho}_c(\mathbf{r})\hat{\rho}_c(\mathbf{r}')\rangle_R - \langle\hat{\rho}_c(\mathbf{r})\rangle_R\langle\hat{\rho}_c(\mathbf{r}')\rangle_R \tag{5.130}$$

is the correlation function of the local charge density of colloidal particles obtained by the averaging over the microstates of the basis system. Taking into account that $\langle\hat{\rho}_c(\mathbf{r})\rangle = 0$, we obtain

$$\begin{aligned}\mathcal{K}(\mathbf{r}-\mathbf{r}') &= \langle\hat{\rho}_c(\mathbf{r})\hat{\rho}_c(\mathbf{r}')\rangle_R \\ &= \sum_{ij}\sum_{\alpha\gamma} q_\alpha q_\gamma \left\langle\left(\delta\left(\mathbf{r}-\mathbf{R}_i-\boldsymbol{\xi}_i^{(\alpha)}\right)-\delta(\mathbf{r}-\mathbf{R}_i)\right)\left(\delta\left(\mathbf{r}-\mathbf{R}_j-\boldsymbol{\xi}_j^{(\gamma)}\right)-\delta(\mathbf{r}-\mathbf{R}_j)\right)\right\rangle_R \\ &= \sum_{ij}\sum_{\alpha\gamma} q_\alpha q_\gamma \int\frac{d\mathbf{k}}{(2\pi)^3}\int\frac{d\mathbf{q}}{(2\pi)^3} e^{i\mathbf{k}\mathbf{r}+i\mathbf{q}\mathbf{r}'}\left\langle e^{-i\mathbf{k}\mathbf{R}_i-i\mathbf{q}\mathbf{R}_j}\right\rangle_{hs}\left\langle\left(e^{-i\mathbf{k}\boldsymbol{\xi}_i^{(\alpha)}}-1\right)\left(e^{-i\mathbf{q}\boldsymbol{\xi}_j^{(\gamma)}}-1\right)\right\rangle_\xi,\end{aligned} \tag{5.131}$$

where we utilize the Fourier-transformation of the delta-function

$$\delta(\mathbf{x}) = \int\frac{d\mathbf{k}}{(2\pi)^3}e^{i\mathbf{k}\mathbf{x}}, \tag{5.132}$$

and the averaging $\langle(...)\rangle_{hs}$ over the configurations of hard spheres and the averaging $\langle(...)\rangle_\xi$ over the configurations $\boldsymbol{\xi}_i^{(\alpha)}$ of peripheral charges. Further, taking into account the identity

$$\left\langle e^{-i\mathbf{k}\mathbf{R}_i-i\mathbf{q}\mathbf{R}_j}\right\rangle_{hs} = \frac{(2\pi)^3}{V}\delta(\mathbf{k}+\mathbf{q})\left\langle e^{-i\mathbf{k}(\mathbf{R}_i-\mathbf{R}_j)}\right\rangle_{hs} \tag{5.133}$$

and using the definition of the structure factor

$$S(\mathbf{k}) = \frac{1}{V}\sum_{ij}\left\langle e^{-i\mathbf{k}(\mathbf{R}_i-\mathbf{R}_j)}\right\rangle_{hs} = n(1+n\tilde{h}_2(\mathbf{k})) \tag{5.134}$$

after algebraic transformations we obtain

$$\mathcal{K}(\mathbf{r}-\mathbf{r}') = \int \frac{d\mathbf{k}}{(2\pi)^3} e^{i\mathbf{k}(\mathbf{r}-\mathbf{r}')}\left(|q(\mathbf{k})|^2 S(\mathbf{k}) + n\sum_{\alpha=1}^{m} q_\alpha^2(1-|g_\alpha(\mathbf{k})|^2)\right), \tag{5.135}$$

where we take into account that $g_\alpha(\mathbf{k}) = \left\langle e^{-i\mathbf{k}\xi_i^{(\alpha)}}\right\rangle_\xi$, and the auxiliary function $q(\mathbf{k})$ is determined by the relation (5.102).

Thus, we arrive at the following expression for the configuration integral within the Gaussian approximation

$$Q = Q_b \int \frac{\mathcal{D}\varphi}{C} \exp\left[-\frac{\beta}{2}(\varphi G^{-1}\varphi)\right], \tag{5.136}$$

where the renormalized inverse Green's function takes the form

$$G^{-1}(\mathbf{r},\mathbf{r}') = G_0^{-1}(\mathbf{r},\mathbf{r}') + \beta\mathcal{K}(\mathbf{r}-\mathbf{r}'). \tag{5.137}$$

Equation (5.137) can be rewritten in Fourier representation as

$$G(\mathbf{k}) = \frac{G_0(\mathbf{k})}{1+\beta\tilde{\mathcal{K}}(\mathbf{k})G_0(\mathbf{k})} = \frac{4\pi}{\varepsilon(k^2+\varkappa^2(\mathbf{k}))}, \tag{5.138}$$

where we define the following screening function [106]

$$\varkappa^2(\mathbf{k}) = \frac{4\pi}{\varepsilon k_B T}\left(|q(\mathbf{k})|^2 S(\mathbf{k}) + n\sum_{\alpha=1}^{m} q_\alpha^2(1-|g_\alpha(\mathbf{k})|^2)\right). \tag{5.139}$$

The excess free energy within the Gaussian approximation has the form

$$F_{ex} = -k_B T \ln Q = F_{hs} + \frac{Vk_BT}{2}\int \frac{d\mathbf{k}}{(2\pi)^3} \ln\frac{G_0(\mathbf{k})}{G(\mathbf{k})}. \tag{5.140}$$

Finally, subtracting the intrinsic electrostatic energy from the final expression, we obtain

$$F_{ex} = F_{hs} + \frac{Vk_BT}{2}\int \frac{d\mathbf{k}}{(2\pi)^3}\left(\ln\left(1+\frac{\varkappa^2(\mathbf{k})}{k^2}\right) - \frac{\varkappa^2(\mathbf{k})}{k^2}\right). \tag{5.141}$$

Note that for the excess free energy F_{hs} of a system of hard spheres we can utilize the Perkus–Yevick or Carnahan–Starling approximation [19]. For colloidal particles with a single peripheral charge ($m = 1$), the screening function (5.139) simplifies to

$$\varkappa^2(\mathbf{k}) = \frac{8\pi n q^2}{\varepsilon k_B T}(1 - g(\mathbf{k}))\left(1 + \frac{1}{2} n\tilde{h}_2(\mathbf{k})(1 - g(\mathbf{k}))\right), \tag{5.142}$$

Using the approximation $n\tilde{h}_2(\mathbf{k}) \approx n\tilde{h}_2(0) = nk_B T\chi - 1 = z_0 - 1$, (where χ denotes the isothermal compressibility of the system of the hard spheres) and the characteristic function $g(\mathbf{k}) = (1 + k^2 l^2/6)^{-1}$, we arrive at the expression (5.315) for the electrostatic free energy, as will be derived below within the fluctuation theory framework.

For the opposite case of a large number of peripheral charges on colloidal particles ($m \gg 1$), the screening function takes the following form

$$\varkappa^2(\mathbf{k}) \simeq \frac{4\pi q^2 n^2}{\varepsilon k_B T}(1 - g(\mathbf{k}))^2 \tilde{h}_2(\mathbf{k}). \tag{5.143}$$

Utilizing the same approximation and the same model function $g(\mathbf{k})$, we obtain

$$\frac{F_{el}}{V k_B T} = -\frac{1}{\sqrt{6}\pi l^3}\left((1 + \tilde{y}_c)\sqrt{1 + \frac{\tilde{y}_c}{4}} - \frac{9\tilde{y}_c}{8} - 1\right), \tag{5.144}$$

where $\tilde{y}_c = y_c z_0$ and $y_c = 2\pi q^2 m^2 n/(3\varepsilon k_B T)$. The limiting regimes following from the expression (5.144) have the following form

$$\frac{F_{el}}{V k_B T} = \begin{cases} -\frac{5\sqrt{6}\pi q^4 l}{96(\varepsilon k_B T)^2} z_0^2 n^2, & y_c \ll 1 \\ -\frac{\kappa^3}{12\pi} z_0^{3/2}, & y_c \gg 1. \end{cases} \tag{5.145}$$

Within the Percus–Yevick approximation, we have

$$z_0 = (1 - \eta)^4/(1 + 2\eta)^2, \tag{5.146}$$

and in the case of Carnahan–Starling one

$$z_0 = (1 - \eta)^4/(1 + 4\eta + 4\eta^2 - 4\eta^3 + \eta^4), \tag{5.147}$$

where $\eta = \pi n d^3/6$. As in the case of the dipole particles with excluded volume, the presence of a hard core in quadrupole particles results in a reduction in the magnitude of the electrostatic interactions contribution to the total free energy of the solution.

5.3 Statistical Field Theory of Salt Solutions of Spherical Micellar Aggregates

Let's now utilize previously established theory of solutions of multipolar colloidal particles to formulate a theory concerning salt solutions of micellar aggregates.

Let's consider a salt solution containing N_+ point cations with charges $q_+ > 0$, N_- point anions with charges $q_- < 0$, and N_M micellar aggregates confined within volume V at temperature T. The solvent is modeled as a continuous dielectric medium with dielectric constant ε. Each micellar aggregate is represented as a system comprising m elementary charges $q = e$, situated at fluctuating distances $\boldsymbol{\xi}_\alpha$ from the center, characterized by a distribution function $g(\boldsymbol{\xi}_\alpha)$. Additionally, there are $N_M m$ point counterions with charges $-e$ in the solution. Let's assume that the micellar solution is dilute enough to neglect all interactions except the electrostatic ones. Considering the excluded volume of micelles can be approached similarly to solutions of neutral multipolar colloidal particles. With these modeling assumptions in mind, we can express the configuration integral of the system as follows

$$Q = \int d\Gamma_i \int d\Gamma_M \exp\left[-\beta U\right], \tag{5.148}$$

where

$$\int d\Gamma_M(\ldots) = \int \cdots \int \prod_{j=1}^{N_M} d\Gamma_j(\ldots) \tag{5.149}$$

is the integration measure over the internal configurations Γ_j of the micelle, i.e.

$$\int d\Gamma_j(\ldots) = \int \frac{d\mathbf{R}_j}{V} \int \cdots \int \prod_{\alpha=1}^{m} d\boldsymbol{\xi}_j^{(\alpha)} g(\xi_j^{(\alpha)})(\ldots), \tag{5.150}$$

where $\boldsymbol{\xi}_j^{(\alpha)}$ represents the vectors of charge displacements from the positions of the centers $\mathbf{R}_j$. The integration measure over the coordinates $\mathbf{r}_k^{(\pm)}$ of the salt ions and counterions is defined as:

$$\int d\Gamma_i(\ldots) = \frac{1}{V^{N_+ + N_- + N_M m}} \int \prod_{k=1}^{N_+} d\mathbf{r}_k^{(+)} \int \prod_{s=1}^{N_-} d\mathbf{r}_s^{(-)} \int \prod_{n=1}^{mN_M} d\mathbf{r}_n^{(c)}(\ldots). \tag{5.151}$$

The Hamiltonian of electrostatic interactions containing self-interaction can be written as follows

$$U = \frac{1}{2} \int d\mathbf{r} \int d\mathbf{r}'\, \hat{\rho}(\mathbf{r}) G_0(\mathbf{r} - \mathbf{r}') \hat{\rho}(\mathbf{r}') = \frac{1}{2} \left(\hat{\rho} G_0 \hat{\rho}\right), \tag{5.152}$$

where $G_0(\mathbf{r} - \mathbf{r}') = 1/(\varepsilon|\mathbf{r} - \mathbf{r}'|)$ is the Green's function of the Poisson equation and

$$\hat{\rho}(\mathbf{r}) = \hat{\rho}_M(\mathbf{r}) + \hat{\rho}_i(\mathbf{r}) + \rho_{ext}(\mathbf{r}) \tag{5.153}$$

represents the local charge density of the system, where

$$\hat{\rho}_M(\mathbf{r}) = q \sum_{\alpha=1}^{m} \sum_{j=1}^{N_M} \delta\left(\mathbf{r} - \mathbf{R}_j - \boldsymbol{\xi}_j^{(\alpha)}\right) \tag{5.154}$$

is the charge density of the micelles;

$$\hat{\rho}_i(\mathbf{r}) = q_+ \sum_{k=1}^{N_+} \delta\left(\mathbf{r} - \mathbf{r}_k^{(+)}\right) + q_- \sum_{k=1}^{N_-} \delta\left(\mathbf{r} - \mathbf{r}_k^{(-)}\right) - q \sum_{n=1}^{mN_M} \delta\left(\mathbf{r} - \mathbf{r}_n^{(c)}\right) \tag{5.155}$$

is the local charge density of the ions. As usual, $\rho_{ext}(\mathbf{r})$, denotes the density of the external charges.

Using the Hubbard–Stratonovich transformation, we arrive at the following representation of the configuration integral in the form of the functional integral

$$Q = \int \frac{\mathcal{D}\varphi}{C_0} \exp\left[-\frac{\beta}{2}(\varphi G_0^{-1}\varphi) + i\beta(\rho_{ext}\varphi)\right] Q_M^{N_M}[\varphi] Q_c^{mN_M}[\varphi] Q_+^{N_+}[\varphi] Q_-^{N_-}[\varphi] \tag{5.156}$$

with one-particle configuration integrals

$$Q_M[\varphi] = \int \frac{d\mathbf{R}}{V} \int d\sigma(\boldsymbol{\xi}_1, \ldots, \boldsymbol{\xi}_m) \exp\left(i\beta q \sum_{\alpha=1}^{m} \varphi(\mathbf{R} + \boldsymbol{\xi}_\alpha)\right), \tag{5.157}$$

$$Q_\pm[\varphi] = \int \frac{d\mathbf{r}}{V} \exp(i\beta q_\pm \varphi(\mathbf{r})), \tag{5.158}$$

and

$$Q_c[\varphi] = \int \frac{d\mathbf{r}}{V} \exp(-iq\beta\varphi(\mathbf{r})) \tag{5.159}$$

with the following normalized integration measure over the displacements of the charges:

$$\int d\sigma(\boldsymbol{\xi}_1, \ldots, \boldsymbol{\xi}_m)(\ldots) = \int d\boldsymbol{\xi}_1 \ldots \int d\boldsymbol{\xi}_m \prod_{\alpha=1}^{m} g(\boldsymbol{\xi}_\alpha)(\ldots), \tag{5.160}$$

i.e. $\int d\sigma = 1$. In the thermodynamic limit $V \to \infty,\ N_{m,\pm} \to \infty,\ N_{m,\pm}/V \to n_{m,\pm}$ we obtain:

$$Q_M^{N_M}[\varphi] \simeq \exp\left[n_M \int d\mathbf{R} \int d\sigma(\boldsymbol{\xi}_1, \dots, \boldsymbol{\xi}_m) \left(\exp\left(i\beta q \sum_{\alpha=1}^{m} \varphi(\mathbf{R} + \boldsymbol{\xi}_\alpha)\right) - 1\right)\right], \tag{5.161}$$

$$Q_\pm^{N_\pm}[\varphi] \simeq \exp\left[n_\pm \int d\mathbf{r}\, (\exp(i\beta q_\pm \varphi(\mathbf{r})) - 1)\right], \tag{5.162}$$

$$Q_c^{mN_M}[\varphi] \simeq \exp\left[mn_M \int d\mathbf{r}\, (\exp(-i\beta q \varphi(\mathbf{r})) - 1)\right]. \tag{5.163}$$

Thus, we arrive at the following functional representation of the configuration integral:

$$Q = \int \frac{\mathcal{D}\varphi}{C_0} \exp\left[-S[\varphi]\right], \tag{5.164}$$

where we introduce the following functional

$$S[\varphi] = \frac{\beta}{2}(\varphi G_0^{-1}\varphi) - i\beta(\rho_{ext}\varphi) - W[\varphi], \tag{5.165}$$

where

$$W[\varphi] = n_M \int d\mathbf{R} \int d\sigma(\boldsymbol{\xi}_1, \dots, \boldsymbol{\xi}_m) \left(e^{i\beta q \sum\limits_{\alpha=1}^{m} \varphi(\mathbf{R}+\boldsymbol{\xi}_\alpha)} - 1\right)$$
$$+ n_+ \int d\mathbf{r}\, (e^{i\beta q_+ \varphi(\mathbf{r})} - 1) + n_- \int d\mathbf{r}\, (e^{i\beta q_- \varphi(\mathbf{r})} - 1) + mn_M \int d\mathbf{r}\, (e^{-i\beta q \varphi(\mathbf{r})} - 1). \tag{5.166}$$

Next, let's derive an expression for the electrostatic free energy of the solution within the Gaussian approximation. Expanding the functional $S[\varphi]$ from (5.165) into a power series around the mean field $\varphi(\mathbf{r}) = i\psi(\mathbf{r})$ satisfying the Euler–Lagrange equation

$$\left.\frac{\delta S[\varphi]}{\delta \varphi(\mathbf{r})}\right|_{\varphi = i\psi} = 0 \tag{5.167}$$

and neglecting the terms above the second order, we obtain

$$Q = \exp\left[-S[i\psi]\right] \int \frac{\mathcal{D}\varphi}{C} \exp\left[-\frac{\beta}{2}\left(\delta\varphi \mathcal{G}^{-1}\delta\varphi\right) + O[\varphi^3]\right], \tag{5.168}$$

where $\delta\varphi = \varphi - i\psi$, and we introduce the following renormalized Green's function

$$\mathcal{G}^{-1}(\mathbf{r}, \mathbf{r}'|\psi) = k_B T \frac{\delta^2 S[i\psi]}{\delta\varphi(\mathbf{r})\delta\varphi(\mathbf{r}')} = G_0^{-1}(\mathbf{r}, \mathbf{r}') + S_i(\mathbf{r}, \mathbf{r}') + S_M(\mathbf{r}, \mathbf{r}'), \tag{5.169}$$

where

$$S_i(\mathbf{r},\mathbf{r}') = \beta\left(q_+^2 n_+ e^{-\beta q_+\psi(\mathbf{r})} + q_-^2 n_- e^{-\beta q_-\psi(\mathbf{r})} + q^2 m n_M e^{\beta q\psi(\mathbf{r})}\right)\delta(\mathbf{r}-\mathbf{r}'), \tag{5.170}$$

$$S_M(\mathbf{r},\mathbf{r}') = \beta n_M q^2 \sum_{\delta\gamma}\left\langle e^{-\beta q\sum_{\alpha=1}^{m}\psi(\mathbf{r}+\boldsymbol{\xi}_\alpha-\boldsymbol{\xi}_\gamma)}\delta(\mathbf{r}-\mathbf{r}'+\boldsymbol{\xi}_\delta-\boldsymbol{\xi}_\gamma)\right\rangle_\xi$$

is the renormalized inverse Green's function with the mean-field electrostatic potential $\psi(\mathbf{r})$ satisfying the following self-consistent field equation obtained from the Euler–Lagrange equation (5.167)

$$\Delta\psi(\mathbf{r}) = -\frac{4\pi}{\varepsilon}\left(\rho_{ext}(\mathbf{r}) + \bar{\rho}_i(\mathbf{r}) + \bar{\rho}_M(\mathbf{r})\right), \tag{5.171}$$

where

$$\bar{\rho}_i(\mathbf{r}) = q_+ n_+ e^{-\beta q_+\psi(\mathbf{r})} + q_- n_- e^{-\beta q_-\psi(\mathbf{r})} - q m n_M e^{\beta e\psi(\mathbf{r})} \tag{5.172}$$

determines the average charge density of the ions and

$$\bar{\rho}_M(\mathbf{r}) = n_M q \sum_{\gamma=1}^{m}\left\langle e^{-\beta q\sum_{\alpha=1}^{m}\psi(\mathbf{r}+\boldsymbol{\xi}_\gamma-\boldsymbol{\xi}_\alpha)}\right\rangle_\xi \tag{5.173}$$

is the average charge density of the micelles. Note that in the expression (5.173) we introduce the following averaging over the configurations of the micelle charges

$$\langle(\ldots)\rangle_\xi = \int d\sigma(\boldsymbol{\xi}_1,\ldots,\boldsymbol{\xi}_m)(\ldots). \tag{5.174}$$

Thus, formally evaluating the Gaussian functional integral, we arrive at the following general expression for the configuration integral within the Gaussian approximation:

$$Q \approx \exp\left\{-S[i\psi] + \frac{1}{2}tr\left(\ln\mathcal{G} - \ln G_0\right)\right\}. \tag{5.175}$$

As previously, when there are no external charges in the solution (i.e., $\rho_{ext}(\mathbf{r}) = 0$), the electrostatic potential of the self-consistent field $\psi(\mathbf{r}) = 0$, so the mean field contribution to the electrostatic free energy is $F_{el}^{(MF)} = k_B T S[0] = 0$. Thus, in this case, the electrostatic free energy is completely determined by thermal fluctuations of the electrostatic potential near its zero value and the renormalized inverse Green's function simplifies to:

$$\mathcal{G}^{-1}(\mathbf{r},\mathbf{r}'|0) = G^{-1}(\mathbf{r},\mathbf{r}') = G_0^{-1}(\mathbf{r},\mathbf{r}') + \frac{2Iq^2}{k_B T}\delta(\mathbf{r}-\mathbf{r}') + \frac{q^2 m n_M}{k_B T}\delta(\mathbf{r}-\mathbf{r}') + \frac{n_M q^2}{k_B T}\sum_{\delta,\gamma}\left\langle\delta(\mathbf{r}-\mathbf{r}'+\boldsymbol{\xi}_\delta-\boldsymbol{\xi}_\gamma)\right\rangle_\xi,$$

while the electrostatic free energy of the ion-dipole mixture within the Gaussian approximation can be calculated as follows:

$$F_{el} \approx \frac{k_B T}{2} tr\,(\ln G_0 - \ln G) = \frac{V k_B T}{2} \int \frac{d\mathbf{k}}{(2\pi)^3} \ln \frac{G_0(\mathbf{k})}{G(\mathbf{k})}, \tag{5.176}$$

where $G_0(\mathbf{k}) = 4\pi/(\varepsilon k^2)$ and $G(\mathbf{k}) = 4\pi/(\varepsilon(k^2 + \varkappa^2(\mathbf{k})))$ represent the Fourier transforms of Green's functions, which become translationally invariant in the thermodynamic limit.

Hence, the electrostatic free energy of the solution within the Gaussian approximation has the form

$$F_{el} = \frac{V k_B T}{2} \int \frac{d\mathbf{k}}{(2\pi)^3} \left(\ln \left(1 + \frac{\varkappa^2(\mathbf{k})}{k^2} \right) - \frac{\varkappa^2(\mathbf{k})}{k^2} \right) \tag{5.177}$$

with the screening function

$$\varkappa^2(\mathbf{k}) = \kappa_s^2 + \frac{8\pi q^2 m n_M}{\varepsilon k_B T} + \frac{4\pi q^2 n_M m(m-1)}{\varepsilon k_B T} |g(\mathbf{k})|^2 \tag{5.178}$$

where $\kappa_s^2 = 8\pi q^2 I/(\varepsilon k_B T)$ is the square of the inverse Debye length, attributed to salt, and we introduce the following characteristic function:

$$g(\mathbf{k}) = \int d\mathbf{r} e^{-i\mathbf{k}\mathbf{r}} g(\mathbf{r}). \tag{5.179}$$

It should be noted that we subtracted the following self-energy of the charges from the final expression for the electrostatic free energy:

$$U_{self} = \frac{V k_B T}{2} \int \frac{d\mathbf{k}}{(2\pi)^3} \frac{\varkappa^2(\mathbf{k})}{k^2}. \tag{5.180}$$

Let's now evaluate the electrostatic contribution to the excess chemical potential of the micelles at infinite dilution, i.e., when $n_M \to 0$. We have:

$$\mu_{el} = \lim_{n_M \to 0} \frac{\partial f_{el}}{\partial n_M}, \tag{5.181}$$

where $f_{el} = F_{el}/V$. The result is as follows [114]

$$\mu_{el} = -\frac{q^2 m \kappa_s^2}{\pi \varepsilon} \int_0^{\infty} dk \frac{(2 + (m-1) g^2(k))}{k^2 + \kappa_s^2}. \tag{5.182}$$

Utilizing the following model characteristic function

$$g(k) = \frac{1}{1 + k^2 R^2}, \tag{5.183}$$

we arrive at the expression

$$\mu_{el} = -\frac{q^2 m \kappa_s}{\varepsilon} \left(1 + \frac{(m-1)(2+\sqrt{y_s})}{(1+\sqrt{y_s})^2} \right). \tag{5.184}$$

Under the condition $m \gg 1$ we obtain the following expression

$$\mu_{el} \simeq -\frac{Q^2 \kappa_s}{4\varepsilon} \frac{2+\sqrt{y_s}}{(1+\sqrt{y_s})^2}, \tag{5.185}$$

where $Q = qm$ represents the total charge of the micelle and $y_s = \kappa_s^2 R^2$. It's noteworthy that when $y_s \ll 1$, indicating that the Debye length attributed to the salt is much larger than the micelle size, we recover the Debye–Hückel approximation

$$\mu_{el} \simeq -\frac{Q^2 \kappa_s}{2\varepsilon}. \tag{5.186}$$

In the opposite regime $y_s \gg 1$ we obtain:

$$\mu_{el} \simeq -\frac{Q^2}{4\varepsilon R}. \tag{5.187}$$

5.4 Nonlocal Statistical Theory of Polar Molecules Solutions with Account of Steric Interactions

In the preceding discussion, we formulated a nonlocal statistical theory of polar molecules in electrolyte solutions, albeit without considering the impact of excluded volumes, focusing solely on low solution concentrations. However, to comprehensively describe the phase behavior across a wide range of concentrations, it becomes necessary to incorporate not only electrostatic interactions but also short-range repulsive ones. In this section, we formulate a general field-theoretic framework based on thermodynamic perturbation theory and the Gaussian approximation to describe the phase behavior of solutions featuring the electrically neutral soft molecules with arbitrary internal electrical structure. As a practical demonstration of this methodology, we apply it to describe the phase behavior of solutions of polar molecules with Gaussian repulsion potential (soft-core).

General statement. In this section, we formulate a field-theoretic methodology based on the thermodynamic perturbation theory and the Gaussian approximation to evaluate the free energy of solution of electrically neutral soft molecules with a complex internal electrical structure. Consider a solution of N electrically neutral molecules, which we model as a set of l charged centers bearing charges q_α ($\alpha = 1, \ldots, l$), ensuring that $\sum_{\alpha=1}^{l} q_\alpha = 0$. We postulate that within each molecule, the positions of these centers, denoted by $\mathbf{r}^{(\alpha)}$, are stochastic variables characterized by the distribution function $g\left(\mathbf{r}^{(1)}, \ldots, \mathbf{r}^{(l)}\right)$, which in general case is determined by the quantum nature of molecules and satisfies the following normalization condition

$$\int \cdots \int d\mathbf{r}^{(\alpha_1)} \ldots d\mathbf{r}^{(\alpha_{l-1})} g\left(\mathbf{r}^{(1)}, \ldots, \mathbf{r}^{(l)}\right) = 1, \tag{5.188}$$

where $\alpha_i \neq \alpha_k$ for $i \neq k$. Similar to previous sections, we assume an implicit solvent represented by a continuous dielectric medium with a constant dielectric permittivity ε. Let's also incorporate soft potentials between molecular centers, described by the $l \times l$ matrix $V_{\alpha\gamma}(\mathbf{r})$, in addition to the Coulomb interaction.

Thus, taking into account these assumptions, we can write the configuration integral of the solution in the form

$$Q = \int d\Gamma \exp\left[-\beta U_0 - \beta U_{el}\right], \tag{5.189}$$

where the integration measure over the positions of molecule centers has the form:

$$\int d\Gamma(\ldots) = \int \ldots \int \prod_{i=1}^{N} \frac{d\mathbf{r}_i^{(1)} \ldots d\mathbf{r}_i^{(l)}}{V} g(\mathbf{r}_i^{(1)}, \ldots, \mathbf{r}_i^{(l)})(\ldots), \tag{5.190}$$

and the Hamiltonian of non-electrostatic interactions is expressed as:

$$U_0 = \frac{1}{2} \int d\mathbf{r} \int d\mathbf{r}' \sum_{\alpha\gamma} \hat{n}_\alpha(\mathbf{r}) V_{\alpha\gamma}(\mathbf{r} - \mathbf{r}') \hat{n}_\gamma(\mathbf{r}') = \frac{1}{2} \left(\hat{n} \mathbf{V} \hat{n}\right), \tag{5.191}$$

where $\hat{n}_\alpha(\mathbf{r}) = \sum_{i=1}^{N} \delta\left(\mathbf{r} - \mathbf{r}_i^{(\alpha)}\right)$ represent the local concentrations of the centers of the αth kind and $V_{\alpha\gamma}(\mathbf{r} - \mathbf{r}')$ denotes the matrix of soft potentials of non-electrostatic interactions between centers. The Hamiltonian of electrostatic interactions has the form

$$U_{el} = \frac{1}{2} \int d\mathbf{r} \int d\mathbf{r}' \, \hat{\rho}(\mathbf{r}) G_0(\mathbf{r} - \mathbf{r}') \hat{\rho}(\mathbf{r}') = \frac{1}{2} \left(\hat{\rho} G_0 \hat{\rho}\right), \tag{5.192}$$

where $\hat{\rho}(\mathbf{r}) = \sum_{\alpha=1}^{l} q_\alpha \hat{n}_\alpha(\mathbf{r})$ is the local charge density of the centres, satisfying the condition of electrical neutrality, i.e. $\int d\mathbf{r} \, \hat{\rho}(\mathbf{r}) = 0$.

Further, using the Hubbard–Stratonovich transformation

$$\exp\left[-\frac{\beta}{2}(\hat{\rho}G_0\hat{\rho})\right] = \int \frac{\mathcal{D}\varphi}{C_0} \exp\left[-\frac{\beta}{2}(\varphi G_0^{-1}\varphi) + i\beta(\hat{\rho}\varphi)\right], \tag{5.193}$$

we arrive at the following configuration integral

$$Q = \int \frac{\mathcal{D}\varphi}{C_0} \exp\left[-\frac{\beta}{2}(\varphi G_0^{-1}\varphi)\right] \int d\Gamma \exp\left[-\beta U_0\right] \exp\left[i\beta(\hat{\rho}\varphi)\right]. \tag{5.194}$$

Now, considering the solution of molecules without Coulomb interactions as a reference system, let's write the configuration integral as:

$$Q = Q_0 \int \frac{\mathcal{D}\varphi}{C_0} \exp\left[-\frac{\beta}{2}(\varphi G_0^{-1}\varphi)\right] \langle \exp\left[i\beta(\hat{\rho}\varphi)\right] \rangle_0, \tag{5.195}$$

where

$$Q_0 = \int d\Gamma \exp\left[-\beta U_0\right] \tag{5.196}$$

represents the configuration integral of the reference system, which we will evaluate further, and the symbol

$$\langle (\ldots) \rangle_0 = \frac{1}{Q_0} \int d\Gamma \exp\left[-\beta U_0\right] (\ldots) \tag{5.197}$$

denotes the averaging over the statistics of the reference system. Using the Gaussian approximation, i.e. truncating the cumulant expansion [2] at the second term, we derive:

$$Q \approx Q_0 \int \frac{\mathcal{D}\varphi}{C_0} \exp\left[-\frac{\beta}{2}(\varphi G_0^{-1}\varphi) - \frac{\beta^2}{2}(\varphi S_c \varphi)\right], \tag{5.198}$$

where

$$S_c(\mathbf{r} - \mathbf{r}') = \langle \hat{\rho}(\mathbf{r})\hat{\rho}(\mathbf{r}') \rangle_0 = \sum_{\alpha\gamma} q_\alpha q_\gamma S_{\alpha\gamma}(\mathbf{r} - \mathbf{r}') \tag{5.199}$$

is a correlator of the local charge density with the structural matrix $S_{\alpha\gamma}(\mathbf{r} - \mathbf{r}')$ of the reference system, whose elements can be calculated by the formula

$$S_{\alpha\gamma}(\mathbf{r} - \mathbf{r}') = \langle \delta\hat{n}_\alpha(\mathbf{r})\delta\hat{n}_\gamma(\mathbf{r}') \rangle_0, \tag{5.200}$$

where we introduce the fluctuations of the local concentrations of the centres $\delta\hat{n}_\alpha(\mathbf{r}) = \hat{n}_\alpha(\mathbf{r}) - \langle \hat{n}_\alpha(\mathbf{r}) \rangle_0$.

Hence, in the framework of GA the configuration integral is defined by the relation

$$Q \approx Q_0 \int \frac{\mathcal{D}\varphi}{C_0} \exp\left[-\frac{\beta}{2}(\varphi G^{-1}\varphi)\right] = Q_0 \exp\left[\frac{1}{2} tr\,(\ln G - \ln G_0)\right], \quad (5.201)$$

where the symbol $tr(A)$ denotes the trace of the integral operator and

$$G^{-1}(\mathbf{r}-\mathbf{r}') = G_0^{-1}(\mathbf{r}-\mathbf{r}') + \beta S_c(\mathbf{r}-\mathbf{r}') \quad (5.202)$$

represents the renormalized inverse Green's function within the GA. It's worth noting that the trace of the operator A with the translationally invariant kernel $A(\mathbf{r}, \mathbf{r}') = A(\mathbf{r}-\mathbf{r}')$ can be determined by the relation

$$tr(A) = V \int \frac{d\mathbf{k}}{(2\pi)^3} \tilde{A}(\mathbf{k}), \quad (5.203)$$

where V denotes the total volume of the system and

$$\tilde{A}(\mathbf{k}) = \int d\mathbf{r} A(\mathbf{r}) e^{-i\mathbf{k}\mathbf{r}} \quad (5.204)$$

is the Fourier transform of the function $A(\mathbf{r})$.

Therefore, taking into account the expression (5.202), the definition of the operator's trace and subtracting the electrostatic self-energy of the molecules from the final expression, we obtain

$$F_{ex} = F_0 + \frac{V k_B T}{2} \int \frac{d\mathbf{k}}{(2\pi)^3} \left(\ln\left(1 + \frac{\varkappa^2(\mathbf{k})}{k^2}\right) - \frac{\varkappa^2(\mathbf{k})}{k^2} \right), \quad (5.205)$$

where

$$\varkappa^2(\mathbf{k}) = \frac{4\pi}{k_B T \varepsilon} \sum_{\alpha\gamma} q_\alpha q_\gamma S_{\alpha\gamma}(\mathbf{k}) \quad (5.206)$$

represents the screening function and $S_{\alpha\gamma}(\mathbf{k})$ are the structural factors of the reference system. It's worth noting that the expression (5.206) was first obtained in [115].

Evaluation of the free energy and structural factors of the reference system within the Gaussian approximation. To evaluate the excess free energy (5.205) of the solution, we need to calculate the excess free energy $F_0 = -k_B T \ln Q_0$ and the structural factors $S_{\alpha\gamma}(\mathbf{k})$ of the reference system. For this purpose, we define the generating functional (see also Chap. 2):

$$Q_0[\Psi] = \int d\Gamma \exp\left[-\frac{\beta}{2}\left(\hat{n}\mathbf{V}\hat{n}\right) - \beta\left(\delta\hat{n}\Psi\right)\right], \quad (5.207)$$

where $\left(\delta\hat{n}\Psi\right) = \int d\mathbf{r} \sum_\alpha \delta\hat{n}_\alpha(\mathbf{r})\Psi_\alpha(\mathbf{r})$, and we introduce the following auxiliary fields $\Psi_\alpha(\mathbf{r})$, satisfying the relation $\int d\mathbf{r}\, \Psi_\alpha(\mathbf{r}) = 0$. Thus, we can determine the structural factors by finding the second functional derivative as follows

$$S_{\alpha\gamma}(\mathbf{r}-\mathbf{r}') = \frac{1}{\beta^2}\frac{\delta^2}{\delta\Psi_\alpha(\mathbf{r})\delta\Psi_\gamma(\mathbf{r}')}\frac{Q_0[\Psi]}{Q_0[0]}\bigg|_{\Psi=0}. \quad (5.208)$$

Further, using the Hubbard–Stratonovich transformation

$$e^{-\frac{\beta}{2}(\hat{n}\mathbf{V}\hat{n})} = \int\frac{\mathcal{D}\Phi}{\mathcal{N}_V}e^{-\frac{\beta}{2}(\Phi\mathbf{V}^{-1}\Phi)+i\beta(\hat{n}\Phi)}, \quad (5.209)$$

where

$$(\Phi\mathbf{V}^{-1}\Phi) = \int d\mathbf{r}\int d\mathbf{r}'\sum_{\alpha\gamma}\Phi_\alpha(\mathbf{r})V^{-1}_{\alpha\gamma}(\mathbf{r}-\mathbf{r}')\Phi_\gamma(\mathbf{r}'), \quad (5.210)$$

$$(\hat{n}\Phi) = \int d\mathbf{r}\sum_\alpha\hat{n}_\alpha(\mathbf{r})\Phi_\alpha(\mathbf{r}) \quad (5.211)$$

and $\mathcal{N}_V = \int\mathcal{D}\Phi\, e^{-\frac{\beta}{2}(\Phi\mathbf{V}^{-1}\Phi)}$ represents the normalization constant of the Gaussian measure, we obtain:

$$Q_0[\Psi] = \int\frac{\mathcal{D}\Phi}{\mathcal{N}_V}e^{-\frac{\beta}{2}(\Phi\mathbf{V}^{-1}\Phi)}\int d\Gamma e^{i\beta(\hat{n}\Phi)-\beta(\delta\hat{n}\Psi)}, \quad (5.212)$$

where we introduce the inverse matrix operator $\mathbf{V}^{-1}$, defined by the integral relation

$$\int d\mathbf{r}''\sum_\lambda V_{\alpha\lambda}(\mathbf{r}-\mathbf{r}'')V^{-1}_{\lambda\gamma}(\mathbf{r}''-\mathbf{r}') = \delta_{\alpha\gamma}\delta(\mathbf{r}-\mathbf{r}'). \quad (5.213)$$

Thus, rewriting the field variable as $\Phi(\mathbf{r}) = \Phi_0 + \Lambda(\mathbf{r})$, where $\Phi_0 = \int d\mathbf{r}\,\Phi(\mathbf{r})/V$ and performing Gaussian integration over Φ_0 in (5.212), we obtain

$$Q_0[\Psi] = Q_0^{(MF)}e^{\beta(n\Psi)}\int\frac{\mathcal{D}\Lambda}{\mathcal{N}_V}e^{-\frac{\beta}{2}(\Lambda\mathbf{V}^{-1}\Lambda)}\int d\Gamma e^{i\beta(\hat{n}[\Lambda+i\Psi])}, \quad (5.214)$$

where

$$Q_0^{(MF)} = \exp\left[-\frac{\beta}{2}(n\mathbf{V}n)\right] \quad (5.215)$$

represents the mean-field approximation for the configuration integral of the reference system with average concentrations of charged centers $n_\alpha(\mathbf{r}) = \langle\hat{n}_\alpha(\mathbf{r})\rangle_0$; $\mathcal{N}_V = \int\mathcal{D}\Lambda\, e^{-\frac{\beta}{2}(\Lambda\mathbf{V}^{-1}\Lambda)}$ is the normalization constant of the Gaussian integration measure over the vector fields $\Lambda_\alpha(\mathbf{r})$. Then, shifting the field variable $\Lambda \to \Lambda - i\Psi$, we arrive at the expression

$$Q_0[\Psi] = Q_0^{(MF)}e^{\beta(n\Psi)+\frac{\beta}{2}(\Psi\mathbf{V}^{-1}\Psi)}\int\frac{\mathcal{D}\Lambda}{\mathcal{N}_V}e^{-\frac{\beta}{2}(\Lambda\mathbf{V}^{-1}\Lambda)+i\beta(\Lambda\mathbf{V}^{-1}\Psi)}Q^N[\Lambda], \quad (5.216)$$

where

$$Q[\Lambda] = \int \cdots \int \frac{d\mathbf{r}^{(1)} \ldots d\mathbf{r}^{(l)}}{V} g(\mathbf{r}^{(1)}, \ldots, \mathbf{r}^{(l)}) \exp\left[i\beta \sum_{\alpha=1}^{l} \Lambda_\alpha(\mathbf{r}^{(\alpha)}) \right] \tag{5.217}$$

is the one-particle configuration integral. In the thermodynamic limit $N \to \infty$, $V \to \infty$, $N/V \to n$ we obtain

$$Q_0[\Psi] = Q_0^{(MF)} e^{\beta(n\Psi) + \frac{\beta}{2}(\Psi \mathbf{V}^{-1} \Psi)} \int \frac{\mathcal{D}\Lambda}{\mathcal{N}_V} e^{-\frac{\beta}{2}(\Lambda \mathbf{V}^{-1} \Lambda) + i\beta(\Lambda \mathbf{V}^{-1} \Psi) + W[\Lambda]}, \tag{5.218}$$

where we introduce the auxiliary functional

$$W[\Lambda] = n \int \cdots \int d\mathbf{r}^{(1)} \ldots d\mathbf{r}^{(l)} g(\mathbf{r}^{(1)}, \ldots, \mathbf{r}^{(l)}) \left(e^{i\beta \sum\limits_{\alpha=1}^{l} \Lambda_\alpha(\mathbf{r}^{(\alpha)})} - 1 \right). \tag{5.219}$$

Given that $\int d\mathbf{r}\, \Lambda(\mathbf{r}) = 0$, expansion of the functional $W[\Lambda]$ into a power series up to the quadratic term leads to

$$\begin{aligned} Q_0[\Psi] &\approx Q_0^{(MF)} e^{\beta(n\Psi) + \frac{\beta}{2}(\Psi \mathbf{V}^{-1} \Psi)} \int \frac{\mathcal{D}\Lambda}{\mathcal{N}_V} e^{-\frac{\beta}{2}(\Lambda \mathbf{D}^{-1} \Lambda) + i\beta(\Lambda \mathbf{V}^{-1} \Psi)} \\ &= Q_0^{(GA)} e^{\beta(n\Psi) + \frac{\beta}{2}(\Psi \mathbf{V}^{-1} \Psi) - \frac{\beta}{2}(\Psi \mathbf{V}^{-1} \mathbf{D} \mathbf{V}^{-1} \Psi)}, \end{aligned} \tag{5.220}$$

where

$$Q_0^{(GA)} = Q_0^{(MF)} \exp\left[\frac{1}{2} Tr\, (\ln \mathbf{D} - \ln \mathbf{V}) \right] \tag{5.221}$$

is the configuration integral of the reference system within the Gaussian approximation, and the operator $\mathbf{D}$ is defined by the following relation:

$$D_{\alpha\gamma}^{-1}(\mathbf{r} - \mathbf{r}') = V_{\alpha\gamma}^{-1}(\mathbf{r} - \mathbf{r}') + \beta W_{\alpha\gamma}(\mathbf{r} - \mathbf{r}'), \tag{5.222}$$

where operator $\mathbf{W}$ has the form

$$W_{\alpha\gamma}(\mathbf{r} - \mathbf{r}') = n \left(\delta_{\alpha\gamma} \delta(\mathbf{r} - \mathbf{r}') + (1 - \delta_{\alpha\gamma}) g_{\alpha\gamma}(\mathbf{r} - \mathbf{r}') \right) \tag{5.223}$$

with the structure functions

$$\begin{aligned} g_{\alpha\gamma}(\mathbf{r} - \mathbf{r}') &= \int \cdots \int d\mathbf{r}^{(1)} \ldots d\mathbf{r}^{(\alpha-1)} d\mathbf{r}^{(\alpha+1)} \ldots d\mathbf{r}^{(\gamma-1)} d\mathbf{r}^{(\gamma+1)} \ldots d\mathbf{r}^{(l)} \\ &\times g(\mathbf{r}^{(1)}, \ldots, \mathbf{r}^{(\alpha-1)}, \mathbf{r}, \mathbf{r}^{(\alpha+1)}, \ldots, \mathbf{r}^{(\gamma-1)}, \mathbf{r}', \mathbf{r}^{(\gamma+1)}, \ldots, \mathbf{r}^{(l)}), \end{aligned} \tag{5.224}$$

which define the distribution functions over the distance between the centers of the αth and γth sorts. The symbol $Tr(\dots)$ denotes the trace of the integral matrix operator according to the definition

$$Tr(\mathbf{A}) = \int d\mathbf{r} \sum_{\alpha} A_{\alpha\alpha}(\mathbf{r}, \mathbf{r}). \tag{5.225}$$

Using the relation (5.208), we arrive at the following expression for the structural operator

$$\mathbf{S} = \frac{1}{\beta}\left(\mathbf{V}^{-1} - \mathbf{V}^{-1}\mathbf{D}\mathbf{V}^{-1}\right), \tag{5.226}$$

which after simple transformations can be rewritten as

$$S^{-1}_{\alpha\gamma}(\mathbf{r} - \mathbf{r}') = W^{-1}_{\alpha\gamma}(\mathbf{r} - \mathbf{r}') + \beta V_{\alpha\gamma}(\mathbf{r} - \mathbf{r}') \tag{5.227}$$

or within the Fourier representation

$$S^{-1}_{\alpha\gamma}(\mathbf{k}) = W^{-1}_{\alpha\gamma}(\mathbf{k}) + \beta \tilde{V}_{\alpha\gamma}(\mathbf{k}). \tag{5.228}$$

Hence, using the definition of integral matrix operator trace (5.225) and considering, that

$$\frac{\det \tilde{D}(\mathbf{k})}{\det \tilde{V}(\mathbf{k})} = \frac{\det S(\mathbf{k})}{\det W(\mathbf{k})}, \tag{5.229}$$

we arrive at the final expression for the excess free energy of the reference system within the GA:

$$F_0 = -k_B T \ln Q_0^{(GA)} = \frac{V}{2} n^2 \sum_{\alpha,\gamma} \tilde{V}_{\alpha\gamma}(0) - \frac{V k_B T}{2} \int \frac{d\mathbf{k}}{(2\pi)^3} \ln \frac{\det S(\mathbf{k})}{\det W(\mathbf{k})}. \tag{5.230}$$

The first term on the right-hand side of Eq. (5.230) defines the mean-field approximation for the excess free energy of the reference system, while the second one defines the fluctuation correction to the mean-field approximation within the GA framework. Physically, the fluctuation correction is determined by the Gaussian thermal fluctuations of molecular fields $\Phi_\alpha(\mathbf{r})$ near their mean values.

Evaluation of the total structural factors by the generating functional method[2]. Now, let's utilize the generating functional method discussed earlier to derive the structural factors, considering the effect of electrostatic interactions. Let's start by expressing the generating functional for the auxiliary fields $\Phi_\alpha(\mathbf{r})$

$$Q[\Phi] = \int d\Gamma\, e^{-\frac{\beta}{2}(\hat{\rho}G_0\hat{\rho}) - \frac{\beta}{2}(\hat{n}\mathbf{V}\hat{n}) + \beta(\delta\hat{n}\Phi)}. \tag{5.231}$$

---

[2] The derivation presented below was developed together with P. E. Brandyshev.

We will utilize this functional to determine the correlation functions using the following functional relation:

$$\hat{S}_{\alpha\beta}(\mathbf{r}-\mathbf{r}') = \frac{1}{\beta^2 Q[0]} \frac{\delta^2}{\delta\Phi_\alpha(\mathbf{r})\delta\Phi_\beta(\mathbf{r}')} Q[\Phi]\bigg|_{\Phi=0}. \tag{5.232}$$

Here we do not specify the form of integration measure $d\Gamma$ over the configurations of charged particles.

Using the HS transformation

$$e^{-\frac{\beta}{2}(\hat{\rho}G_0\hat{\rho})} = \int \frac{\mathcal{D}\varphi}{C_0} e^{-\frac{\beta}{2}(\varphi G_0^{-1}\varphi)+i\beta(\hat{\rho}\varphi)}, \tag{5.233}$$

we can rewrite the expression (5.231) in the following form

$$Q[\Phi] = \int \frac{\mathcal{D}\varphi}{C_0} e^{-\frac{\beta}{2}(\varphi G_0^{-1}\varphi)} \int d\Gamma e^{-\frac{\beta}{2}(\hat{n}\mathbf{V}\hat{n})+i\beta(\hat{\rho}\varphi)+\beta(\delta\hat{n}\Phi)}. \tag{5.234}$$

Introducing the configuration integral of the reference system

$$Q_0 = \int d\Gamma e^{-\frac{\beta}{2}(\hat{n}\mathbf{V}\hat{n})}, \tag{5.235}$$

we can transform the relation (5.234) to the form

$$Q[\Phi] = Q_0 \int \frac{\mathcal{D}\varphi}{C_0} e^{-\frac{\beta}{2}(\varphi G_0^{-1}\varphi)} \left\langle e^{i\beta(\hat{\rho}\varphi)+\beta(\delta\hat{n}\Phi)} \right\rangle_0. \tag{5.236}$$

where, as before, we introduce averaging over the statistics of the reference system. Truncating the cumulant series at the second order, we obtain

$$\left\langle \exp(i\beta(\hat{\rho}\varphi)+\beta(\delta\hat{n}\Phi)) \right\rangle_0 = \exp\left(\left\langle i\beta(\hat{\rho}\varphi)+\beta(\delta\hat{n}\Phi)+\frac{\beta^2}{2}(i(\hat{\rho}\varphi)+(\delta\hat{n}\Phi))^2 \right\rangle_c\right). \tag{5.237}$$

Taking into account that the average charge density is zero, we obtain

$$\left\langle \exp(i\beta(\hat{\rho}\varphi)+\beta(\delta\hat{n}\Phi)) \right\rangle_0 = \exp\left(\frac{\beta^2}{2}\left(-(\varphi S_c\varphi)+\frac{2i}{\beta}(\varphi J)+(\Phi\mathbf{S}\Phi)\right)\right), \tag{5.238}$$

where we introduce the following shorthand notations

$$(\Phi\mathbf{S}\Phi) = \int d\mathbf{r} \int d\mathbf{r}' \sum_{\mu\nu} \Phi_\mu(\mathbf{r}) S_{\mu\nu}(\mathbf{r}-\mathbf{r}')\Phi_\nu(\mathbf{r}'), \tag{5.239}$$

$$(\varphi J) = \int d\mathbf{r}\, \varphi(\mathbf{r}) J(\mathbf{r}) \tag{5.240}$$

and

$$J(\mathbf{r}) = \beta \int d\mathbf{r}' \sum_{\alpha\beta} q_\alpha S_{\alpha\beta}(\mathbf{r}-\mathbf{r}')\Phi_\beta(\mathbf{r}'), \tag{5.241}$$

with familiar correlation functions of the reference system

$$S_{\alpha\beta}(\mathbf{r}-\mathbf{r}') = \left\langle \delta\hat{n}_\alpha(\mathbf{r})\delta\hat{n}_\beta(\mathbf{r}')\right\rangle_0 \tag{5.242}$$

and already known the correlation function of the local charge density fluctuations, obtained by averaging over the configurations of the reference system

$$S_c(\mathbf{r}-\mathbf{r}') = \sum_{\alpha\beta} q_\alpha q_\beta S_{\alpha\beta}(\mathbf{r}-\mathbf{r}'). \tag{5.243}$$

Thus, substituting the latter expression into the expression (5.236), we obtain

$$Q[\Phi] = Q_0 e^{\frac{\beta^2}{2}(\Phi S\Phi)} \int \frac{\mathcal{D}\varphi}{C_0} e^{-\frac{\beta}{2}(\varphi G_0^{-1}\varphi)-\frac{\beta^2}{2}(\varphi S_c\varphi)+i\beta(\varphi J)}. \tag{5.244}$$

Further we have

$$Q[\Phi] = Q_0 e^{\frac{\beta^2}{2}(\Phi \mathbf{S}\Phi)} \int \frac{\mathcal{D}\varphi}{C_0} e^{-\frac{\beta}{2}(\varphi G^{-1}\varphi)+i\beta(\varphi J)}. \tag{5.245}$$

After evaluating the Gaussian integral we obtain

$$Q[\Phi] = Q_0 e^{\frac{\beta^2}{2}(\Phi \mathbf{S}\Phi)-\frac{\beta}{2}(JGJ)} e^{\frac{1}{2}tr(\ln G - \ln G_0)}. \tag{5.246}$$

Using (5.241) and taking into account (5.232), we obtain

$$\hat{S}_{\mu\nu}(\mathbf{r}-\mathbf{r}') = S_{\mu\nu}(\mathbf{r}-\mathbf{r}') - \Sigma_{\mu\nu}(\mathbf{r}-\mathbf{r}'), \tag{5.247}$$

where we introduce the following functions

$$\Sigma_{\mu\nu}(\mathbf{r}-\mathbf{r}') = \beta \sum_{\alpha\beta} q_\alpha q_\beta \int d\mathbf{x} \int d\mathbf{y}\, S_{\alpha\mu}(\mathbf{r}-\mathbf{x})G(\mathbf{x}-\mathbf{y})S_{\beta\nu}(\mathbf{y}-\mathbf{r}'). \tag{5.248}$$

Considering that the operator G^{-1} in the Fourier representation has the form

$$G^{-1}(\mathbf{k}) = G_0^{-1}(\mathbf{k}) + \beta S_c(\mathbf{k}), \quad G_0^{-1}(\mathbf{k}) = \frac{\varepsilon k^2}{4\pi} \tag{5.249}$$

or

$$G^{-1}(\mathbf{k}) = \frac{\varepsilon}{4\pi}(k^2 + \varkappa^2(\mathbf{k})) \tag{5.250}$$

where we used already known screening function $\varkappa^2(\mathbf{k})$. Hence, we obtain

$$G(\mathbf{k}) = \frac{4\pi}{\varepsilon(k^2 + \varkappa^2(\mathbf{k}))}, \tag{5.251}$$

so that the full structural factors within the GA have the form

$$\hat{S}_{\mu\nu}(\mathbf{k}) = S_{\mu\nu}(\mathbf{k}) - \frac{4\pi}{\varepsilon k_B T} \frac{\sum_{\alpha\beta} q_\alpha S_{\alpha\mu}(\mathbf{k}) q_\beta S_{\beta\nu}(\mathbf{k})}{k^2 + \varkappa^2(\mathbf{k})}. \tag{5.252}$$

The charge structure factor has the form

$$\hat{S}_c(\mathbf{k}) = \sum_{\mu\nu} q_\mu q_\nu \hat{S}_{\mu\nu}(\mathbf{k}) = S_c(\mathbf{k}) \frac{k^2}{k^2 + \varkappa^2(\mathbf{k})}. \tag{5.253}$$

The expressions (5.252) and (5.253) were first derived in Ref. [115] within the framework of fluctuation theory. These expressions were also derived in later papers [116–118].

Soft polar molecules. Now we can construct a general statistical theory of solutions of polar molecules. Let's assume that each polar molecule represents a dimer of charged centers with charges $\pm q$ located at a fluctuating distance. As before, we introduce the distribution function $g(\mathbf{r})$ of the distance between the charged centers. Thus, the relation for the structure factor of polar molecules within the Gaussian approximation takes the following form:

$$S^{-1}(\mathbf{k}) = W^{-1}(\mathbf{k}) + \beta \tilde{V}(\mathbf{k}) = \begin{pmatrix} \frac{1}{n(1-g^2(\mathbf{k}))} + \beta \tilde{V}_{11}(\mathbf{k}) & -\frac{g(\mathbf{k})}{n(1-g^2(\mathbf{k}))} + \beta \tilde{V}_{12}(\mathbf{k}) \\ -\frac{g(\mathbf{k})}{n(1-g^2(\mathbf{k}))} + \beta \tilde{V}_{12}(\mathbf{k}) & \frac{1}{n(1-g^2(\mathbf{k}))} + \beta \tilde{V}_{22}(\mathbf{k}) \end{pmatrix}, \tag{5.254}$$

where

$$\tilde{V}(\mathbf{k}) = \begin{pmatrix} \tilde{V}_{11}(\mathbf{k}) & \tilde{V}_{12}(\mathbf{k}) \\ \tilde{V}_{12}(\mathbf{k}) & \tilde{V}_{22}(\mathbf{k}) \end{pmatrix} \tag{5.255}$$

is the matrix of the Fourier transforms of the soft potentials of non-electrostatic interactions; the structural matrix for dipole particles has the form

$$W(\mathbf{k}) = \begin{pmatrix} n & ng(\mathbf{k}) \\ ng(\mathbf{k}) & n \end{pmatrix}, \tag{5.256}$$

where

$$g(\mathbf{k}) = \int d\mathbf{r} g(\mathbf{r}) e^{-i\mathbf{k}\mathbf{r}} \tag{5.257}$$

represents the characteristic function of the probability distribution over the distance between the charged centers.

Further, utilizing the relations (5.206) and (5.254), we arrive at an analytical expression for the screening function of a solution of soft polar molecules [119]:

$$\varkappa^2(\mathbf{k}) = \kappa^2\,(1 - g(\mathbf{k}))\,Q(\mathbf{k}), \tag{5.258}$$

where we introduce the following auxiliary function defining the deviation of the screening function from that of a solution of polar molecules without excluded volume interactions

$$Q(\mathbf{k}) = \frac{1 + \frac{\beta n}{2}\left(\tilde{V}_{11}(\mathbf{k}) + \tilde{V}_{22}(\mathbf{k}) + 2\tilde{V}_{12}(\mathbf{k})\right)(1 + g(\mathbf{k}))}{1 + \beta n\left(\tilde{V}_{11}(\mathbf{k}) + \tilde{V}_{22}(\mathbf{k}) + 2\tilde{V}_{12}(\mathbf{k})g(\mathbf{k})\right) + \beta^2 n^2\left(1 - g^2(\mathbf{k})\right)\Delta(\mathbf{k})}, \tag{5.259}$$

where $\Delta(\mathbf{k}) = \tilde{V}_{11}(\mathbf{k})\tilde{V}_{22}(\mathbf{k}) - \tilde{V}_{12}^2(\mathbf{k})$. Here, $\kappa^2 = r_D^{-2} = 8\pi q^2 n/(\varepsilon k_B T)$. It is worth noting that in the absence of nonelectrostatic interactions ($\tilde{V}_{\alpha\gamma}(\mathbf{k}) = 0$), the screening function (5.258) reduces to the previously obtained expression (5.30).

Solution of polar molecules with the Gaussian core. Unfortunately, the formula (5.258) does not allow us to obtain an analytical expression for the electrostatic free energy of the solution in the general case. However, it can be achieved for a simplified model of soft polar molecules. Specifically, consider a reference system with $\tilde{V}_{11}(\mathbf{k}) = \tilde{V}(\mathbf{k}) > 0$ and $\tilde{V}_{22}(\mathbf{k}) = \tilde{V}_{12}(\mathbf{k}) = 0$. Essentially, such a reference system describes a solution of soft particles with an "imaginary" point particle on a spring attached to the center of mass of the particles. This model can be applied to a coarse-grained description of macromolecules, e.g. polar polypeptides or betaines dissolved in water. Additionally, it could describe zwitterionic liquids whose molecules represent a pair of covalently bonded large organic cation and small anion.

Thus, in such a case we arrive at the following relation for the screening function:

$$\varkappa^2(\mathbf{k}) = \kappa_D^2\,(1 - g(\mathbf{k}))\,Q(\mathbf{k}), \tag{5.260}$$

where

$$Q(\mathbf{k}) = 1 - \frac{1}{2}n\tilde{h}_2(\mathbf{k})\,(1 - g(\mathbf{k}))\,, \tag{5.261}$$

with

$$\tilde{h}_2(\mathbf{k}) = \frac{\beta\tilde{V}(\mathbf{k})}{1 + \beta n\tilde{V}(\mathbf{k})}. \tag{5.262}$$

Further, as above, we assume the following distribution function

$$g(\mathbf{r}) = \frac{3}{2\pi l^2 r}\exp\left[-\sqrt{6}r/l\right], \tag{5.263}$$

which defines the characteristic function

$$g(\mathbf{k}) = \frac{1}{1 + \frac{k^2 l^2}{6}}. \tag{5.264}$$

Using the expression (5.264), assuming the following approximation

$$Q(\mathbf{k}) \approx 1 - \frac{1}{2} n \tilde{h}_2(0) \left(1 - g(\mathbf{k})\right) = 1 - \frac{\alpha}{2} \left(1 - g(\mathbf{k})\right) \tag{5.265}$$

and integrating over the inverse space in (5.205), we arrive at the analytic expression for the electrostatic free energy [119]

$$F_{el} = -\frac{V k_B T}{l^3} \theta(y, \alpha), \tag{5.266}$$

where we introduce the following rather cumbersome auxiliary function:

$$\theta(y,\alpha) = \left(1 - \frac{3\alpha}{4}\right)\sigma(y) + \frac{3\sqrt{6}\exp\left[\frac{\alpha y}{4(1+\sqrt{1+y})^2}\right]\left(y\left(4+\sqrt{1+y}(\alpha+2)\right) + 2y^2 + 4(1+\sqrt{1+y})\right)}{4\pi(1+\sqrt{1+y})} - \frac{3\sqrt{6}\left((\alpha+4)(y^2+2y+y\sqrt{1+y}) + 8(1+\sqrt{1+y})\right)}{8\pi\left(1+\sqrt{1+y}\right)}. \tag{5.267}$$

Here the parameter $\alpha = \beta n \tilde{V}(0)/(1 + \beta n \tilde{V}(0)) < 1$ defines the "strength" of steric interactions, and $y = l^2/6r_D^2$; we also introduce the already known auxiliary function

$$\sigma(y) = \frac{\sqrt{6}}{4\pi}\left(2(1+y)^{3/2} - 2 - 3y\right). \tag{5.268}$$

As in the case of a solution of polar particles without excluded volume interactions, the following limiting expressions can be obtained

$$\frac{F_{el}}{V k_B T} = \begin{cases} -\frac{\sqrt{6}\pi q^4 l n^2}{3(\varepsilon k_B T)^2}\left(1 - \frac{3\alpha}{4} + \frac{5\alpha^2}{32}\right), & y \ll 1 \\ -\frac{1}{12\pi {r_D}^3}\left(1 - \frac{3}{2}\alpha - 3\left(1 - e^{\alpha/4}\right)\right), & y \gg 1. \end{cases} \tag{5.269}$$

The first regime ($r_D \gg l$) corresponds to the situation when electrostatic interactions are described by the Keesom interactions of dipoles in a dilute dipole gas [108], taking into account steric interactions. The second regime ($r_D \ll l$) describes the case when charged groups of polar particles stop "feeling" their connectivity and thus behave as free ions in an electrolyte solution. As seen from Eq. (5.269), in both limiting regimes, the presence of repulsive interaction between polar molecules leads to a decrease in the absolute value of the electrostatic free energy. This behavior can be explained by the fact that accounting for the finite volume of polar molecules prevents their mutual overlap, which weakens the electrostatic interactions. It is worth noting that within the five percent error interval, we can use the following approximation for the electrostatic free energy of a solution of Gaussian dipole particles

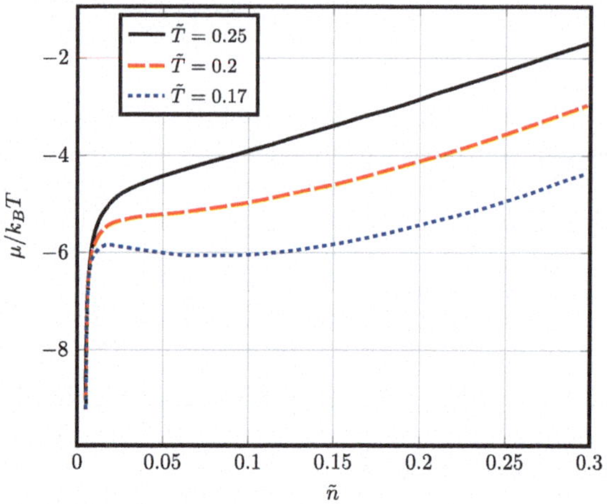

Fig. 5.3 Dependences of the chemical potential $\mu = k_B T \ln(nR^3) + \partial f_{ex}/\partial n$ (f_{ex} is the excess free energy density) on reduced concentration $\tilde{n}$ at different temperatures: $\tilde{T} = 0.25$ (solid line), $\tilde{T} = 0.2$ (dashed line), and $\tilde{T} = 0.17$ (dotted line). The data are shown for $\tilde{V}_0 = 1$ and $\tilde{l} = 1$

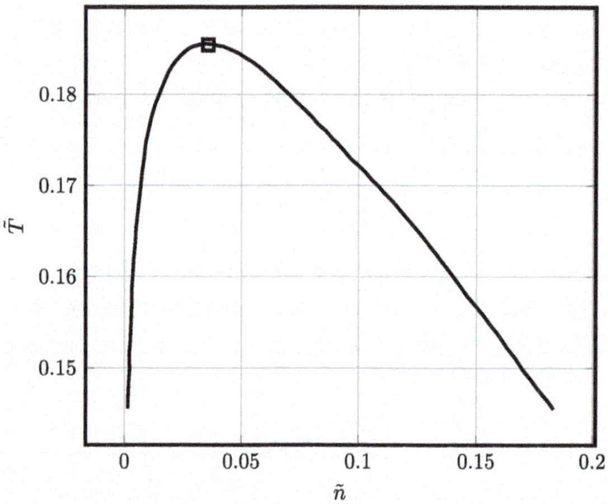

Fig. 5.4 Typical binodal for GCDM. The data are shown for $\tilde{V}_0 = 1$ and $\tilde{l} = 1$. Critical parameters in this case are $\tilde{T}_c = 0.186$ and $\tilde{n}_c = 0.035$

model electrolyte [120, 121]. Notably, the critical parameters of the GCDM are of the same order of magnitude as those realized for the restricted primitive model electrolyte. Moreover, similar critical behavior is also realized in systems of polar dumbbells [122].

Finally, let's discuss how repulsive interactions affect the electrostatic contribution to the excess free energy of GCDM. In Fig. 5.5, we depict the dependencies of the reduced electrostatic free energy $\tilde{f}_{el} = \beta f_{el} R^3$ on the reduced concentration $\tilde{n}$ at various values of the parameter $\tilde{V}_0$. It's evident that increasing $\tilde{V}_0$ results in a significant rise in the absolute value of the electrostatic free energy at sufficiently high concentrations. This behavior arises from the weakening of electrostatic interactions due to the reduction of Gaussian cores overlap as the repulsion parameter V_0 increases.

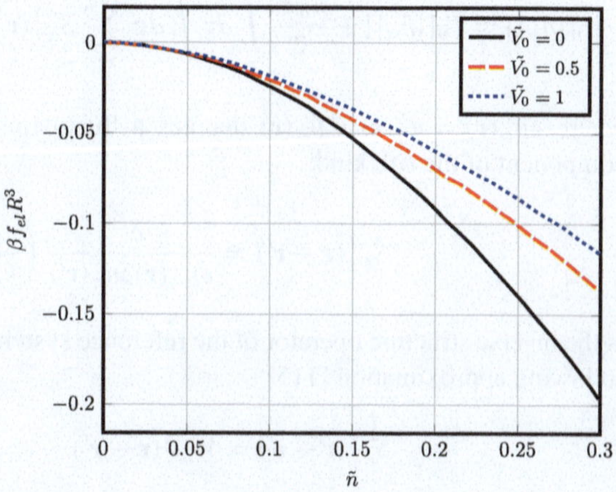

Fig. 5.5 Reduced electrostatic free energy density $\tilde{f}_{el} = \beta f_{el} R^3$ as a function of the reduced concentration $\tilde{n}$ at fixed temperature $\tilde{T} = 1$ and different repulsion parameters: $\tilde{V}_0 = 0$ (solid line), $\tilde{V}_0 = 0.5$ (dashed line) and $\tilde{V}_0 = 1$ (dashed line). The data are shown for $\tilde{l} = 1$

5.5 Erukhimovich Fluctuation Theory of Molecular Liquids

In this section, we provide a brief overview of the fluctuation theory of molecular liquids, where molecules are regarded as sets of spatially correlated charged sites with charges q_α, developed in the works of Erukhimovich and co-authors [115]. The simplest example of such a fluid is the polar fluid, which we discuss in detail. Specifically, we derive an analytical expression for the excess free energy of a system of polar hard spheres within the framework of the Gaussian approximation.

Let's write the partition function of the fluid in the form of a functional integral over the fluctuations of the local densities $n_\alpha(\mathbf{r})$ of the charged centers [115]:

$$Z = \mathcal{N} \int \prod_\alpha \mathcal{D}n_\alpha \exp\left[-\beta F_0[\{n_\alpha\}] - \beta U_{cl}[\{n_\alpha\}]\right], \tag{5.275}$$

where $F_0[n_\alpha]$ denotes the free energy functional of the reference fluid model without the electrostatic interactions, while $\mathcal{N}$ stands for the normalization constant, to be defined subsequently;

$$U_{cl}[\{n_\alpha\}] = \frac{1}{2} \int d\mathbf{r} \int d\mathbf{r}' \, \frac{\rho_c(\mathbf{r})\rho_c(\mathbf{r}')}{|\mathbf{r} - \mathbf{r}'|} \tag{5.276}$$

represents the energy of the Coulomb interactions of the charged sites, $\rho_c(\mathbf{r}) = \sum_\alpha q_\alpha n_\alpha(\mathbf{r})$ is the local fluctuating charge density.

Further, we expand the free energy functional of the reference system into a functional power series around the average density

$$F_0[\{n_\alpha\}] \approx F_0[\{\bar{n}_\alpha\}] + \frac{k_B T}{2} \int d\mathbf{r} \int d\mathbf{r}' \sum_{\alpha\gamma} S_{\alpha\gamma}^{-1}(\mathbf{r}-\mathbf{r}')\delta n_\alpha(\mathbf{r})\delta n_\gamma(\mathbf{r}'), \quad (5.277)$$

where $\delta n_\alpha(\mathbf{r}) = n_\alpha(\mathbf{r}) - \bar{n}_\alpha(\mathbf{r})$ denotes a fluctuation of the local density of the component of the αth kind;

$$S_{\alpha\gamma}^{-1}(\mathbf{r}-\mathbf{r}') = \frac{\delta^2}{\delta n_\alpha(\mathbf{r})\delta n_\gamma(\mathbf{r}')} \left(\frac{F_0}{k_B T}\right)\Bigg|_{n=\bar{n}} \quad (5.278)$$

is the inverse structure operator of the reference system [19], for which we adopt the following approximation [115]

$$S_{\alpha\gamma}^{-1}(\mathbf{r}-\mathbf{r}') = W_{\alpha\gamma}^{-1}(\mathbf{r}-\mathbf{r}') - c_{\alpha\gamma}(\mathbf{r}-\mathbf{r}'), \quad (5.279)$$

where $c_{\alpha\gamma}(\mathbf{r}-\mathbf{r}')$ is the matrix of direct correlation functions of the nonbonded centers and

$$W_{\alpha\gamma}(\mathbf{r}-\mathbf{r}') = n\delta_{\alpha\gamma}\delta(\mathbf{r}-\mathbf{r}') + n(1-\delta_{\alpha\gamma})g_{\alpha\gamma}(\mathbf{r}-\mathbf{r}') \quad (5.280)$$

is the already defined matrix of internal structure factors of molecules defined above; $g_{\alpha\gamma}(\mathbf{r}-\mathbf{r}')$ denotes the distance distribution function between the sites of αth and γth kinds within one molecule; $n = \bar{n}_\alpha$ is the average particle density. This approximation assumes that the pure entropy effects of molecule structure (which is described by $W_{\alpha\gamma}(\mathbf{r})$) and the interactions between sites (stemming from $c_{\alpha\gamma}(\mathbf{r})$) are treated independently.

Thus, we have the following relation

$$Z \approx \exp[-\beta F_0[\{\bar{n}_\alpha\}]]\mathcal{N} \int \prod_\alpha \mathcal{D}n_\alpha \exp\left[-\frac{1}{2}\int d\mathbf{r} \int d\mathbf{r}' \sum_{\alpha\gamma} \hat{S}_{\alpha\gamma}^{-1}(\mathbf{r}-\mathbf{r}')\delta n_\alpha(\mathbf{r})\delta n_\gamma(\mathbf{r}')\right], \quad (5.281)$$

where

$$\hat{S}_{\alpha\gamma}^{-1}(\mathbf{r}-\mathbf{r}') = S_{\alpha\gamma}^{-1}(\mathbf{r}-\mathbf{r}') + \frac{q_\alpha q_\gamma}{\varepsilon k_B T|\mathbf{r}-\mathbf{r}'|} \quad (5.282)$$

is the inverse structural operator within the Gaussian approximation. Then, we select the normalization constant $\mathcal{N}$ such that the electrostatic contribution to the total free energy is zero when $q_\alpha = 0$. By calculating the Gaussian functional integral, we obtain:

$$Z \approx \exp[-\beta F_0[\{\bar{n}_\alpha\}]]\sqrt{\frac{Det G}{Det S}} = \exp\left[-\beta F_0[\{\bar{n}_\alpha\}] + \frac{1}{2}Tr\ln(SG^{-1})\right], \quad (5.283)$$

where we take into account the identity

$$\sqrt{\frac{Det G}{Det S}} = \sqrt{Det(GS^{-1})} = \exp\left[\frac{1}{2}Tr\ln(GS^{-1})\right] \quad (5.284)$$

and that $Tr(B)$ represents the trace of the integral matrix operator $B_{\alpha\gamma}(\mathbf{r}, \mathbf{r}')$, i.e.

$$Tr(B) = \int d\mathbf{r} \sum_{\alpha} B_{\alpha\alpha}(\mathbf{r}, \mathbf{r}). \tag{5.285}$$

For translationally invariant kernels $B_{\alpha\gamma}(\mathbf{r}, \mathbf{r}') = B_{\alpha\gamma}(\mathbf{r} - \mathbf{r}')$ we have:

$$Tr(B) = V \int \frac{d\mathbf{k}}{(2\pi)^3} \sum_{\alpha} \tilde{B}_{\alpha\alpha}(\mathbf{k}), \tag{5.286}$$

where we introduce the Fourier transforms

$$\tilde{A}_{\alpha\gamma}(\mathbf{k}) = \int d\mathbf{r} e^{-i\mathbf{k}\mathbf{r}} A_{\alpha\gamma}(\mathbf{r}) \tag{5.287}$$

and V denotes the total volume of the system. Thus, we obtain

$$\hat{S} = (S^{-1} + A)^{-1} = (S^{-1}(I + SA))^{-1} = (I + SA)^{-1} S, \tag{5.288}$$

which in its turn leads to

$$\hat{S} S^{-1} = (I + K)^{-1}, \tag{5.289}$$

where I is the unit operator with kernel

$$I_{\alpha\gamma}(\mathbf{r} - \mathbf{r}') = \delta_{\alpha\gamma} \delta(\mathbf{r} - \mathbf{r}') \tag{5.290}$$

and operator $K = SA$ is determined by convolution

$$K_{\alpha\gamma}(\mathbf{r} - \mathbf{r}') = \sum_{\lambda} \int d\mathbf{r}'' \, S_{\alpha\lambda}(\mathbf{r} - \mathbf{r}'') A_{\lambda\gamma}(\mathbf{r}'' - \mathbf{r}'), \tag{5.291}$$

where

$$A_{\alpha\lambda}(\mathbf{r} - \mathbf{r}') = \frac{q_\alpha q_\lambda}{\varepsilon k_B T |\mathbf{r} - \mathbf{r}'|}. \tag{5.292}$$

Thereby, we get

$$Tr \ln(S \hat{S}^{-1}) = -Tr \ln(I + K) = -V \int \frac{d\mathbf{k}}{(2\pi)^3} \sum_{\alpha} \tilde{B}_{\alpha\alpha}(\mathbf{k}), \tag{5.293}$$

where

$$B = \ln(I + K) = \sum_{n=1}^{\infty} \frac{(-1)^{n+1}}{n} K^n. \tag{5.294}$$

Thus, we have

$$Tr\ln(I+K)=V\sum_{n=1}^{\infty}\frac{(-1)^{n+1}}{n}\sum_{\alpha}\int\frac{d\mathbf{k}}{(2\pi)^3}\left(K^n\right)_{\alpha\alpha}(\mathbf{k}), \tag{5.295}$$

with

$$\left(K^n\right)_{\alpha\alpha}(\mathbf{k})=\sum_{\alpha_1..\alpha_{n-1}}\tilde{K}_{\alpha\alpha_1}(\mathbf{k})\tilde{K}_{\alpha_1\alpha_2}(\mathbf{k})..\tilde{K}_{\alpha_{n-2}\alpha_{n-1}}(\mathbf{k})\tilde{K}_{\alpha_{n-1}\alpha}(\mathbf{k}). \tag{5.296}$$

Hence we get

$$\sum_{\alpha}\left(K^n\right)_{\alpha\alpha}(\mathbf{k})=\left(\frac{4\pi\sum\limits_{\alpha\gamma}q_\alpha q_\gamma S_{\alpha\gamma}(\mathbf{k})}{k_BT\varepsilon\mathbf{k}^2}\right)^n, \tag{5.297}$$

where we take into account (see (5.291)) that

$$\tilde{K}_{\alpha\gamma}(\mathbf{k})=\frac{4\pi}{k_BT\varepsilon\mathbf{k}^2}\sum_{\lambda}q_\alpha q_\lambda S_{\lambda\gamma}(\mathbf{k}). \tag{5.298}$$

Finally, we obtain

$$tr\ln(S\hat{S}^{-1})=V\int\frac{d\mathbf{k}}{(2\pi)^3}\ln\left(1+\frac{4\pi\sum\limits_{\alpha\gamma}q_\alpha q_\gamma S_{\alpha\gamma}(\mathbf{k})}{k_BT\varepsilon\mathbf{k}^2}\right). \tag{5.299}$$

Hence, the expression for the total free energy takes the form:

$$F=F_0+F_{el}, \tag{5.300}$$

where

$$F_{el}=\frac{Vk_BT}{2}\int\frac{d\mathbf{k}}{(2\pi)^3}\left(\ln\left(1+\frac{\varkappa^2(\mathbf{k})}{k^2}\right)-\frac{\varkappa^2(\mathbf{k})}{k^2}\right) \tag{5.301}$$

is the electrostatic free energy,

$$\varkappa^2(\mathbf{k})=\frac{4\pi}{\varepsilon k_BT}\sum_{\alpha\gamma}q_\alpha q_\gamma S_{\alpha\gamma}(\mathbf{k}) \tag{5.302}$$

represents the screening function, as introduced previously. It's important to note that we have subtracted the intrinsic electrostatic energy of the liquid molecules from the final expression; $S_{\alpha\gamma}(\mathbf{k})$ denotes the Fourier images of structural factors, which are determined from the matrix equation

$$S^{-1}_{\alpha\gamma}(\mathbf{k})=W^{-1}_{\alpha\gamma}(\mathbf{k})-c_{\alpha\gamma}(\mathbf{k}). \tag{5.303}$$

It's important to note that if we adopt the mean-spherical approximation, where $c_{\alpha\gamma}(\mathbf{k}) = -\beta\tilde{V}_{\alpha\gamma}(\mathbf{k})$, with $\tilde{V}_{\alpha\gamma}(\mathbf{k})$ representing the Fourier transforms of the potentials of non-Coulomb interactions between the centers, then we arrive at the well-known theory. The correlation functions of the reference system can be obtained independently through the methods of fluid theory [19] or from the computer experiments.

5.6 System of Polar Hard Spheres Within the Fluctuation Theory of Liquids

Now, we can apply the formulated general theory to describe polar molecules. In this case, we consider polar molecules as ionic pairs of charges $\pm q$. The relation for the structure operator of polar particles takes the following form

$$S^{-1}(\mathbf{k}) = W^{-1}(\mathbf{k}) - C(\mathbf{k}) = \begin{pmatrix} \frac{1}{n(1-g^2(\mathbf{k}))} - c_{11}(\mathbf{k}) & -\frac{g(\mathbf{k})}{n(1-g^2(\mathbf{k}))} - c_{12}(\mathbf{k}) \\ -\frac{g(\mathbf{k})}{n(1-g^2(\mathbf{k}))} - c_{12}(\mathbf{k}) & \frac{1}{n(1-g^2(\mathbf{k}))} - c_{22}(\mathbf{k}) \end{pmatrix}, \tag{5.304}$$

where

$$C(\mathbf{k}) = \begin{pmatrix} c_{11}(\mathbf{k}) & c_{12}(\mathbf{k}) \\ c_{12}(\mathbf{k}) & c_{22}(\mathbf{k}) \end{pmatrix} \tag{5.305}$$

is the matrix of direct correlation functions [19] of the reference system, while the structure matrix of polar molecules has the following form

$$W(\mathbf{k}) = \begin{pmatrix} n & ng(\mathbf{k}) \\ ng(\mathbf{k}) & n \end{pmatrix}, \tag{5.306}$$

where

$$g(\mathbf{k}) = \int d\mathbf{r} g(\mathbf{r}) e^{-i\mathbf{k}\mathbf{r}} \tag{5.307}$$

is the Fourier transform of the distribution function $g(\mathbf{r})$, and n is the particle number density. The screening function is

$$\varkappa^2(\mathbf{k}) = \frac{4\pi}{\varepsilon k_B T} \sum_{\alpha\gamma} q_\alpha q_\gamma S_{\alpha\gamma}(\mathbf{k}) = \frac{8\pi n q^2}{\varepsilon k_B T} (1 - g(\mathbf{k}))\, Q(\mathbf{k}), \tag{5.308}$$

with

$$Q(\mathbf{k}) = \frac{1 - \frac{n}{2}(c_{11}(\mathbf{k}) + c_{22}(\mathbf{k}) + 2c_{12}(\mathbf{k}))(1 + g(\mathbf{k}))}{1 - n(c_{11}(\mathbf{k}) + c_{22}(\mathbf{k}) + 2c_{12}(\mathbf{k})g(\mathbf{k})) + n^2(1 - g^2(\mathbf{k}))\Delta(\mathbf{k})} \tag{5.309}$$

and

$$\Delta(\mathbf{k}) = c_{11}(\mathbf{k})c_{22}(\mathbf{k}) - c_{12}^2(\mathbf{k}). \tag{5.310}$$

Let's consider a reference system with $c_{11}(\mathbf{k}) = c(\mathbf{k})$ and $c_{22}(\mathbf{k}) = c_{12}(\mathbf{k}) = 0$, where $c(\mathbf{k})$ represents the direct correlation function of hard spheres. Thus, one of the centers (center with number 1) is a hard sphere center, while the other center (with number 2) is an imaginary point center that is not correlated with the other centers. Consequently, we obtain the expression previously derived by another method (compare with Eq. (5.142)).

$$Q(\mathbf{k}) = 1 + \frac{n}{2}\tilde{h}_2(\mathbf{k})\,(1 - g(\mathbf{k}))\,, \tag{5.311}$$

where

$$\tilde{h}_2(\mathbf{k}) = \frac{\tilde{c}_2(\mathbf{k})}{1 - n\tilde{c}_2(\mathbf{k})} \tag{5.312}$$

is the Fourier transform of the pair correlation function of the reference system. Further, using the following expression for the Fourier transform of the distribution function

$$g(\mathbf{k}) = \frac{1}{1 + \frac{k^2 l^2}{6}}, \tag{5.313}$$

approximation

$$Q(\mathbf{k}) \approx 1 + \frac{n}{2}\tilde{h}(0)\,(1 - g(\mathbf{k})) = 1 + \frac{z_0 - 1}{2}\,(1 - g(\mathbf{k}))\,, \tag{5.314}$$

and evaluating the integral (5.301), we obtain the following relation for the electrostatic free energy

$$F_{el} = -\frac{Vk_BT}{l^3}\theta(y, \alpha), \tag{5.315}$$

where l is the dipole length; we also introduce the following auxiliary functions

$$\theta(y, \alpha) = \left(1 - \frac{3\alpha}{4}\right)\sigma(y) - \frac{3\sqrt{6}\left((\alpha + 4)(y^2 + 2y + y\sqrt{1 + y}) + 8(1 + \sqrt{1 + y})\right)}{8\pi\left(1 + \sqrt{1 + y}\right)}$$
$$+ \frac{3\sqrt{6}e^{\left[\frac{\alpha y}{4(1+\sqrt{1+y})^2}\right]}\left(y\left(4 + \sqrt{1 + y}(\alpha + 2)\right) + 2y^2 + 4(1 + \sqrt{1 + y})\right)}{4\pi(1 + \sqrt{1 + y})} \tag{5.316}$$

and

$$\sigma(y) = \frac{\sqrt{6}}{4\pi}\left[2(1 + y)^{3/2} - 2 - 3y\right]. \tag{5.317}$$

Here, $\alpha = 1 - z_0 = 1 - nk_BT\chi$, χ is the isothermal compressibility of hard spheres; $y = 4\pi q^2 l^2 n / 3\varepsilon k_B T$. Instead of the aforementioned cumbersome expression, one can utilize the interpolation formula proposed in [123], which provides a rather accurate approximation

$$F_{el} = -\frac{Vk_BT}{l^3}\left(\left(1-\frac{3\alpha}{4}\right)\sigma(y) + \frac{15}{256\pi}\frac{\alpha^2 y^2}{1+\sqrt{1+y}}\right). \tag{5.318}$$

Within the Percus–Yevick approximation [19] we have the following expression for the free energy of the reference system:

$$F_0 = Nk_BT\left(\ln(\Lambda^3 n) - 1 - \ln(1-\eta) + \frac{3\eta}{1-\eta} + \frac{3\eta^2}{2(1-\eta)^2}\right), \tag{5.319}$$

where N denotes the number of particles. The equation of state has the form:

$$P_0 = nk_BT\frac{1+\eta+\eta^2}{(1-\eta)^3}. \tag{5.320}$$

Thus, utilizing the expression for the isothermal compressibility $\chi = n^{-1}\partial n/\partial P_0$, we arrive at the relation for the parameter α:

$$\alpha = \frac{\eta(4-\eta)(2+\eta^2)}{(1+2\eta)^2}. \tag{5.321}$$

Chapter 6
Further Developments in Modified Poisson-Boltzmann Equations Accounting for Ionic Specificity

In this chapter, we formulate modified Poisson-Boltzmann equations that take into account not only the excluded volume, but also the ionic chemical specificity like static polarizability and dipole moment and structural effects. We then delve into practical applications of the theory across a range of diverse physical scenarios.

6.1 Incorporation of Short-Range Ion Correlations

As demonstrated earlier, the modified PB equation predicts a monotonic behavior of local ion concentrations within the electric double layer at both the metal/electrolyte interface and nanoscale pores, even at high bulk ion concentrations. However, it is well-established that concentrated electrolyte solutions and ionic liquids exhibit non-monotonic behavior of local ion concentrations in EDLs, along with the oscillations of disjoining pressure within nanopores as pore thickness varies. These oscillations primarily stem from short-range ion correlations (or packing effects), which the mean-field theory outlined above can't encompass. In this section, we delve into the simplest form of self-consistent field theory, aiming to incorporate the short-range correlations of ions, which arise from their molecular structure, albeit at a phenomenological level.

Self-consistent field equations and stress tensor. Let's explore an ionic fluid model, which captures the molecular structure of ions at a phenomenological level by employing a bilinear form over concentration gradients in the grand thermodynamic potential density [83, 93]. Thus, we can express the grand thermodynamic potential as:

$$\Omega = F_{el} + \Omega_{liq}, \tag{6.1}$$

Y. A. Budkov and N. N. Kalikin, *Statistical Field Theory of Ion-Molecular Fluids*,
https://doi.org/10.1007/978-3-031-68364-0_6

where

$$F_{el} = \int d\mathbf{r} \left(-\frac{\varepsilon (\nabla \psi)^2}{8\pi} + \rho \psi \right) \tag{6.2}$$

determines the contribution of the electrostatic interactions within the mean-field approximation to the free energy; $\rho(\mathbf{r}) = \sum_\alpha q_\alpha n_\alpha(\mathbf{r})$ represents the charge density of the ions, with q_α denoting the charge of the α-th type of ion; ε is the dielectric permittivity of the medium; and $n_\alpha(\mathbf{r})$ represents the local ion concentrations. The grand thermodynamic potential of the ionic fluid takes the form:

$$\Omega_{liq} = \int d\mathbf{r} \left(f(\{n_\alpha\}) + \sum_\alpha n_\alpha u_\alpha + \frac{1}{2} \sum_{\alpha\gamma} \varkappa_{\alpha\gamma} \nabla n_\alpha \cdot \nabla n_\gamma - \sum_\alpha \mu_\alpha n_\alpha \right), \tag{6.3}$$

where the first term in the integrand represents the free energy density of the fluid, dependent on local ion concentrations n_α; the second term determines the density of the potential energy of ions in external fields with energies $u_\alpha(\mathbf{r})$; and the third term accounts for the ion structure using the Cahn–Hilliard approach with "structural" constants $\varkappa_{\alpha\gamma}$ (as discussed in Chap. 2 and [83]), and, finally, μ_α are the chemical potentials of the components. It is important to note that structural interactions, in general, incorporate the effects related to the molecular structure of ions. They may include steric effects arising from the asymmetric shape of ions, their conformational flexibility (in the case of ionic liquids), and solvent effects such as the deformation of solvation shells due to their overlapping [30].

Thus, the grand thermodynamic potential has the form

$$\Omega = \int d\mathbf{r}\, \omega(\mathbf{r}), \tag{6.4}$$

where as previously we define the GTP density as

$$\omega = -\frac{\varepsilon (\nabla \psi)^2}{8\pi} + \rho \psi + \frac{1}{2} \sum_{\alpha\gamma} \varkappa_{\alpha\gamma} \nabla n_\alpha \cdot \nabla n_\gamma + f(\{n_\alpha\}) + \sum_\alpha n_\alpha u_\alpha - \sum_\alpha \mu_\alpha n_\alpha. \tag{6.5}$$

Note that the summation in (6.5) is performed over all ionic species. Additionally, the integration in (6.4) is performed over the volume of the ionic fluid confined between the surfaces of conducting or dielectric bodies. Finally, the Euler-Lagrange equations

$$\frac{\delta \Omega}{\delta \psi} = \frac{\partial \omega}{\partial \psi} - \partial_i \left(\frac{\partial \omega}{\partial \psi_{,i}} \right) = 0, \quad \frac{\delta \Omega}{\delta n_\alpha} = \frac{\partial \omega}{\partial n_\alpha} - \partial_i \left(\frac{\partial \omega}{\partial n_{\alpha,i}} \right) = 0, \tag{6.6}$$

yield

$$\nabla^2 \psi = -\frac{4\pi}{\varepsilon} \sum_\alpha q_\alpha n_\alpha, \quad \sum_\gamma \varkappa_{\alpha\gamma} \nabla^2 n_\alpha = v_\alpha, \tag{6.7}$$

where $v_\alpha = q_\alpha\psi + u_\alpha + \bar{\mu}_\alpha - \mu_\alpha$ and $\bar{\mu}_\alpha = \partial f/\partial n_\alpha$ are the intrinsic chemical potentials. The first equation represents the standard Poisson equation for the electrostatic potential, while the second one governs the thermodynamic equilibrium of the α-th ions. These self-consistent field equations can be solved with appropriate boundary conditions for the potential and concentrations. It is important to note that the boundary conditions are determined by the nature of the macroscopic body immersed in the ionic fluid. For the ion concentrations, we assume $n_\alpha = 0$ at the surface of the bodies, reflecting the strong repulsion of ions from the surface molecules upon direct contact. Any specific attraction of ions to the surface can be incorporated through the potential energies of external fields $u_\alpha(\mathbf{r})$.

Thus, using (6.7), we arrive at

$$\frac{\partial \omega}{\partial x_i} = \frac{\partial}{\partial x_k}\left(\psi_{,i}\frac{\partial \omega}{\partial \psi_{,k}}\right) + \frac{\partial}{\partial x_k}\left(\sum_\alpha n_{\alpha,i}\frac{\partial \omega}{\partial n_{\alpha,k}}\right) + \sum_\alpha n_\alpha \frac{\partial u_\alpha}{\partial x_i}. \tag{6.8}$$

or

$$\frac{\partial \sigma_{ik}}{\partial x_k} - \sum_\alpha n_\alpha \frac{\partial u_\alpha}{\partial x_i} = 0, \tag{6.9}$$

where $n_{\alpha,i} = \partial_i n_\alpha$, $\psi_{,i} = \partial_i \psi$, $\partial_i = \partial/\partial x_i$ and

$$\sigma_{ik} = \omega\delta_{ik} - \psi_{,i}\frac{\partial \omega}{\partial \psi_{,k}} - \sum_\alpha n_{\alpha,i}\frac{\partial \omega}{\partial n_{\alpha,k}} \tag{6.10}$$

is the total stress tensor. Using the Eq. (6.7), stress tensor can be represented in the form

$$\sigma_{ik} = \sigma_{ik}^{(M)} + \sigma_{ik}^{(h)} + \sigma_{ik}^{(s)}, \tag{6.11}$$

where

$$\sigma_{ik}^{(M)} = \frac{\varepsilon}{4\pi}\left(\mathcal{E}_i\mathcal{E}_k - \frac{1}{2}\mathcal{E}^2\delta_{ik}\right) \tag{6.12}$$

is previously obtained Maxwell stress tensor with $\mathcal{E}_i = -\partial_i\psi$;

$$\sigma_{ik}^{(h)} = -P\delta_{ik} \tag{6.13}$$

is the hydrostatic stress tensor with the local osmotic pressure of the ions, $P = \sum_\alpha n_\alpha \partial f/\partial n_\alpha - f$, whereas

$$\sigma_{ik}^{(s)} = \sum_{\alpha\gamma} \varkappa_{\alpha\gamma}\left[\left(\frac{1}{2}\nabla n_\alpha \cdot \nabla n_\gamma + n_\alpha \nabla^2 n_\gamma\right)\delta_{ik} - \partial_i n_\alpha \partial_k n_\gamma\right] \tag{6.14}$$

is the contribution of structural interactions to the total stress tensor, so-called *structural* stress tensor [93].

Expression of disjoining pressure in terms of the stress tensor. Now, using the obtained expression for the stress tensor, let's derive the expression for the disjoining pressure for the case of the ionic fluid confined in a nanosized pore. The disjoining pressure can be evaluated using the formula:

$$\Pi = -\frac{\partial(\Omega/\mathcal{A})}{\partial H} - P_b, \tag{6.15}$$

where P_b denotes the pressure in the bulk fluid and

$$\Omega/\mathcal{A} = \int_0^H dz\, \omega(z) \tag{6.16}$$

represents the grand thermodynamic potential per unit area of the wall. Thus, we obtain:

$$\frac{1}{A}\frac{\partial \Omega}{\partial H} = \frac{1}{A}\int_0^H dz \left(\frac{\delta \Omega}{\delta \psi(z)}\frac{\partial \psi(z)}{\partial H} + \sum_\alpha \frac{\delta \Omega}{\delta n_\alpha(z)}\frac{\partial n_\alpha(z)}{\partial H}\right) + \sigma_{zz}(H) + \int_0^H dz \sum_\alpha \frac{\partial \omega}{\partial u_\alpha}\frac{\partial u_\alpha}{\partial H}$$
$$= \int_0^H dz \sum_\alpha n_\alpha(z)\frac{\partial u_\alpha(z)}{\partial H} + \sigma_{zz}(H), \tag{6.17}$$

where we take into account that $\partial\omega/\partial u_\alpha = n_\alpha$ and $\delta\Omega/\delta\psi(z) = 0$, $\delta\Omega/\delta n_\alpha(z) = 0$. Thus, we arrive at

$$\Pi = -\sum_\alpha \int_0^H dz n_\alpha(z)\frac{\partial u_\alpha(z)}{\partial H} - \sigma_{zz}(H) - P_b, \tag{6.18}$$

where $\sigma_{zz}(H) = (\omega - \psi'\partial\omega/\partial\psi' - \sum_\alpha n'_\alpha \partial\omega/\partial n'_\alpha)|_{z=H}$ is the normal stress at $z = H$.

Now let's consider an important case from a practical standpoint, where the external potentials are generated by identical walls, i.e.

$$u_\alpha(z) = \phi_\alpha(z) + \phi_\alpha(H - z), \tag{6.19}$$

where ϕ_α is a potential of single wall. Thus, taking into account that $n_\alpha(z) = n_\alpha(H - z)$, we arrive at the expression

$$\Pi = -\sum_\alpha \int_0^H dz\, n_\alpha(z)\phi'_\alpha(z) - \sigma_{zz}(H) - P_b. \tag{6.20}$$

The expression (6.20) can be rewritten in a more convenient form for computations. Utilizing the equation of mechanical equilibrium

$$\frac{d\sigma_{zz}(z)}{dz} - \sum_{\alpha} n_{\alpha}(z)u'_{\alpha}(z) = 0, \tag{6.21}$$

after integration from $z = H/2$ to $z = H$, we obtain

$$\sigma_{zz}(H) = \sigma_{zz}\left(\frac{H}{2}\right) + \sum_{\alpha} \int_{H/2}^{H} dz\, n_{\alpha}(z)\phi'_{\alpha}(z) - \sum_{\alpha} \int_{0}^{H/2} dz\, n_{\alpha}(z)\phi'_{\alpha}(z), \tag{6.22}$$

where we take into account (6.19) and the condition discussed above: $n_{\alpha}(z) = n_{\alpha}(H - z)$. Substituting (6.22) for $\sigma_{zz}(H)$ into (6.20), after some manipulations, we obtain

$$\Pi = P_n - P_b - 2\sum_{\alpha} \int_{H/2}^{H} dz\, n_{\alpha}(z)\phi'_{\alpha}(z), \tag{6.23}$$

where

$$P_n = -\sigma_{zz}\left(\frac{H}{2}\right) = P_m - \sum_{\alpha\gamma} \varkappa_{\alpha\gamma} n_{\alpha}\left(\frac{H}{2}\right) n''_{\gamma}\left(\frac{H}{2}\right) \tag{6.24}$$

is the normal pressure in the midpoint of the pore, where we take into account $\psi'(H/2) = n'_{\alpha}(H/2) = 0$ and introduce the osmotic pressure of the ions in the midpoint of the pore, $P_m = P\,(H/2)$. Using the Eq. (6.7), the expression (6.24) can be simplified to

$$P_n = P_m - \sum_{\alpha} n_{\alpha,m} v_{\alpha,m}, \tag{6.25}$$

where $n_{\alpha,m} = n_{\alpha}\,(H/2)$ and $v_{\alpha,m} = v_{\alpha}\,(H/2)$. The second term on the right-hand side of expression (6.25) describes the contribution of the structural interactions of the ions to the normal stress in the midpoint of the pore.

Asymmetric lattice gas as a reference model. Before analyzing the numerical results, let's decide on the model to use to describe the ionic fluid confined in a slit-like pore with identically charged walls and a fixed surface electrostatic potential ψ_0. As a reference model, we propose to use the asymmetric lattice gas model, which allows us to consider the asymmetry of ion sizes. The free energy density of an asymmetric lattice gas is [82]

$$f = \frac{k_B T}{v}\left(\frac{\phi_+}{N_+}\ln\phi_+ + \frac{\phi_-}{N_-}\ln\phi_- + (1 - \phi_+ - \phi_-)\ln(1 - \phi_+ - \phi_-)\right), \tag{6.26}$$

where v denotes the lattice gas cell volume, $\phi_\pm = N_\pm n_\pm v$ are the local volume fractions of the ions, N_α is the number of cells, occupied by the ions of α-th kind ($\alpha = \pm$); k_B is the Boltzmann constant, T is temperature. The intrinsic chemical potentials in such a case are

$$\bar{\mu}_\pm = \frac{\partial f}{\partial n_\pm} = k_B T \left(\ln \phi_\pm + 1 - N_\pm \ln (1 - \phi_+ - \phi_-) - N_\pm\right), \tag{6.27}$$

and the bulk chemical potentials are

$$\mu_\pm = k_B T \left(\ln \phi_\pm^{(b)} + 1 - N_\pm (\ln (1 - \phi_+^{(b)} - \phi_-^{(b)}) + 1)\right), \tag{6.28}$$

where $\phi_\pm^{(b)} = N_\pm n_\pm^{(b)} v$ are the bulk volume fractions of the ions. Supposing that $q_\pm = \pm q$, from the electroneutrality condition in the bulk phase, $q_+ n_+^{(b)} + q_- n_-^{(b)} = 0$, we obtain that $n_\pm^{(b)} = n$. To describe the interactions of the ions with the pore walls we use the following Gaussian short-range potentials

$$u_\pm(z) = -\varepsilon_\pm \exp\left[-\frac{z^2}{2\sigma_\pm^2}\right], \tag{6.29}$$

where $\varepsilon_\pm$ denotes positive energetic parameters (the depth of the potential), and $\sigma_\pm$ is the effective width of the potential.

"Structural" attraction of the charged walls. Figure 6.1 illustrates the impact of changing pore width and surface potential on the concentration profiles of cations and anions, as well as the electrostatic potential profiles. At small pore widths, both types of ions predominantly concentrate in the center of the pore, resulting in the absence of an electric double layer at the ionic fluid/wall interface. However, with increasing surface potential, cation concentration rises while anion concentration decreases due to electrostatic attraction and repulsion, respectively. As the pore width increases, EDLs begin to form near the walls. Around a width of 5 nm, the EDLs become nearly isolated, forming an electrically neutral bulk phase of ionic fluid in the center of the pore. It's noteworthy that positive surface potentials lead to similar concentration profile behavior.

Next, we explore the behavior of the disjoining pressure. Figure 6.2a illustrates the disjoining pressure as a function of pore width at various negative surface potentials. For sufficiently small pore thickness, the disjoining pressure exhibits a non-monotonic trend, indicating an effective attraction between the walls ($\Pi < 0$). This minimum becomes more pronounced with increasing absolute value of the surface potential. To comprehend the physical basis of this behavior, we examine the ion concentration profiles at pore widths before ($H = 1$ nm), at ($H = 1.75$ nm), and after ($H = 4$ nm) the position of the minimum (refer to Fig. 6.3). In the case of a narrow 1 nm wide pore, unimodal concentration profiles are observed, with ions concentrated at the pore's center. At $H = 1.75$ nm, we observe maxima in the cation concentration profiles near the walls and a minimum in the pore's center. This suggests that

$$F_{el} = -\frac{Vk_BT}{l^3}\left(\left(1-\frac{3\alpha}{4}\right)\sigma(y)+\frac{15}{256\pi}\frac{\alpha^2 y^2}{1+\sqrt{1+y}}\right). \quad (5.270)$$

Further, employing the general relation (5.230), we arrive at an expression for the excess free energy of the reference system, which characterizes the excess free energy of a one-component fluid with a central pairwise interaction potential (see Chap. 2) having a positive Fourier transform [33, 34, 38]

$$F_0 = \frac{Vn^2\tilde{V}(0)}{2} + \frac{Vk_BT}{2}\int\frac{d\mathbf{k}}{(2\pi)^3}\left(\ln\left(1+\beta n\tilde{V}(\mathbf{k})\right)-\beta n\tilde{V}(\mathbf{k})\right). \quad (5.271)$$

Note that we have subtracted the self-energy of the particles to obtain the final result. Thus, using the Gaussian core model (GCM) [37] as a reference system, i.e., using the Gaussian potential

$$V(r) = V_0 \exp\left[-r^2/R^2\right] \quad (5.272)$$

to describe the nonelectrostatic interactions of the polar particles and taking into account that $\tilde{V}(\mathbf{k}) = \pi^{3/2}R^3V_0\exp\left[-k^2R^2/4\right]$, we obtain the expression for the excess free energy of the GCM (see also Chap. 2)

$$\frac{F_0}{Nk_BT} = \frac{\xi}{2}\left(1-\beta V_0\mathcal{F}(\xi)\right), \quad (5.273)$$

where $\xi = \pi^{3/2}\beta n V_0 R^3$, R denotes the effective size of the Gaussian core and V_0 is an energetic parameter, determining the strength of repulsion between the polar molecules; moreover, we introduce the following auxiliary function

$$\mathcal{F}(\xi) = \frac{4}{3\sqrt{\pi}}\int\limits_0^1\frac{dt(-\ln t)^{3/2}t}{1+\xi t}. \quad (5.274)$$

As mentioned in Chap. 2, the expression (5.273) accurately characterizes the thermodynamic behavior of the GCM within the liquid-phase region of the phase diagram, particularly for sufficiently small values of V_0 [37]. Henceforth, we will refer to the system of Gaussian polar molecules as the Gaussian core dipolar model (GCDM).

For numerical analysis, we introduce following dimensionless parameters: $\tilde{n} = nR^3$, $\tilde{T} = \varepsilon Rk_BT/q^2$, $\tilde{V}_0 = \varepsilon RV_0/q^2$, $\tilde{l} = l/R$. We assume $\tilde{l} = 1$, which is quite reasonable for the case of macromolecules mentioned above.

Figure 5.3 demonstrates that at sufficiently high temperatures, the chemical potential behaves in a standard manner, monotonically increasing with concentration growth [12]. However, below the critical temperature, a van der Waals loop appears on the chemical potential curve, indicating a liquid-liquid phase separation. The typical phase equilibrium curve (binodal) for GCDM, shown in Fig. 5.4, was obtained using Maxwell's rule for finding concentrations in coexisting phases [12]. This binodal exhibits strong asymmetry, similar to that observed for the restricted primitive

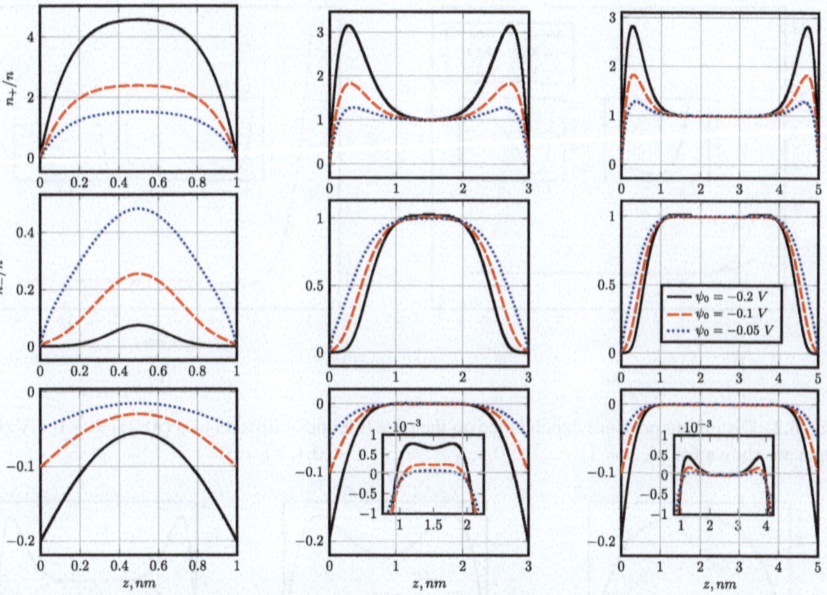

Fig. 6.1 Dimensionless ionic concentrations, $n_\pm/n$, corresponding to different pore thicknesses plotted for different negative surface potentials, ψ_0. Data are shown for $\chi_{++} = 1$, $\chi_{--} = 0.5$, $\chi_{+-} = \chi_{-+} = 0.1$, $\varepsilon_\pm = 0$

more cations migrate from the pore's center to the walls to form an EDLs. Consequently, three layers emerge: two layers of cations near the pore walls and a layer of mixed cations and anions between them. The formation of this 'sandwich' structure manifests as an effective "structural" attraction between the walls. For pore widths of 4 nm, a similar scenario unfolds, with isolated EDLs near the walls and a bulk fluid phase in the pore's middle. In this region, the disjoining pressure exponentially decreases (see the right inset in Fig. 6.2b).

It's important to underline that the model described above does not incorporate the layering effect observed in both ionic liquids and concentrated electrolyte solutions, situated at electrified interfaces [124–126]. This effect results in pronounced oscillations of the disjoining pressure at small pore thicknesses. To accurately depict these phenomena, it's necessary to consider higher-order derivatives in the grand thermodynamic potential [127] or to employ nonlocal theories. However, despite these limitations, the presented simple statistical theory remains valuable for modeling ionic fluids at moderate concentrations where the layering of ions is not a significant factor.

The effect of specific adsorption of the ions. Additionally, it is informative to explore the impact of specific ion adsorption, described by Gaussian potential wells (6.29), on local ion concentrations and disjoining pressure. Figure 6.4 illustrates various ion concentration profiles for different pore widths while varying the characteristic energy parameter of cation adsorption, ε_+. As a result, it can be seen

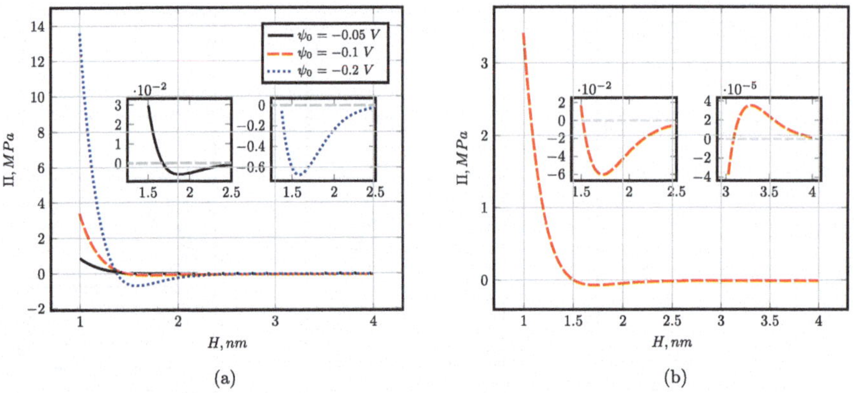

Fig. 6.2 Disjoining pressure dependence on the pore thickness: different ψ_0 (**a**), $\psi_0 = -0.1$ V (**b**). Data are shown for $\chi_{++} = 1$, $\chi_{--} = 0.5$, $\chi_{+-} = \chi_{-+} = 0.1$, $\varepsilon_\pm = 0$

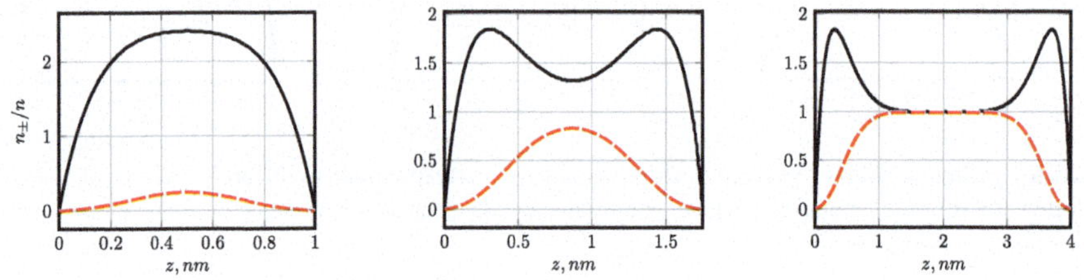

Fig. 6.3 Ion concentration profiles at pore thicknesses up to ($H = 1$ nm), at ($H = 1.75$ nm), and after ($H = 4$ nm) disjoining pressure minimum. Data are shown for $\chi_{++} = 1$, $\chi_{--} = 0.5$, $\chi_{+-} = \chi_{-+} = 0.1$, $\varepsilon_\pm = 0$, $\psi_0 = -0.1$ V

that for narrow pores ($H = 1$ nm), increasing ε_+ leads to a predictable increase of the local cation concentration. However, this also results in a notable elevation of local anion concentration due to Coulomb attraction between cations and anions. With increasing H, higher values of ε_+ lead to sharper peaks in the cation concentration profile, indicating the formation of a dense Stern layer within the EDLs. Moreover, the deepening of the minimum on the disjoining pressure profiles and its shift towards narrower pore regions is observed with increased ε_+. Such behavior can be attributed to the specific adsorption of cations, which facilitates the formation of EDL at smaller pore widths (H) than possible without specific adsorption. This is evidenced by the pronounced peaks in the cation concentration profile depicted in Fig. 6.5 at $H = 1$ nm and $\varepsilon_+ = 10k_BT$.

Asymmetry of the ion sizes. Now, let's examine the scenario where ion sizes differ, i.e., when $N_+ \neq N_-$, and analyze how this asymmetry affects the disjoining pressure behavior. Figure 6.6 presents typical disjoining pressure profiles for various cation sizes: $N_+ = 1$, $N_+ = 2$, and $N_+ = 3$, while maintaining a fixed anion size $N_- = 1$, for negatively charged pore walls ($\psi_0 = -0.1$ V). As depicted in the figure, increasing the cation size relative to the anion size nearly abolishes the minimum on

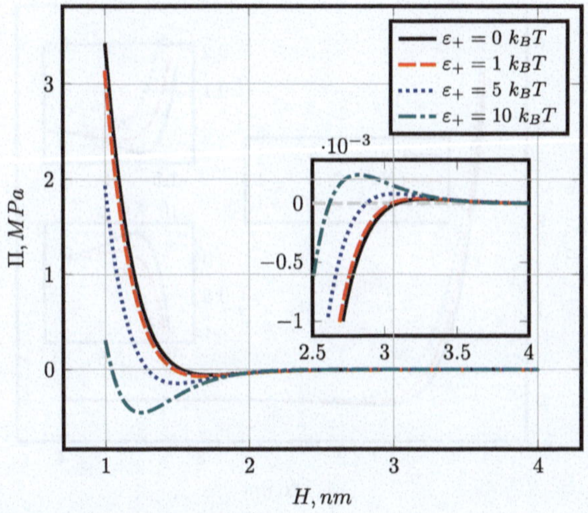

Fig. 6.4 Disjoining pressure dependences on the pore thickness plotted for different specific adsorption parameter, ε_+. Data are show for $\chi_{++} = 1$, $\chi_{--} = 0.5$, $\chi_{+-} = \chi_{-+} = 0.1$, $\varepsilon_- = 0$, $\sigma_+/v^{1/3} = 1$, $\psi_0 = -0.1$ V

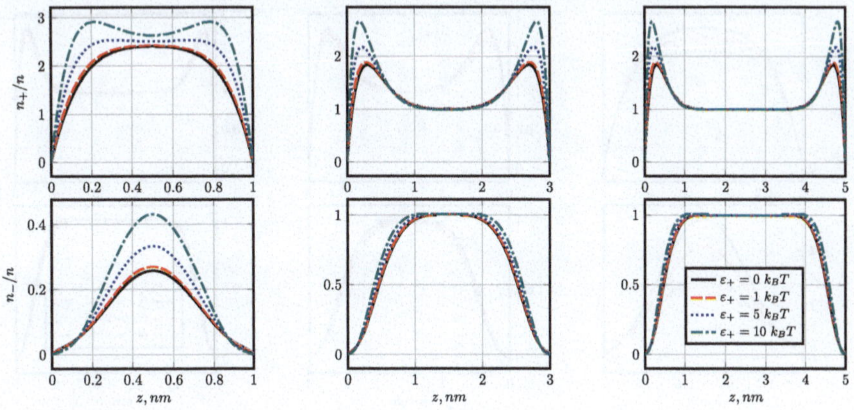

Fig. 6.5 Ion concentration profiles corresponding to different pore thicknesses plotted for different specific adsorption parameter, ε_+. Data are shown for $\chi_{++} = 1$, $\chi_{--} = 0.5$, $\chi_{+-} = \chi_{-+} = 0.1$, $\varepsilon_- = 0$, $\sigma_+/v^{1/3} = 1$, $\psi_0 = -0.1$ V

the disjoining pressure curve. This phenomenon can be elucidated by the fact that larger cation sizes result in increased steric interactions, leading to a reduction in the local cation concentration within the pore volume (refer to Fig. 6.7). Consequently, this reduction "compensates" the structural attraction between the walls.

When examining walls with positive charge ($\psi_0 = 0.1$ V), the cation size influences the disjoining pressure differently, leading to a more pronounced minimum (refer to Fig. 6.8). Initially, this conclusion might appear counterintuitive, but it

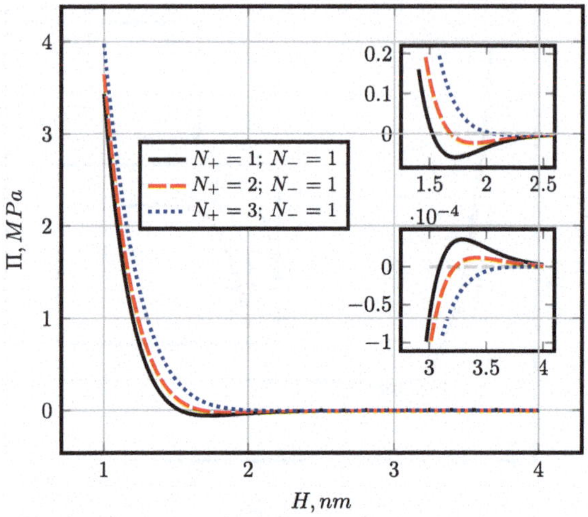

Fig. 6.6 Disjoining pressure dependences on the pore thickness plotted for different ionic size asymmetry $N_+ = [1, 2, 3]$ and $N_- = 1$. Data are shown for $\chi_{++} = 1$, $\chi_{--} = 0.5$, $\chi_{+-} = \chi_{-+} = 0.1$, $\varepsilon_\pm = 0$, $\sigma_+/v^{1/3} = 1$, $\psi_0 = -0.1$ V

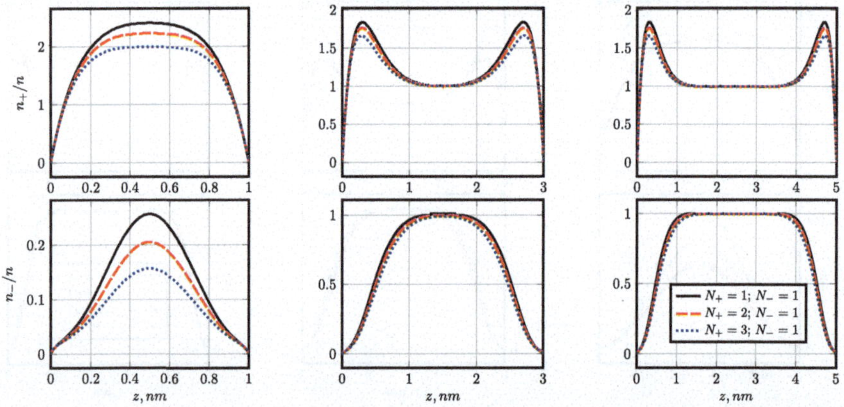

Fig. 6.7 Ion concentration profiles corresponding to different pore thicknesses plotted for different ionic size asymmetry $N_+ = [1, 2, 3]$ and $N_- = 1$. Data are shown for $\chi_{++} = 1$, $\chi_{--} = 0.5$, $\chi_{+-} = \chi_{-+} = 0.1$, $\varepsilon_\pm = 0$, $\sigma_+/v^{1/3} = 1$, $\psi_0 = -0.1$ V

can be elucidated by the well-known phenomenon of Asakura-Oosawa depletion forces [21]. When cations are in close proximity to positively charged walls, a depletion of cations occurs in this near-wall region. Consequently, both cations and anions are displaced from the vicinity of the wall, resulting in an "empty space" there. This, in turn, induces an "entropic" attraction of the displaced cation towards the wall. The described behavior is evident in the concentration profiles depicted in Fig. 6.9—larger cation sizes correspond to higher local concentrations of cations and

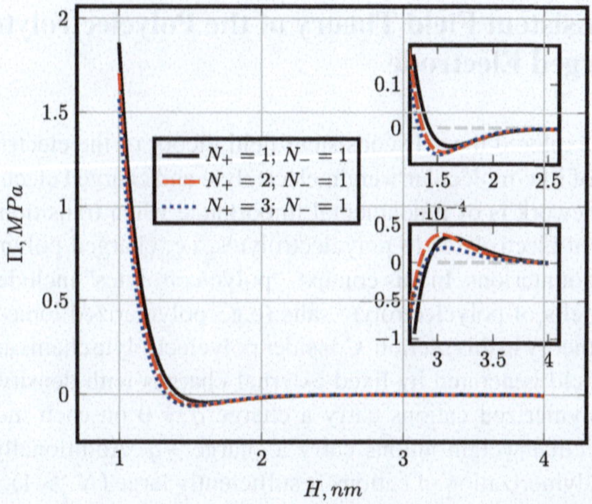

Fig. 6.8 Disjoining pressure dependences on the pore thickness plotted for different ionic size asymmetry $N_+ = [1, 2, 3]$ and $N_- = 1$. Data are shown for $\chi_{++} = 1$, $\chi_{--} = 0.5$, $\chi_{+-} = \chi_{-+} = 0.1$, $\varepsilon_\pm = 0$, $\sigma_+/v^{1/3} = 1$, $\psi_0 = 0.1$ V

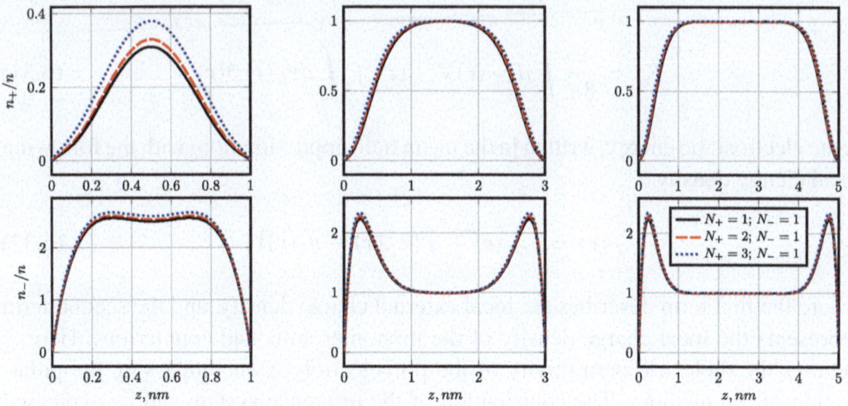

Fig. 6.9 Ion concentration profiles corresponding to different pore thicknesses plotted for different ionic size asymmetry $N_+ = [1, 2, 3]$ and $N_- = 1$. Data are shown for $\chi_{++} = 1$, $\chi_{--} = 0.5$, $\chi_{+-} = \chi_{-+} = 0.1$, $\varepsilon_\pm = 0$, $\sigma_+/v^{1/3} = 1$, $\psi_0 = 0.1$ V

anions. It's important to note that the modified Poisson-Boltzmann equations, which do not incorporate structural effects, omit the effects of depletion forces. However, by considering structural interactions in the grand thermodynamic potential within the framework of the Cahn-Hilliard-type theory, we can qualitatively describe these phenomena.

6.2 Self-consistent Field Theory of the Polyelectrolyte Fluid at Charged Electrode

Previously, we explored the self-consistent field theory of the electric double layer at the interface of low-molecular weight electrolyte and charged electrode. A similar theoretical framework is of fundamental importance when transitioning from low-molecular weight electrolytes to polyelectrolytes, i.e. charged polymer chains and low-molecular counterions. In this context, "polyelectrolytes" include both polymer solutions and melts of polyelectrolyte salts (e.g., polymerized ionic liquids). Let's formulate such theory in this section. Consider polyelectrolyte chains and counterions in an external field generated by fixed external charges with density $\rho_{ext}(\mathbf{r})$. Let's assume that polymerized cations carry a charge $q > 0$ on each monomeric unit, while low-molecular-weight anions carry a charge $-q$. Additionally, assume that the degree of polymerization of cations is sufficiently large ($N \gg 1$), allowing us to neglect the translational entropy of the polymer chains' centers of mass. Then, we can express the grand thermodynamic potential of the polyelectrolyte as[128, 129]:

$$\Omega = F_{el} + \Omega_{ref}, \tag{6.30}$$

where

$$F_{el} = -\frac{\varepsilon}{8\pi}\int d\mathbf{r}\psi(\mathbf{r})\nabla^2\psi(\mathbf{r}) + \int d\mathbf{r}\rho(\mathbf{r})\psi(\mathbf{r}) \tag{6.31}$$

is the electrostatic energy, written in the mean field approximation, with the following local charge density

$$\rho(\mathbf{r}) = \rho_{ext}(\mathbf{r}) + q\left(n_p(\mathbf{r}) - n_c(\mathbf{r})\right), \tag{6.32}$$

where the first term describes the local external charge density, and the second term represents the local charge density of the monomer units and counterions. Here, ε denotes the dielectric permittivity of the polyelectrolyte, encompassing the polarization of the medium. The contribution of the reference system can be expressed as:

$$\Omega_{ref} = F_{conf} + F_{vol} - \int d\mathbf{r}\left(\mu_p n_p(\mathbf{r}) + \mu_c n_c(\mathbf{r})\right), \tag{6.33}$$

where

$$F_{conf}[n_p] = \frac{k_B T b^2}{6}\int d\mathbf{r}\,(\nabla n_p^{1/2}(\mathbf{r}))^2 \tag{6.34}$$

is the Lifshitz conformation free energy [26, 28] of flexible polymer chains with a bond length b. The contribution of steric interactions of the monomer units and counterions can be expressed as:

$$F_{vol}[n_p, n_c] = \int d\mathbf{r} f(n_p(\mathbf{r}), n_c(\mathbf{r})) \tag{6.35}$$

where $f = f(n_p(\mathbf{r}), n_c(\mathbf{r}))$ is the free energy density as a function of the local concentrations, $n_{p,c}(\mathbf{r})$, of the monomer units and counterions. In this case, we can utilize the the following expression for the lattice model:

$$f = \frac{k_B T}{v}\left(\phi_c \ln \phi_c + \frac{1}{N}\phi_p \ln \phi_p + (1-\phi_c-\phi_p)\ln(1-\phi_c-\phi_p)\right), \tag{6.36}$$

where $\phi_c = n_c v$ and $\phi_p = n_p v$ are the local volume fractions of the counterions and monomer units, respectively, with v being the unit cell volume determined through the bond length via the standard relation for the lattice model, $v = b^3$. μ_p and μ_c are the bulk chemical potentials of the monomer units and the counterions. Let's write the Euler-Lagrange equations, $\delta\Omega/\delta n_{p,c}(\mathbf{r}) = 0$ and $\delta\Omega/\delta\psi(\mathbf{r}) = 0$, which lead to the following self-consistent field equations for ion concentrations and electrostatic potential:

$$\begin{cases} \bar{\mu}_c(\mathbf{r}) - q\psi(\mathbf{r}) = \mu_c \\ \bar{\mu}_p(\mathbf{r}) - \frac{k_B T b^2}{6 n_p^{1/2}(\mathbf{r})}\Delta n_p^{1/2}(\mathbf{r}) + q\psi(\mathbf{r}) = \mu_p \\ -\frac{\varepsilon}{4\pi}\Delta\psi(\mathbf{r}) = \rho_{ext}(\mathbf{r}) + q\left(n_p(\mathbf{r}) - n_c(\mathbf{r})\right), \end{cases} \tag{6.37}$$

where

$$\bar{\mu}_c = \frac{\partial f}{\partial n_c} = k_B T \ln\left(\frac{\phi_c}{1-\phi_c-\phi_p}\right), \tag{6.38}$$

$$\bar{\mu}_p = \frac{\partial f}{\partial n_p} = -k_B T \ln(1-\phi_c-\phi_p) \tag{6.39}$$

are the intrinsic chemical potentials of monomer units and counterions, respectively, where we neglect the term related to the entropy of translational motion of the chains as a whole, due to $N \gg 1$. Taking into account that in the bulk phase of the solution, where $\psi = 0$, the condition of local electroneutrality is fulfilled, $n_p = n_c = n_0$, we can obtain the following expressions for the bulk chemical potentials of particles:

$$\mu_c = k_B T \ln\left(\frac{\phi_0}{1-2\phi_0}\right), \tag{6.40}$$

$$\mu_p = -k_B T \ln(1-2\phi_0), \tag{6.41}$$

where $\phi_0 = n_0 v$ is the volume fraction of the monomer units and counterions in the bulk solution.

Linear approximation. The system of self-consistent field equations (6.37) for the local concentrations of monomer units and counterions, $n_{p,c}(\mathbf{r})$, and the electrostatic potential, $\psi(\mathbf{r})$, obtained above cannot be solved analytically in the general case. However, by analogy with the classical Debye-Hückel theory for solutions of strong electrolytes, such a system of equations can be analyzed in the limit of weak electrostatic interactions, when the local concentrations of monomeric units and counterions differ weakly from their values in the bulk solution, n_0, i.e.

$$n_p = n_0 + \bar{n}_p, \quad n_c = n_0 + \bar{n}_c. \tag{6.42}$$

where we introduce small perturbations of concentrations, $|\bar{n}_{p,c}| \ll n_0$. Thus, introducing dimensionless parameters $x_p = \bar{n}_p/n_0$, $x_c = \bar{n}_c/n_0$, $\bar{\psi} = q\psi/k_B T$, we can linearize the self-consistent field equations (6.37) and arrive at the following system of linear equations:

$$\begin{cases} \Delta^2 x_p - \lambda^2 \Delta x_p + \frac{\chi^4}{4} x_p = -\frac{48\pi q}{\varepsilon k_B T b^2 (1-\phi_0)} \rho_{ext} \\ \bar{\psi} = \frac{b^2}{12}(1-\phi_0)\Delta x_p - \phi_0 x_p \\ x_c = \frac{(1-2\phi_0)\bar{\psi} - \phi_0 x_p}{1-\phi_0}, \end{cases} \tag{6.43}$$

where we also introduce following auxiliary parameters

$$\lambda^2 = \frac{12\phi_0}{b^2(1-\phi_0)} + \frac{1-2\phi_0}{r_D^2(1-\phi_0)}, \; \chi^4 = \frac{48}{r_D^2 b^2} \frac{1+2\phi_0}{1-\phi_0}, \tag{6.44}$$

where $r_D = \left(4\pi q^2 n_0/\varepsilon k_B T\right)^{-1/2}$ is the Debye radius corresponding to the presence of mobile counterions in the system. Let us now analyze the solution of the set of Eq. (6.43) for two physically relevant cases: the case of a point charge, and the case of a flat charged electrode placed in a polyelectrolyte solution (or polymerized ionic liquid).

Point charge case. Let us find the solution of the system (6.43) for the case when the polyelectrolyte is perturbed by the presence of a fixed point sample charge q_0, placed at the origin. Taking into account that the density of the test point charge is

$$\rho_{ext}(\mathbf{r}) = q_0 \delta(\mathbf{r}), \tag{6.45}$$

where $\delta(\mathbf{r})$ is the Dirac delta-function, we obtain

$$\Delta^2 x_p - \lambda^2 \Delta x_p + \frac{\chi^4}{4} x_p = -\frac{48\pi q q_0}{\varepsilon k_B T b^2 (1-\phi_0)} \delta(\mathbf{r}). \tag{6.46}$$

Using integral Fourier transformation

$$x_p(\mathbf{r}) = \int \frac{d\mathbf{k}}{(2\pi)^3} e^{i\mathbf{k}\mathbf{r}} \tilde{x}_p(\mathbf{k}), \tag{6.47}$$

we arrive at the following expression for the Fourier transform

$$\tilde{x}_p(\mathbf{k}) = -\frac{48\pi q q_0}{\varepsilon k_B T b^2(1-\phi_0)} \frac{1}{k^4 + \lambda^2 k^2 + \frac{\chi^4}{4}}. \tag{6.48}$$

We consider only the physically relevant case when $\lambda < \chi$, which can be reformulated as the following condition on the Debye radius and the chain segment length:

$$\frac{\sqrt{(1+2\phi_0)(1-\phi_0)} - 1}{2\sqrt{3}\phi_0} < \frac{r_D}{b} < \frac{\sqrt{(1+2\phi_0)(1-\phi_0)} + 1}{2\sqrt{3}\phi_0}. \tag{6.49}$$

Condition (6.49) covers the cases of a polymerized ionic liquid melt and polyelectrolyte solution in a polar solvent (the approximate physical parameters used in the model are given below). In this case, the inverse Fourier transform leads to:

$$x_p = a\frac{\exp\left[-\varkappa r\right]}{r} \sin kr, \tag{6.50}$$

where we introduce the constants $\varkappa = \sqrt{\chi^2 + \lambda^2}/2$, $k = \sqrt{\chi^2 - \lambda^2}/2$ (we discuss their physical meaning later) and the coefficient

$$a = -\frac{6 q q_0}{\varepsilon k_B T b^2 \varkappa k(1-\phi_0)}. \tag{6.51}$$

Dimensionless electrostatic potential has the following form

$$\bar{\psi} = \frac{\exp\left[-\varkappa r\right]}{r}(b_1 \sin kr + b_2 \cos kr), \tag{6.52}$$

where

$$b_1 = a\left(\theta - \phi_0\right),\ b_2 = -a\eta, \tag{6.53}$$

with $\theta = \chi^2 b^2(1-\phi_0)/24$ and $\eta = \varkappa k b^2(1-\phi_0)/6$.

Perturbation of the counterions concentration is determined by the expression

$$x_c = \frac{\exp\left[-\varkappa r\right]}{r}(c_1 \sin kr + c_2 \cos kr) \tag{6.54}$$

with the following coefficients

$$c_1 = \frac{(1-2\phi_0)b_1 - \phi_0 a}{1-\phi_0},\ c_2 = \frac{1-2\phi_0}{1-\phi_0} b_2. \tag{6.55}$$

It's interesting to note that for the case of $\phi_0 = 0.5$, indicating absence of the voids in the bulk liquid structure, we find $x_p = -x_c$, or $n_p + n_c = 2n_0$. Additionally, a similar expression to (6.52) was derived within field-theoretic approaches studying polyelectrolyte solutions in works [115, 130].

It's worth noting that the length $\varkappa^{-1}$ characterizes the spatial decay of perturbations in the concentrations of monomeric units and counterions, as well as the electrostatic potential, relative to their values in the bulk phase. This length is analogous to the Debye radius from the theory of simple electrolyte solutions. Additionally, the length k^{-1} can be interpreted as the degree of "periodicity" of the predominantly positively or negatively charged regions, arising from the intricate interplay between the conformational entropy of the polymer chains, electrostatic interactions, and the excluded volume interactions of the monomeric units.

Case of the flat charged electrode. Now let's examine the outcome of the linear theory for the scenario of a flat infinite uniformly charged electrode with a surface charge density σ. In this case the charge density of the electrode is represented as:

$$\rho_{ext}(z) = \sigma\delta(z), \tag{6.56}$$

which leads to the following expressions for the perturbation of the monomer units concentration

$$x_p^{(IV)}(z) - \lambda^2 x_p''(z) + \frac{\chi^4}{4} x_p(z) = -\frac{48\pi q\sigma}{\varepsilon k_B T b^2(1-\phi_0)}\delta(z), \tag{6.57}$$

where the upper index (IV) denotes the forth order derivative. The boundary conditions can be written in the form

$$x_p'(0) = 0,\ x_p'''(0) = -\frac{48q\sigma}{\varepsilon k_B T b^2(1-\phi_0)},\ x_p(\infty) = x_p'(\infty) = 0,$$

where we assume the absence of specific adsorption of polymer chains onto the electrode surface. Below, we will discuss the possible extension of the theory to accommodate the specific adsorption of monomeric chains onto the electrode. The solution to Eq. (6.57) with the specified boundary conditions leads to

$$x_p(z) = e^{-\varkappa z}\left(f_1\cos(kz) + f_2\sin(kz)\right), \tag{6.58}$$

with the following coefficients

$$f_1 = -\frac{6q\sigma}{\varkappa(\varkappa^2+k^2)\varepsilon k_B T b^2(1-\phi_0)},\ f_2 = -\frac{6q\sigma}{k(\varkappa^2+k^2)\varepsilon k_B T b^2(1-\phi_0)}. \tag{6.59}$$

Dimensionless electrostatic potential is

$$\bar{\psi} = e^{-\varkappa z}\left(g_1\cos(kz) + g_2\sin(kz)\right), \tag{6.60}$$

where

$$g_1 = (\theta-\phi_0)f_1 - \eta f_2,\ g_2 = (\theta-\phi_0)f_2 + \eta f_1. \tag{6.61}$$

Small perturbation of the counterion concentration is expressed as follows

$$x_c = e^{-\varkappa z} \left(U_1 \cos(kz) + U_2 \sin(kz)\right), \tag{6.62}$$

while the coefficients U_1, U_2 are linearly related to the previously mentioned coefficients

$$U_1 = \frac{(1-2\phi_0)g_1 - \phi_0 f_1}{1-\phi_0}, \; U_2 = \frac{(1-2\phi_0)g_2 - \phi_0 f_2}{1-\phi_0}. \tag{6.63}$$

It is noteworthy that, similar to the scenario with the point test charge, when $\phi_0 = 0.5$ we have $x_p = -x_c$, or equivalently, $n_c(z) + n_p(z) = 2n_0$.

Setting a voltage at the electrode, $\psi_0 = \psi(0)$, and utilizing Eq. (5.104), after some algebra, we arrive at the following expression for the surface charge density:

$$\sigma = C_0 \psi_0, \tag{6.64}$$

with a linear differential capacitance of EDL

$$C_0 = \frac{\varepsilon}{4\pi d}, \tag{6.65}$$

where

$$d = \frac{1 + \frac{\phi_0}{\theta}}{\sqrt{\lambda^2 + \chi^2}} \tag{6.66}$$

can be interpreted as the effective thickness of the diffuse part of EDL. This thickness is proportional to the effective screening length, $\varkappa^{-1}$, as defined earlier. Comparing the obtained expressions, one can clearly discern the difference between the results for polyelectrolytes and for low molecular weight ionic liquids, studied theoretically within the symmetric lattice gas model [60]. Specifically, the differential capacitance, C_0, of a polyelectrolyte in the linear approximation depends on the molecular parameters of the ionic species: the bond length, b, and the effective excluded volume of species, v. In the limit of negligibly small excluded volume interactions ($v \to 0$) at constant bond length b, we can obtain the expression

$$C_0 = C_{GC}\sqrt{1 + \frac{4\sqrt{3}r_D}{b}}, \tag{6.67}$$

which is similar to the well-known Debye capacitance of the diffuse layer, $C_{GC} = \varepsilon\varepsilon_0/r_D$, for the case of low molecular weight electrolyte solutions and monomeric ionic liquids. It's important to note that Eq. (6.67) is accurate only for sufficiently dilute polyelectrolyte solutions at a charged electrode in the θ-solvent regime, where the polymer chains behave as ideal Gaussian coils. In the region where $r_D \ll b$, Eq. (6.67) reduces to the standard Gouy-Chapman expression, $d \approx r_D$. However, when $r_D \gg b$, the thickness of the EDL is determined by $d \approx 48^{-1/4}(r_D b)^{1/2}$. At the same time, decreasing the bond length, b, as can be seen from Eq. (6.67), leads to an increase in differential capacitance. This behavior can be explained by the fact that a decrease in bond length results in a higher linear charge density of the polymer chains, and hence a larger accumulated charge in the EDL.

Polyelectrolyte at charged electrode: nonlinear case. Now, let's examine the scenario of a polyelectrolyte at a flat charged electrode with a constant surface potential, ψ_0. In this case, the self-consistent field equations simplify to the following form:

$$\begin{cases} \bar{\mu}_p - \frac{k_B T b^2}{6\phi_p^{1/2}}(\phi_p^{1/2})'' + q\psi = \mu_p \\ \psi'' = -\frac{4\pi q}{\varepsilon v}\left(\phi_p - \phi_c\right) \\ \phi_c = (1-\phi_p)\frac{e^{\frac{\mu_c + q\psi}{k_B T}}}{1+e^{\frac{\mu_c + q\psi}{k_B T}}}. \end{cases} \tag{6.68}$$

The solve the Eq. (6.68) we need to specify the boundary conditions. Standard boundary conditions for the electrostatic potential are $\psi(0) = \psi_0$ and $\psi'(\infty) = 0$. Additionally, general boundary conditions for the local polymer volume fraction can be expressed as $\phi'_p(0)/\phi_p(0) = -2/\xi$ and $\phi_p(\infty) = \phi_0$. Here, the so-called extrapolation length, ξ, is inversely proportional to the energy of short-range specific interactions between the monomeric units and the surface [131]. This length is typically positive and on the order of the molecular size when there is attraction between the polymer and the surface. ξ becomes negative if the surface repels the chains and is zero for the case of impermeable solid wall ($\phi_p(0) = 0$). Below, we consider two cases: the case of a non-interacting surface, where $1/\xi = 0$, implying $\phi'_p(0) = 0$. This scenario resembles Kornyshev's theory for a monomeric ionic liquid. The second case models a solid wall, when $\phi_p(0) = 0$. This case corresponds to infinitely strong specific polymer-electrode repulsion at small distances. By solving the system of self-consistent field equations (6.68) at different surface potentials, ψ_0, for these two cases, we can compute the surface charge density as $\sigma = -\varepsilon\psi'(0)/4\pi$, and subsequently determine the differential capacitance, $C = d\sigma/d\psi_0$.

Case of "indifferent" electrode. By numerically solving the self-consistent field equations (6.68) with the boundary conditions $\phi'_p(0) = 0$, we can compute the differential capacitance of the EDL as a function of the applied voltage. Figure 6.10a provides a comparison between the numerically calculated differential capacitance profiles within the proposed theory and the capacitance profile for monomeric IL determined by Kornyshev's formula (4.84).

As observed, similar to the case of monomeric IL, the profile of the differential capacitance for polyelectrolytes exhibits a rather symmetrical shape. Notably, at sufficiently high voltage values, the differential capacitance decreases following a law of $C \sim |\psi_0|^{-0.5}$, akin to the behavior observed in simple ILs. Additionally, at relatively large positive voltage values, the DC profiles coincide with those predicted by Kornyshev's theory, indicating complete depletion of monomeric units from the electrode by counterions (see Fig. 6.11). Consequently, the local structure of the EDL resembles that of monomeric ILs, thereby explaining the consistency with the results

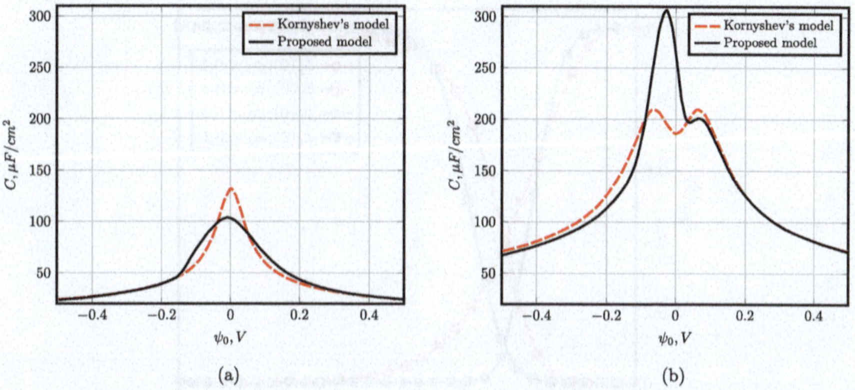

Fig. 6.10 Differential capacitance profile obtained numerically by present theory for PIL and differential capacitance predicted by the analytical mean-field theory for a monomeric IL (see Eq. (4.84))—(**a**). The data are shown for $\phi_0 = 0.4$, $b = v^{1/3} = 0.5$ nm, $T = 298$ K, $\varepsilon = 5$, $q = 1.6 \times 10^{-19}$ C. Differential capacitance profiles predicted by the present theory for a polyelectrolyte solution and predicted by Kornyshev's mean-field theory for a simple electrolyte solution (Eq. (4.84))—(**b**). Here, the following changes in parameters were adopted: $\phi_0 = 0.1$ and $\varepsilon = 40$

of Kornyshev's theory. Furthermore, at sufficiently large negative surface potentials, values of the differential capacitance are also comparable to those obtained within the mean-field theory for simple ILs. This can be attributed to the significant volume fractions of monomeric units ($\phi_p \sim 1$) near the electrode at such relatively high negative potential values, where from an entropy standpoint, there is a minimal distinction between connected charged monomeric units and freely moving ions, as translational entropy is negligible in both cases (see also Fig. 6.11).

It is intriguing to analyze the differential capacitance profile when considering a polyelectrolyte dissolved in a highly polar organic solvent with a dielectric permittivity of approximately $\varepsilon \approx 40$, assuming a volume fraction of monomeric units $\phi_0 = 0.1$. Figure 6.10b illustrates a comparison between the DC profiles predicted by the proposed theory and Kornyshev's mean-field theory for the cases of polyelectrolyte and simple electrolyte solutions, respectively. Unlike the symmetrical bell-shaped capacitance profile observed in the polymerized ionic liquid melt case discussed earlier, the DC profile for the polyelectrolyte solution exhibits two strongly asymmetrical peaks. The higher peak is situated in the negative potential region, while the less pronounced peak appears in the positive region. This behavior significantly differs from that observed in solutions of simple electrolytes predicted by Kornyshev's theory for identical physical parameters. Particularly at negative voltages, where the number of positively charged monomeric units near the electrode surface exceeds that of counterions, we observe considerably larger values of differential capacitance compared to those observed in low molecular weight electrolyte solutions. This peculiar behavior of the differential capacitance can be attributed to the drastic increase in the local volume fraction of monomeric units at the electrode and

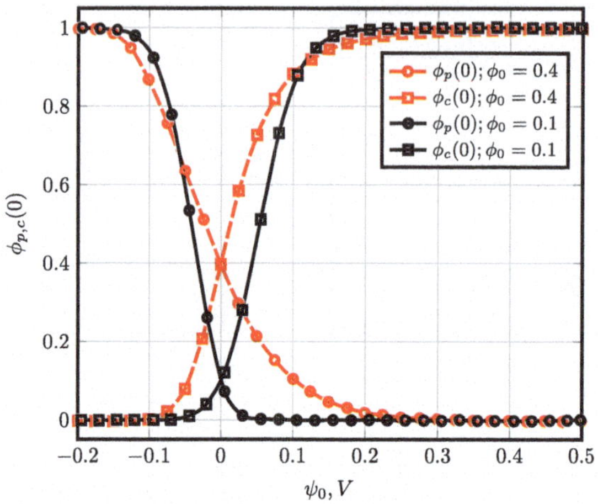

Fig. 6.11 Local volume fractions of monomeric units and counterions on the electrode as functions of the potential drop for the cases of a pure polymeric ionic liquid and and a polymeric ionic liquid solution. The parameters for the pure polymeric ionic liquid (blue lines) are $\phi_0 = 0.4$, $b = v^{1/3} = 0.5$ nm, $T = 298$ K, $\varepsilon = 5$. The parameters for the polymeric ionic liquid solution (black lines) are $\phi_0 = 0.1$, $b = v^{1/3} = 0.5$ nm, $T = 298$ K, $\varepsilon = 40$, $q = 1.6 \times 10^{-19}$ C

the corresponding decrease in the volume fraction of counterions as negative voltage values increase. Conversely, positive potential values lead to a slower increase (decrease) in the local volume fraction of counterions (monomeric units). This discrepancy arises because long polymer chains with very high degrees of polymerization possess zero translational entropy and are thus more strongly attracted to negatively charged electrodes compared to mobile counterions to positively charged ones. Furthermore, it is noteworthy that in both the case of a polyelectrolyte solution and a polymerized ionic liquid melt, the DC at sufficiently large potential values is described by the same power law, $C \sim |\psi_0|^{-0.5}$.

It is insightful to consider separately how variation of the bond length of the polymer chain, $b = v^{1/3}$, influences the behavior of the differential capacitance. As depicted in Fig. 6.12, increasing the chain length results in a monotonic decrease in the differential capacitance across all potential values. This trend can be rationalized by considering that the EDL accumulates less charge due to significant steric interactions between the ions resulting from the elongation of the chain bond length and the increased excluded volume of particles. It's noteworthy to mention that a similar trend in capacitance behavior as a function of bond length is also observed for the case of the polyelectrolyte solution discussed earlier.

Now, let's discuss the profiles of the electrostatic potential and local ion concentrations obtained by numerically solving the self-consistent field equations (6.68) and compare them with predictions of the linear theory. As shown in Fig. 6.13, the potential profile exhibits a minimum at positive voltages, which becomes more pronounced with increasing voltage. Similarly, for negative voltages (see Fig. 6.14),

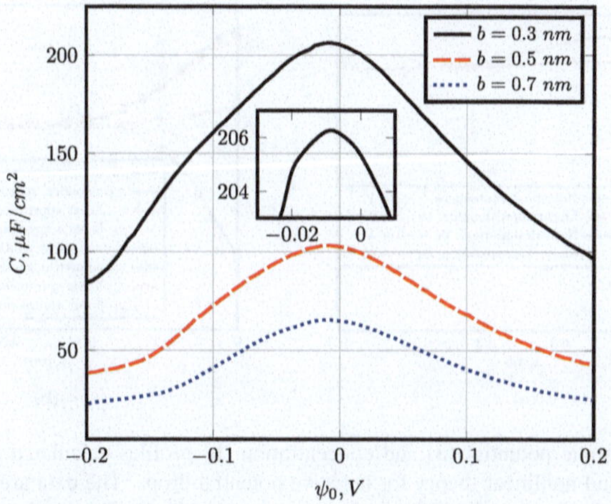

Fig. 6.12 Differential capacitance profiles of a pure PIL on a charged electrode plotted for different bond lengths, $b = v^{1/3}$. The horizontal dashed lines determine the linear capacitance (see Eq. (4.80)). The data are shown for $\phi_0 = 0.4$, $T = 298$ K, $\varepsilon = 5$, $q = 1.6 \times 10^{-19}$ C

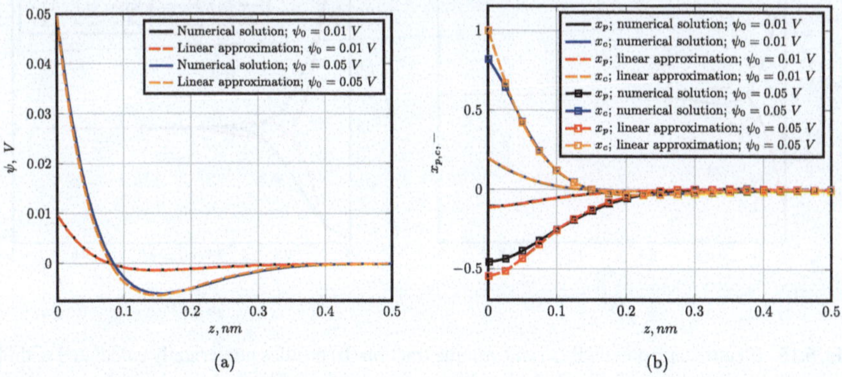

Fig. 6.13 Electrostatic potential (**a**) and concentration (**b**) profiles calculated within the linear analytical theory and nonlinear theory for positive potential drops. The data are shown for $\phi_0 = 0.4$, $b = v^{1/3} = 0.5$ nm, $T = 298$ K, $\varepsilon = 5$, $q = 1.6 \times 10^{-19}$ C. The local concentrations are determined via $x_{p,c}(z) = (n_{p,c}(z) - n_0)/n_0$

the potential profile follows the same trend. This behavior of the potential arises from the interplay between electrostatic interactions and the conformational entropy of macromolecules, akin to the behavior observed for the electrostatic potential of a point charge immersed in a polymerized ionic liquid (PIL) melt (see Eq. (6.52)) or in a polyelectrolyte solution [115, 130]. It's noteworthy that the potential profiles predicted by the linear theory nearly coincide with the numerical results for voltages in the range $|\psi_0| \leq 0.05V$. However, for voltages $|\psi_0| > 0.05V$, the linear theory already exhibits significant deviation from the values of the local electrostatic

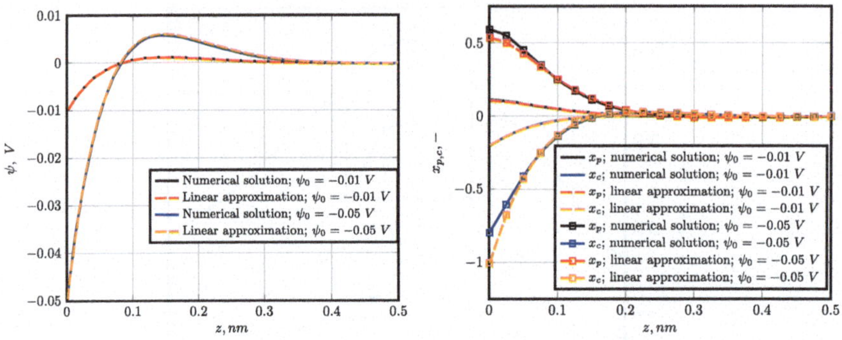

(a) (b)

Fig. 6.14 Electrostatic potential (**a**) and concentration (**b**) profiles calculated within the linear analytical theory and nonlinear theory for negative potential drops. The data are shown for $\phi_0 = 0.4$, $b = v^{1/3} = 0.5$ nm, $T = 298$ K, $\varepsilon = 5$, $q = 1.6 \times 10^{-19}$ C. The local concentrations are determined via $x_{p,c}(z) = (n_{p,c}(z) - n_0)/n_0$

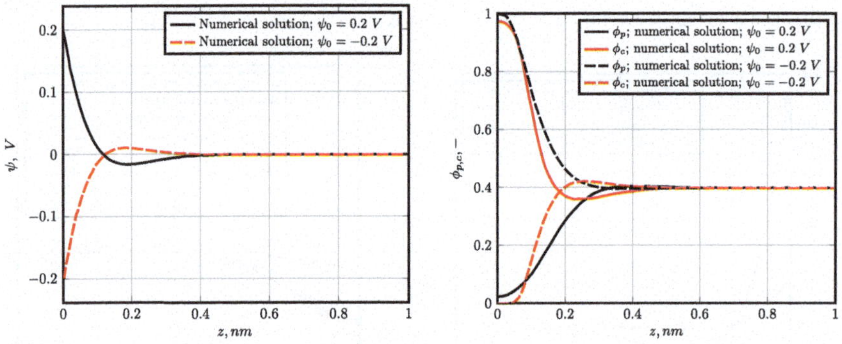

(a) (b)

Fig. 6.15 Electrostatic potential (**a**) and volume fraction (**b**) profiles numerically calculated within the nonlinear theory for sufficiently high voltages $\psi_0 = \pm 0.2$ V. The data are shown for $\phi_0 = 0.4$, $b = v^{1/3} = 0.5$ nm, $T = 298$ K, $\varepsilon = 5$, $q = 1.6 \times 10^{-19}$ C

potential. Moreover, the range of voltages for which the linear theory provides a satisfactory approximation for concentration profiles is even narrower. At voltages around $\psi_0 \approx \pm 0.05$ V, the local ion concentrations predicted by the linear theory slightly deviate from the numerical solution near the electrode surface ($z = 0$), and this deviation increases with higher potential values. Nevertheless, it's worth noting that the expressions (6.58), (6.60), and (6.62) derived in the linear theory determine the asymptotic behavior of the potential profiles and local ion concentrations at sufficiently large distances from the electrode for any applied voltages. Interestingly, the effect of oscillations in the concentration profiles is minimal, and they qualitatively align with those predicted by mean-field theory for conventional ionic liquids.

Finally, let's examine the behavior of electrostatic potential and ion concentration profiles at sufficiently large values of the electrode potential, where the linear approximation becomes inapplicable. In Fig. 6.15, the figure (a) illustrates the electrostatic potential profile, while the figure (b) depicts the local ion concentration profiles for a potential of $\psi_0 = \pm 0.2\ V$. For the case of a polymerized ionic liquid melt pronounced oscillations are absent on the electrostatic potential and ion concentration profiles, contrasting with the behavior of salt-free polyelectrolyte solutions, where such oscillations are commonly observed [132]. Instead, rather pronounced maxima or minima are observed at a certain distance from the electrode. Notably, the potential profiles at $\psi_0 = \pm 0.2\ V$ are almost symmetric. This symmetry can be attributed to the fact that at sufficiently high voltages, the local volume fraction of monomer units (at negative potentials) or counterions (at positive ones) near the electrode is close to unity. Thus, whether the particles are connected in chains or move freely becomes less relevant, as the translational and conformational contributions to entropy are small in both cases.

Case of "hard-wall". It is also instructive to examine another limiting regime, characterized by a "hard-wall" boundary condition where $\phi_p(0) = 0$. This condition describes a strong repulsion between the polymer and the electrode at close distances. In this context, previously discussed linear theory becomes inapplicable even at small potential values due to the significant deviation of the local polymer volume fraction near the electrode from its bulk value, arising from the depletion of monomeric units from the vicinity of the electrode [128]. As illustrated in Fig. 6.16, the differential capacitance curves exhibit qualitative differences compared to those observed for a neutral/indifferent electrode. While a slightly shifted bell-shaped profile of differential capacitance characterizes the case of a polymerized ionic liquid melt at a neutral electrode, a "camel-shaped" profile emerges in the hard-wall regime, featuring two distinctly asymmetric peaks (see 6.16a). One peak, taller and narrower, is located

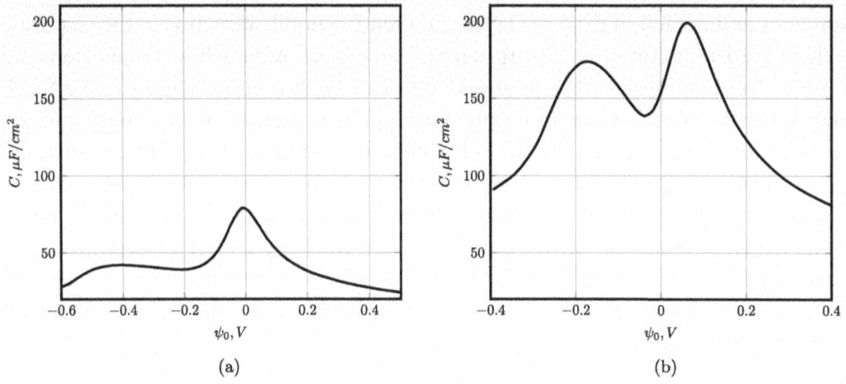

Fig. 6.16 Differential capacitance profiles for a pure PIL a and the same PIL dissolved in an organic solvent b plotted for the hard wall regime. The physical parameters are the same as those taken for Fig. 6.10a, b, respectively

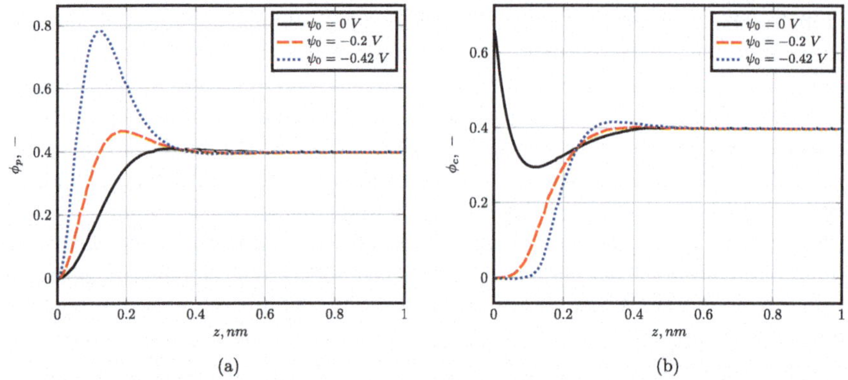

Fig. 6.17 Local volume fraction profiles of monomer units (**a**) and counterions (**b**) for a pure PIL calculated in the hard wall regime. The data are shown for $\phi_0 = 0.4, b = v^{1/3} = 0.5$ nm, $T = 298$ K, $\varepsilon = 5, q = 1.6 \times 10^{-19}\ C$

around zero potential with a slight shift toward the negative potentials, while the other peak emerges in the region of considerably large negative potentials. In the case of a polyelectrolyte solution, the differential capacitance curve also exhibits a standard "camel-shaped" profile, albeit with peaks of roughly similar shapes. The broader and lower peak observed in the case of the polymerized ionic liquid melt can be elucidated by examining the behavior of the local volume fractions of monomer units and counterions, as depicted in Fig. 6.17. At zero potential, monomer units are depleted from a narrow layer at the vicinity of electrode surface, while counterions are found to be in excess compared to the bulk concentration in the same layer. However, as the negative potential increases beyond a certain threshold ($\psi_0 \approx -0.4\ V$), the depleted layer narrows further, and the maximum on the polymer local volume fraction profile increases. In essence, at sufficiently large negative potentials, the interplay between monomer unit depletion from the layer and their Coulomb attraction to the electrode leads to the formation of an adsorption polymer layer. Meanwhile, counterions are depleted from the near-surface region due to Coulomb repulsion and steric effects. Such formation of an adsorption polymer layer is associated with a maximum on the differential capacitance profile in the negative potential region. In contrast, the polyelectrolyte solution scenario resembles that with the boundary conditions corresponding to an indifferent electrode, albeit with a much lower peak in the negative potential region. This difference arises from weaker polymer adsorption due to short-range polymer-surface repulsion. Notably, as observed in the case of the neutral electrode, the same asymptotic behavior for capacitance, $C \sim |\psi_0|^{-0.5}$, is evident at sufficiently high voltages.

6.3 Mechanical Stress in Inhomogeneous Polyelectrolyte Solutions

Conformation stress tensor. Let's now consider a model similar to the one proposed in the previous section, focusing on a polyelectrolyte solution consisting of polymerized cations (macroions) with a high polymerization and low-molecular-weight anions (counterions). GTP of the system can be written as

$$\Omega = \int d\mathbf{r}\,\omega(\mathbf{r}), \tag{6.69}$$

where we introduce the following GTP density

$$\omega = -\frac{\varepsilon(\nabla\psi)^2}{8\pi} + \rho\psi + \frac{k_BTb^2}{6}(\nabla n_p^{1/2})^2 + f - \mu_p n_p - \mu_c n_c. \tag{6.70}$$

Recall that the first and second terms represent the electrostatic energy density in the mean-field approximation, incorporating the local charge density $\rho(\mathbf{r}) = q\left(n_p(\mathbf{r}) - n_c(\mathbf{r})\right)$ and the electrostatic potential $\psi(\mathbf{r})$. Here, ε denotes the dielectric constant of the solvent. The third term accounts for the conformational free energy density [26, 28] of flexible polymer chains with a bond length b. The fourth term, $f = f(n_p, n_c)$, characterizes the contribution of monomer units and counterions volume interactions to the total free energy density with $n_{p,c}(\mathbf{r})$ being the local concentrations of monomer units and counterions. To describe this contribution we use the lattice model (without attraction) [60, 86, 133] in the form: $f = k_BTv^{-1}\left(\phi_c\ln\phi_c + \left(1-\phi_c-\phi_p\right)\ln\left(1-\phi_c-\phi_p\right)\right)$. Here, $\phi_{p,c} = n_{p,c}v$ denotes the local volume fractions of counterions and monomer units, with v representing the unit cell volume related to the bond length by the condition $v = b^3$. Additionally, $\mu_{p,c}$ are the bulk chemical potentials of the monomer units and counterions. The self-consistent field equations, essentially the Euler-Lagrange equations for the introduced functional (6.69), have the following form:

$$\frac{\partial\omega}{\partial\psi} = \partial_i\frac{\partial\omega}{\partial(\partial_i\psi)}, \quad \frac{\partial\omega}{\partial n_p^{1/2}} = \partial_i\frac{\partial\omega}{\partial(\partial_i n_p^{1/2})}, \quad \frac{\partial\omega}{\partial n_c} = 0, \tag{6.71}$$

where $\partial_i = \partial/\partial x_i$ denotes the partial derivative with respect to the Cartesian coordinates, x_i $(i = 1, 2, 3)$. Let us recall that the Einstein sum rule over repeated indices is assumed. Utilizing the defined GTP density (6.70), we derive the following system of equations

$$\begin{cases} \bar{\mu}_c(\mathbf{r}) - q\psi(\mathbf{r}) = \mu_c \\ \bar{\mu}_p(\mathbf{r}) - \frac{k_BTb^2}{6n_p^{1/2}(\mathbf{r})}\nabla^2 n_p^{1/2}(\mathbf{r}) + q\psi(\mathbf{r}) = \mu_p \\ \nabla^2\psi(\mathbf{r}) = -\frac{4\pi q}{\varepsilon}\left(n_p(\mathbf{r}) - n_c(\mathbf{r})\right), \end{cases} \tag{6.72}$$

where the intrinsic chemical potentials of the monomer units and counterions are

$$\bar{\mu}_c = \frac{\partial f}{\partial n_c} = k_B T \ln\left(\frac{\phi_c}{1-\phi_c-\phi_p}\right),$$
$$\bar{\mu}_p = \frac{\partial f}{\partial n_p} = -k_B T \ln\left(1-\phi_c-\phi_p\right). \tag{6.73}$$

Considering the local electroneutrality condition, $n_p = n_c = n_0$, which is fulfilled in the bulk solution when $\psi = 0$, we derive the following expressions for the bulk chemical potentials of the species: $\mu_c = k_B T \ln(\phi_0/(1-2\phi_0))$, $\mu_p = -k_B T \ln(1-2\phi_0)$, where $\phi_0 = n_0 v$ represents the volume fraction of monomer units and counterions. The boundary conditions on the polymer concentration and electrostatic potential are [87, 128] $n_p|_s = 0$, $\psi|_s = \psi_0$, where the symbol $|_s$ denotes that the variables are calculated at the surfaces of macroscopic conductors immersed in the solution. These boundary conditions model a strong repulsion between the monomer units and conducting wall with a fixed surface potential ψ_0. It's worth noting that, for simplicity, we neglect the specific adsorption of counterions here. However, this effect can be incorporated straightforwardly into the model [85].

Applying the approach outlined above to derive the expression for the stress tensor to the GTP (6.69), we obtain the following tensor, which includes three terms:

$$\sigma_{ik} = \sigma_{ik}^{(h)} + \sigma_{ik}^{(M)} + \sigma_{ik}^{(c)}, \tag{6.74}$$

where the first two terms correspond to the hydrostatic and Maxwell stress tensors, which are already known:

$$\sigma_{ik}^{(h)} = -P\delta_{ik}, \tag{6.75}$$

with $P = n_p\bar{\mu}_p + n_c\bar{\mu}_c - f = -k_B T v^{-1}\left(\ln\left(1-\phi_p-\phi_c\right)+\phi_p\right)$, and

$$\sigma_{ik}^{(M)} = \frac{\varepsilon}{4\pi}\left(\mathcal{E}_i\mathcal{E}_k - \frac{1}{2}\mathcal{E}^2\delta_{ik}\right), \tag{6.76}$$

with $\mathcal{E}_i = -\partial_i\psi$, and

$$\sigma_{ik}^{(c)} = \frac{k_B T b^2}{3}\left(\frac{1}{2}\nabla\cdot\left(n_p^{1/2}\nabla n_p^{1/2}\right)\delta_{ik} - \partial_i n_p^{1/2}\partial_k n_p^{1/2}\right) \tag{6.77}$$

determines the contribution of conformational entropy of the polymer chains to the total stress tensor, so-called *conformational* stress tensor.

Polyelectrolyte solution in a slit charged nanopore. Now, let's examine the scenario of a polyelectrolyte solution confined within a slit nanopore with charged conducting walls positioned at $z = 0$ and $z = H$, where H is the pore width. Assuming equilibrium between the confined polyelectrolyte solution and its bulk phase, we can express the self-consistent field equations as follows:

$$\begin{cases} \bar{\mu}_p(z) - \frac{k_B T b^2}{6\phi_p^{1/2}(z)} (\phi_p^{1/2}(z))'' + q\psi(z) = \mu_p \\ \psi''(z) = -\frac{4\pi q}{\varepsilon v} \left(\phi_p(z) - \phi_c(z)\right), \end{cases} \tag{6.78}$$

with the boundary conditions $\psi(0) = \psi(H) = \psi_0$, $\phi_p(0) = \phi_p(H) = 0$; as before, we can analytically express the formula for the volume fraction of counterions in terms of the electrostatic potential and the volume fraction of monomeric units as $\phi_c = (1 - \phi_p)e^{\frac{\mu_c + q\psi}{k_B T}} / \left(1 + e^{\frac{\mu_c + q\psi}{k_B T}}\right)$. Utilizing obtained stress tensor (6.74) we can evaluate the disjoining pressure in the pore. Local mechanical equilibrium can be expressed as

$$\frac{d\sigma_{zz}}{dz} = 0, \tag{6.79}$$

which leads to

$$-\sigma_{zz}(z) = P_b + \Pi = const, \tag{6.80}$$

where σ_{zz} denotes the normal stress, P_b is the bulk pressure, and Π is the disjoining pressure. Thus, determining σ_{zz} at $z = H/2$, where $n'_p(H/2) = \mathcal{E}(H/2) = 0$, we arrive at $\Pi = -\sigma_{zz}(H/2) - P_b = -k_B T b^2 n''_p(H/2)/12 + P_m - P_b$, where we introduce the notation for the pressure in the midpoint of the pore, $P_m = P(H/2)$. The latter equation can then be rewritten in a more convenient form for applications by excluding the second derivative of the polymer concentration. Using the first of the two Eq. (6.78) at $z = H/2$, i.e. $-k_B T b^2 n''_p(H/2)/12 = n_{pm}\left(\mu_p - \mu_{pm} - q\psi_m\right)$, we obtain

$$\Pi = n_{pm}\left(\mu_p - \mu_{pm} - q\psi_m\right) + P_m - P_b, \tag{6.81}$$

where $\psi_m = \psi(H/2)$, $n_{pm} = n_p(H/2)$, $\mu_{pm} = \bar{\mu}_p(H/2)$. As a result, to calculate the disjoining pressure using Eq. (6.81), we first need to solve the self-consistent equations (6.78), and then compute the necessary variables at the midpoint of the pore.

The Fig. 6.18 illustrates the dependency of the disjoining pressure of a polyelectrolyte solution in a polar solvent with a dielectric constant $\varepsilon = 40$ (such as an organic solvent like dimethyl sulfoxide) on the pore width. The surface potential on the walls is fixed at $\psi_0 = 0.1$ V, bulk volume fraction is $\phi_0 = 0.1$. Other model parameters include $b = v^{1/3} = 0.5$ nm, $T = 300$ K, and $q = e$ (where e denotes the electron charge). Insets showing profiles of monomeric units and counterions concentrations at specific pore widths help clarify the reasons behind the pronounced non-monotonic behavior of the disjoining pressure curve. Initially, at narrow widths, there is a repulsion between the pore walls due to the osmotic pressure of counterions and the overlapping of electric double layers formed at the charged walls. However, as the width grows, this transforms into attraction ($P_m < P_b$ or $\Pi < 0$) as the average concentration of counterions in the pore becomes lower than in the bulk solution. As the pore width increases further, polymer chains start to penetrate the volume from the bulk solution (break on the disjoining pressure curve), leading to a abrupt increase in the disjoining pressure due to a significant rise in the osmotic pressure of both

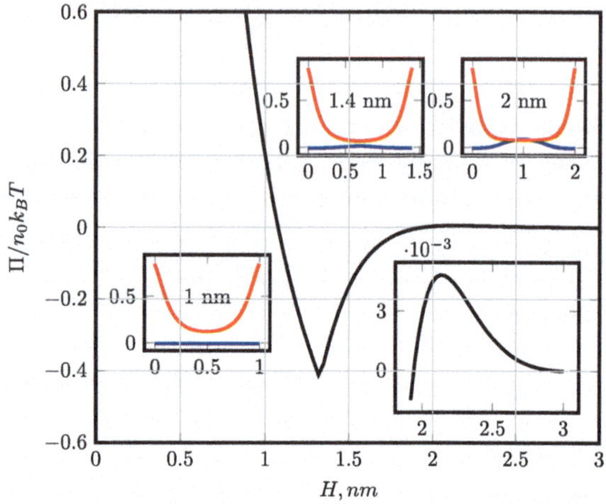

Fig. 6.18 Disjoining pressure as a function of the distance between the walls supplemented by the concentration profiles for monomeric units and counterions. The data are shown for $\phi_0 = 0.1$, $\psi_0 = 0.1\ V$, $\varepsilon = 40$, $b = v^{1/3} = 0.5$ nm, $T = 300$ K and $q = 1.6 \times 10^{-19}\ C$

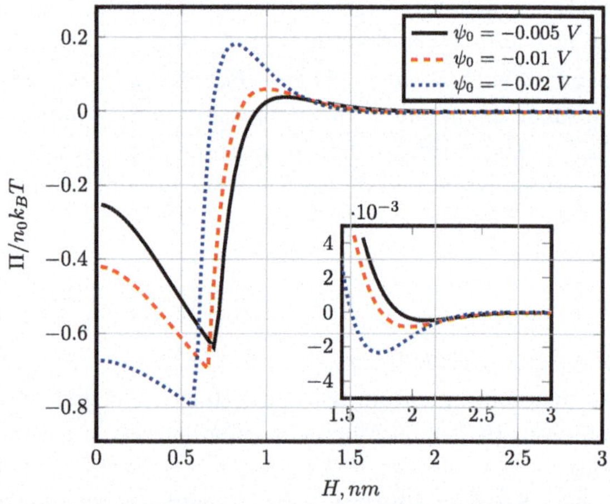

Fig. 6.19 Disjoining pressure of polyelectrolyte solution as a function of distance between the walls for different values of negatively charged surface potential ψ_0. The data are shown for $\phi_0 = 0.1$, $\varepsilon = 40$, $b = v^{1/3} = 0.5$ nm, $T = 300$ K and $q = 1.6 \times 10^{-19}\ C$

counterions and monomeric units. Eventually, at wide enough pores, the disjoining pressure decays exponentially to zero (see the inset at Fig. 6.18).

Interestingly, the same non-monotonic behavior of the disjoining pressure occurs when the monomeric units and the walls are oppositely charged. In such a scenario (Fig. 6.19), when the pore width is small, the concentration of counterions in the pore significantly decreases compared to the bulk solution due to electrostatic repulsion between them and the walls. This eventually results in an effective attraction between the pore walls.

The pore width at which polymer penetration begins can be regulated by adjusting the bulk volume fraction, surface potential, and dielectric constant of the solvent [134]. Thus, increasing the bulk volume fraction of monomeric units, despite the same charge sign on the walls and monomeric units, shifts the minimum to a region of smaller pore width, owing to steric effects. Furthermore, for narrow pores, a higher volume fraction enhances surface potential screening, resulting in smaller pressure values and a deeper minimum. Conversely, with increasing positive surface potential values, the minimum shifts to wider pores due to increasing electrostatic repulsion between polymer chains and walls. Higher dielectric constant of the solvent reduces surface potential screening, thereby boosting counterion concentration at the pore midpoint. Consequently, stronger repulsion between the walls arises from increased osmotic pressure. Increased counterion concentration impedes monomeric units penetration into the pore volume, shifting the minimum on the disjoining pressure curves towards larger pore widths.

The effect observed for the disjoining pressure also reflects in the behavior of the differential capacitance, $C = d\sigma/d\psi_0$, as a function of the surface potential, where $\sigma = -\varepsilon\psi'(0)/4\pi$ denotes the surface charge density on the pore walls. At very small pore widths (see Fig. 6.20a, with parameters identical to those in the figures above), the differential capacitance curve exhibits a similarly abrupt non-monotonic behavior. Then, with a gradual increase in the pore width, a satellite peak first appears, which then smooths out, and finally, the profile transforms to that obtained for an isolated electric double layer, occurring at the interface between the polyelectrolyte solution and the charged electrode. The pronounced non-monotonic behavior of the differential capacitance curve for pores of size ~1 nm can be analyzed in the same way as the situation with disjoining pressure (see insets in Fig. 6.20b). A sharp drop in the differential capacitance is associated with the exclusion of polymer from the pore volume due to electrostatic repulsion at a certain positive value of the surface potential. Further growth of the capacitance is determined by the increase in the concentration of counterions in the pore with increasing surface potential.

Similar to the case of disjoining pressure, the non-monotonic behavior of the differential capacitance can be altered by varying the system parameters [134]. At

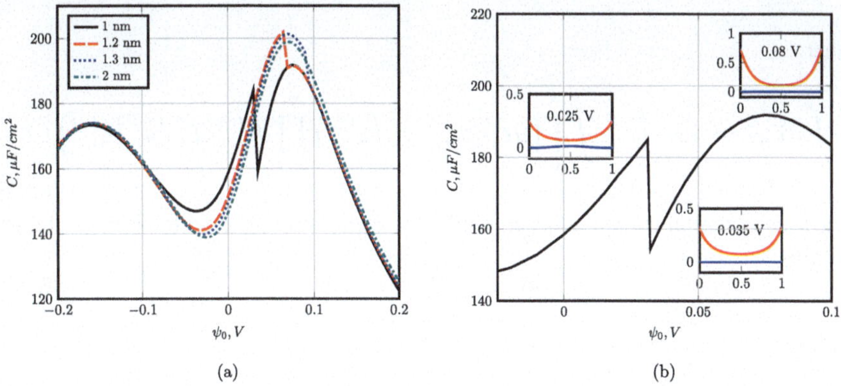

Fig. 6.20 Differential capacitance as a function of potential drop for **a**—1 nm pore supplemented by the concentration profiles at certain surface potentials and **b**—different widths of the pore.The data are shown for $\phi_0 = 0.1$, $\varepsilon = 40$, $b = v^{1/3} = 0.5$ nm, $T = 300$ K and $q = 1.6 \times 10^{-19}$ C

negative surface potentials, an increase in volume fraction simply leads to more particles penetrating the pore, resulting in higher charge accumulation and hence an increase in capacitance. With a change in the sign of the surface potential, a sharp drop in capacitance values can be observed due to the insufficient screening of the surface potential by counterions, which is not enough to keep the polymer in the pore due to electrostatic repulsion. The general idea is as follows: the smaller the volume fraction, the less counterions screen the potential in the pore, and, thus, a smaller positive value of the surface potential is needed to push the polymer out of the pore volume. The capacitance profiles also exhibit sensitivity to changes in the dielectric constant of the solvent [134]. The interpretation is akin to the analysis of the disjoining pressure behavior: as the dielectric constant increases, the screening of the surface potential diminishes. Consequently, for negative surface potentials, a decrease in screening leads to a sharp increase in the polymer concentration in the pore, resulting in a significant increase in capacitance values. Conversely, in the region of positive surface potentials, the more polar the solvent, the smaller the value of the potential needed to push the polymer out of the pore volume.

6.4 Modified Poisson-Boltzmann Equations with Account of Polarizability of the Ions

Consider an ionic fluid composed of ions with charges $q_\pm + e_\pm$ at the center and an additional "peripheral" charge $-e_\pm$ located at a fluctuating distance from the center, resulting in a total charge of $q_\pm$ for the ions. The fluctuating distance, $\boldsymbol{\xi}$, is described by the distribution functions $g_\pm(\boldsymbol{\xi})$. The configuration integral of such a system is given by

$$Q_{N_+N_-} = \int d\Gamma_{N_+N_-} \exp\left[-\beta U_{N_+N_-}\right], \tag{6.82}$$

where

$$\int d\Gamma_{N_+N_-}(\cdot) = \int \cdots \int \prod_{j=1}^{N_+} d\mathbf{r}_j^{(+)} d\boldsymbol{\xi}_j^{(+)} g_+\left(\boldsymbol{\xi}_j^{(+)}\right) \prod_{k=1}^{N_-} d\mathbf{r}_k^{(-)} d\boldsymbol{\xi}_k^{(-)} g_-\left(\boldsymbol{\xi}_k^{(-)}\right)(\cdot). \tag{6.83}$$

The total potential energy is described by the expression (4.4) with the local charge density

$$\hat{\rho}(\mathbf{r}) = \hat{\rho}_i(\mathbf{r}) + \rho_{ext}(\mathbf{r}), \tag{6.84}$$

where

$$\hat{\rho}_i(\mathbf{r}) = \sum_\alpha q_\alpha \hat{c}_\alpha(\mathbf{r}) + \sum_{\alpha=\pm} \sum_{j_\alpha=1}^{N_\alpha} e_\alpha \left(\delta\left(\mathbf{r} - \mathbf{r}_{j_\alpha}^{(\alpha)}\right) - \delta\left(\mathbf{r} - \mathbf{r}_{j_\alpha}^{(\alpha)} - \boldsymbol{\xi}_{j_\alpha}^{(\alpha)}\right)\right) \tag{6.85}$$

is the microscopic charge density of the ions, taking into account their polarizability.

Performing the same transformations as in Chap. 3, we obtain

$$\Xi = \int \frac{\mathcal{D}\varphi}{C_0} e^{-\frac{\beta}{2}(\varphi G_0^{-1}\varphi) + i\beta(\rho_{ext}\varphi)} \Xi_R, \tag{6.86}$$

where we introduce the grand partition function of the reference system

$$\Xi_R = \int \frac{\mathcal{D}\Phi}{N_U} \exp\left[-\frac{\beta}{2}\sum_{\alpha\gamma}(\Phi_\alpha U_{\alpha\gamma}^{-1}\Phi_\gamma) + W[\Phi;\chi]\right] \tag{6.87}$$

with the functional $W[\Phi;\chi]$, determined by the Eq. (4.13) with

$$\chi_\alpha(\mathbf{r}) = q_\alpha\varphi(\mathbf{r}) + iw_\alpha(\mathbf{r}) + k_BT\ln\left[\int d\boldsymbol{\xi}\, g_\alpha(\boldsymbol{\xi}) e^{i\beta e_\alpha(\varphi(\mathbf{r})-\varphi(\mathbf{r}+\boldsymbol{\xi}))}\right]. \tag{6.88}$$

Thus, performing the same evaluations as in Chap. 3, we derive the following expression for the grand thermodynamic potential

$$\Omega[\psi] = \int d\mathbf{r}\left(-\frac{\varepsilon(\nabla\psi)^2}{8\pi} - P(\bar{\mu}_+,\bar{\mu}_-)\right), \tag{6.89}$$

where

$$\bar{\mu}_\alpha(\mathbf{r}) = \mu_\alpha - q_\alpha\psi(\mathbf{r}) + k_BT\ln\left[\int d\boldsymbol{\xi}\, g_\alpha(\boldsymbol{\xi}) e^{-\beta e_\alpha(\psi(\mathbf{r})-\psi(\mathbf{r}+\boldsymbol{\xi}))}\right]. \tag{6.90}$$

Now we need to specify the expression for the distribution function $g_\alpha(\boldsymbol{\xi})$. Let's describe it with the following Gaussian function

$$g_\alpha(\boldsymbol{\xi}) = (2\pi\sigma_\alpha^2)^{-3/2}\int \frac{d\mathbf{n}_\alpha}{4\pi}\exp\left[-\frac{(\boldsymbol{\xi}-\mathbf{s}_\alpha)^2}{2\sigma_\alpha^2}\right], \tag{6.91}$$

averaged over the orientations of the unit vector $\mathbf{n}_\alpha = \mathbf{s}_\alpha/|\mathbf{s}_\alpha|$, where $\mathbf{s}_\alpha$ denotes the vector of the average distance between the charges $\pm e_\alpha$. In this case we obtain

$$\int d\boldsymbol{\xi}\, g_\alpha(\boldsymbol{\xi}) e^{-\beta e_\alpha(\psi(\mathbf{r})-\psi(\mathbf{r}+\boldsymbol{\xi}))} = \int \frac{d\mathbf{n}_\alpha}{4\pi}\int \frac{d\boldsymbol{\xi}}{(2\pi\sigma_\alpha^2)^{3/2}} e^{-\frac{(\boldsymbol{\xi}-\mathbf{s}_\alpha)^2}{2\sigma_\alpha^2}} e^{-\beta e_\alpha(\psi(\mathbf{r})-\psi(\mathbf{r}+\boldsymbol{\xi}))}$$
$$\approx \int \frac{d\mathbf{n}_\alpha}{4\pi}\int \frac{d\boldsymbol{\xi}}{(2\pi\sigma_\alpha^2)^{3/2}} e^{-\frac{(\boldsymbol{\xi}-\mathbf{s}_\alpha)^2}{2\sigma_\alpha^2}} e^{-\beta e_\alpha\boldsymbol{\xi}\nabla\psi(\mathbf{r})} = \frac{\sinh(\beta p_\alpha|\nabla\psi(\mathbf{r})|)}{\beta p_\alpha|\nabla\psi(\mathbf{r})|} e^{\frac{\gamma_\alpha(\nabla\psi(\mathbf{r}))^2}{2k_BT}}, \tag{6.92}$$

where we use the approximation $\psi(\mathbf{r}+\boldsymbol{\xi}) \approx \psi(\mathbf{r}) + \boldsymbol{\xi}\nabla\psi(\mathbf{r})$, and introduce constant dipole moments, $p_\alpha = e_\alpha|\mathbf{s}_\alpha|$, and static polarizabilities, $\gamma_\alpha = \beta\sigma_\alpha^2 e_\alpha^2$, of the ions. Hence, we obtain the following expressions for the intrinsic chemical potentials

$$\bar{\mu}_\alpha(\mathbf{r}) = \mu_\alpha - q_\alpha\psi(\mathbf{r}) + \frac{\gamma_\alpha(\nabla\psi(\mathbf{r}))^2}{2} + k_BT\ln\left[\frac{\sinh(\beta p_\alpha|\nabla\psi(\mathbf{r})|)}{\beta p_\alpha|\nabla\psi(\mathbf{r})|}\right]. \quad (6.93)$$

Varying the functional (6.89) with respect to the electrostatic potential, $\psi(\mathbf{r})$, while taking into account Eq. (6.93) leads to the following self-consistent field equation

$$-\nabla\cdot(\epsilon(\mathbf{r})\nabla\psi(\mathbf{r})) = 4\pi(q_+\bar{n}_+(\mathbf{r}) + q_-\bar{n}_-(\mathbf{r})), \quad (6.94)$$

where we introduce the effective dielectric permittivity of the medium

$$\epsilon(\mathbf{r}) = \varepsilon + 4\pi\sum_{\alpha=\pm}\left(\gamma_\alpha + \frac{p_\alpha^2}{k_BT}\frac{L(\beta p_\alpha\mathcal{E}(\mathbf{r}))}{\beta p_\alpha\mathcal{E}(\mathbf{r})}\right)\bar{n}_\alpha(\mathbf{r}) \quad (6.95)$$

with $\mathcal{E}(\mathbf{r}) = |\nabla\psi(\mathbf{r})|$ representing the absolute value of the local electric field, and $L(x) = \coth x - 1/x$ is the Langevin function.

In conclusion of this section let us discuss what is the total stress tensor can be obtained within the model with poilarizable ions. Using Eq. (4.53), we obtain the expression for the stress tensor

$$\sigma_{ik} = \epsilon\mathcal{E}_i\mathcal{E}_k - \delta_{ik}\left(P_0 + \frac{\varepsilon_b\mathcal{E}^2}{2}\right). \quad (6.96)$$

Note that in the case of unpolarizable ions ($p_\pm = \gamma_\pm = 0$, $\epsilon = \varepsilon$), Eq. (6.96) transforms into Eq. (4.54).

6.5 Modified Poisson-Boltzmann Equations for the Ionic Fluids with the Additions of Polarizable Molecules

Now let's discuss the case of an ionic fluid with an admixture of electrically neutral but polarizable molecules. This model can be used to describe molten salts and ionic liquids with an admixture of a polar liquid or an electrolyte solution with an admixture of polar molecules such as amino acids. As above, we suppose that the ions have charges $q_\pm + e_\pm$ at the center and charges $-e_\pm$ at a fluctuating distance from it. Neutral molecules also have charge e_0 at the center and charge $-e_0$ at a fluctuating distance from it. The distribution functions $g_\alpha(\boldsymbol{\xi})$ $(\alpha = \pm, 0)$ have the same form (see 6.91). The grand partition function takes the following form

$$\Xi = \sum_{N_+=0}^{\infty}\sum_{N_-=0}^{\infty}\sum_{N_0=0}^{\infty}\frac{z_+^{N_+}z_-^{N_-}z_0^{N_0}}{N_+!N_-!N_0!}Q_{N_+N_-N_0} \quad (6.97)$$

with activities $z_\alpha = \Lambda_\alpha^{-3}e^{\beta\mu_\alpha}\theta_\alpha$, $(\alpha = \pm, 0)$ and configuration integrals

$$Q_{N_+N_-N_0} = \int d\Gamma_{N_+N_-N_0} \exp\left[-\beta U_{N_+N_-N_0}\right], \tag{6.98}$$

with

$$\int d\Gamma_{N_+N_-N_0}(\ldots)$$
$$= \int \cdots \int \prod_{j=1}^{N_+} d\mathbf{r}_j^{(+)} d\boldsymbol{\xi}_j^{(+)} g_+\left(\boldsymbol{\xi}_j^{(+)}\right) \prod_{k=1}^{N_-} d\mathbf{r}_k^{(-)} d\boldsymbol{\xi}_k^{(-)} g_-\left(\boldsymbol{\xi}_k^{(-)}\right) \prod_{l=1}^{N_0} d\mathbf{r}_l^{(0)} d\boldsymbol{\xi}_l^{(0)} g_0\left(\boldsymbol{\xi}_l^{(0)}\right)(\ldots), \tag{6.99}$$

where the total potential energy is defined by the expression (4.4). The total local charge density is

$$\hat{\rho}(\mathbf{r}) = \hat{\rho}_i(\mathbf{r}) + \hat{\rho}_d(\mathbf{r}) + \rho_{ext}(\mathbf{r}), \tag{6.100}$$

with $\hat{\rho}_i(\mathbf{r})$ being defined above by the expression (6.85) and

$$\hat{\rho}_d(\mathbf{r}) = \sum_{j=1}^{N_0} e_0\left(\delta\left(\mathbf{r} - \mathbf{r}_j^{(0)}\right) - \delta\left(\mathbf{r} - \mathbf{r}_j^{(0)} - \boldsymbol{\xi}_j^{(0)}\right)\right) \tag{6.101}$$

representing the microscopic bound charge density of the molecules. Thus, performing similar transformations to those from Chap. 3, we arrive at the expression for the grand partition function

$$\Xi = \int \frac{\mathcal{D}\varphi}{C_0} e^{-\frac{\beta}{2}(\varphi G_0^{-1}\varphi) + i\beta(\rho_{ext}\varphi)} \Xi_R[\varphi], \tag{6.102}$$

where, as above, we introduce the grand partition function of the reference system

$$\Xi_R = \int \frac{\mathcal{D}\Phi}{\mathcal{N}_U} \exp\left[-\frac{\beta}{2}\sum_{\alpha\gamma}(\Phi_\alpha U_{\alpha\gamma}^{-1}\Phi_\gamma) + W[\Phi;\chi]\right] \tag{6.103}$$

with the functional

$$W[\Phi;\chi] = \sum_{\alpha=\pm,0} \bar{z}_\alpha \int d\mathbf{r}\, e^{i\beta\Phi_\alpha + i\beta\chi_\alpha}, \tag{6.104}$$

where

$$\chi_\alpha(\mathbf{r}) = q_\alpha\varphi(\mathbf{r}) + i w_\alpha(\mathbf{r}) + k_B T \ln\left[\int d\boldsymbol{\xi}\, g_\alpha(\boldsymbol{\xi}) e^{i\beta e_\alpha(\varphi(\mathbf{r}) - \varphi(\mathbf{r}+\boldsymbol{\xi}))}\right], \quad \alpha = \pm, \tag{6.105}$$

$$\chi_0(\mathbf{r}) = i w_0(\mathbf{r}) + k_B T \ln\left[\int d\boldsymbol{\xi}\, g_0(\boldsymbol{\xi}) e^{i\beta e_0(\varphi(\mathbf{r}) - \varphi(\mathbf{r}+\boldsymbol{\xi}))}\right]. \tag{6.106}$$

Adopting the same approximations as above, we arrive at the expression for the grand thermodynamic potential

$$\Omega[\psi] = \int d\mathbf{r}\left(-\frac{\varepsilon(\nabla\psi)^2}{8\pi} - P(\bar{\mu}_+, \bar{\mu}_-, \bar{\mu}_0)\right) \tag{6.107}$$

and self-consistent field equation

$$-\frac{1}{4\pi}\nabla\cdot(\epsilon(\mathbf{r})\nabla\psi(\mathbf{r})) = q_+\bar{n}_+(\mathbf{r}) + q_-\bar{n}_-(\mathbf{r}), \tag{6.108}$$

with

$$\epsilon(\mathbf{r}) = \varepsilon + 4\pi\sum_{\alpha=\pm,0}\left(\gamma_\alpha + \frac{p_\alpha^2}{k_BT}\frac{L(\beta p_\alpha\mathcal{E}(\mathbf{r}))}{\beta p_\alpha\mathcal{E}(\mathbf{r})}\right)\bar{n}_\alpha(\mathbf{r}) \tag{6.109}$$

and

$$\bar{\mu}_\alpha(\mathbf{r}) = \mu_\alpha - q_\alpha\psi(\mathbf{r}) + \frac{\gamma_\alpha(\nabla\psi(\mathbf{r}))^2}{2} + k_BT\ln\left[\frac{\sinh(\beta p_\alpha|\nabla\psi(\mathbf{r})|)}{\beta p_\alpha|\nabla\psi(\mathbf{r})|}\right], \quad \alpha = \pm, \tag{6.110}$$

$$\bar{\mu}_0(\mathbf{r}) = \mu_0 + \frac{\gamma_0(\nabla\psi(\mathbf{r}))^2}{2} + k_BT\ln\left[\frac{\sinh(\beta p_0|\nabla\psi(\mathbf{r})|)}{\beta p_0|\nabla\psi(\mathbf{r})|}\right], \tag{6.111}$$

where $p_0 = e_0|\mathbf{s}_0|$ and $\gamma_0 = \beta\sigma_0^2e_0^2$ represents the constant dipole moment and the static polarizability of the impurity molecules. The local concentrations of the components are:

$$\bar{n}_\alpha(\mathbf{r}) = \frac{\partial P}{\partial\bar{\mu}_\alpha(\mathbf{r})}, \quad \alpha = \pm, 0. \tag{6.112}$$

It is worth noting that for the case $\gamma_\pm = \gamma_0 = p_\pm = 0$ and $p_0 = p \neq 0$ and

$$P(\bar{\mu}_+, \bar{\mu}_-, \bar{\mu}_0) = \frac{k_BT}{\Lambda_+^3}e^{\beta\bar{\mu}_+} + \frac{k_BT}{\Lambda_-^3}e^{\beta\bar{\mu}_-} + \frac{k_BT}{\Lambda_0^3}e^{\beta\bar{\mu}_0}, \tag{6.113}$$

we arrive at already discussed in Chap. 5 Poisson-Boltzmann-Langevin equation (see also [56, 135])

$$-\frac{1}{4\pi}\nabla\cdot(\epsilon(\mathbf{r})\nabla\psi(\mathbf{r})) = q_+n_+^{(b)}e^{-\beta q_+\psi(\mathbf{r})} + q_-n_-^{(b)}e^{-\beta q_-\psi(\mathbf{r})}, \tag{6.114}$$

with

$$\epsilon(\mathbf{r}) = \varepsilon + \frac{4\pi n_0^{(b)}p^2}{k_BT}\mathcal{G}(\beta p\mathcal{E}(\mathbf{r})), \tag{6.115}$$

where $\mathcal{G}(x) = \sinh x L(x)/x^2$, $n_0^{(b)}$ is the concentration of the molecules in the bulk phase. We used the following expressions for the bulk chemical potentials of ions and molecules: $\mu_{\pm} = k_B T \ln(n_{\pm}^{(b)} \Lambda_{\pm}^3)$, $\mu_0 = k_B T \ln(n_0^{(b)} \Lambda_0^3)$. To incorporate the excluded volume effects of the components, we can utilize a reference system of a mixture of hard spheres and the corresponding relations (4.46)–(4.47), where $n_0 = \bar{n}_+ + \bar{n}_- + \bar{n}_0$, $n_1 = 1/2\, d_+ \bar{n}_+ + 1/2\, d_- \bar{n}_- + 1/2\, d_0 \bar{n}_0$, $n_2 = \pi\, d_+^2 \bar{n}_+ + \pi\, d_-^2 \bar{n}_- + \pi\, d_0^2 \bar{n}_0$, $n_3 = \pi/6\, d_+^3 \bar{n}_+ + \pi/6\, d_-^3 \bar{n}_- + \pi/6\, d_0^3 \bar{n}_0$ (d_0 denotes the effective diameter of the molecules).

It's also important to discuss scenarios involving the short-range specific interactions between the ions and polar molecules. Similar to Chap. 3, we derive the following expression for the grand thermodynamic potential

$$\Omega = \int d\mathbf{r} \left(-\frac{\varepsilon (\nabla \psi)^2}{8\pi} - \frac{1}{2} \sum_{\alpha\gamma} g_{\alpha\gamma} \phi_{\alpha} \phi_{\gamma} - P_R(\bar{\mu}_+, \bar{\mu}_-, \bar{\mu}_0) \right), \tag{6.116}$$

where $g_{\alpha\gamma} = A_{\alpha\gamma}^{-1}$ represents the matrix, inverse to the matrix of the parameters of specific interactions ($\alpha, \gamma = 0, \pm$).

$$\bar{\mu}_{\pm}(\mathbf{r}) = \mu_{\pm} - q_{\pm} \psi(\mathbf{r}) + \frac{\gamma_{\pm} (\nabla \psi(\mathbf{r}))^2}{2} + k_B T \ln \left[\frac{\sinh(\beta p_{\pm} |\nabla \psi(\mathbf{r})|)}{\beta p_{\pm} |\nabla \psi(\mathbf{r})|} \right] - \phi_{\pm}, \tag{6.117}$$

$$\bar{\mu}_0(\mathbf{r}) = \mu_0 + \frac{\gamma_0 (\nabla \psi(\mathbf{r}))^2}{2} + k_B T \ln \left[\frac{\sinh(\beta p_0 |\nabla \psi(\mathbf{r})|)}{\beta p_0 |\nabla \psi(\mathbf{r})|} \right] - \phi_0. \tag{6.118}$$

The self-consistent field equations, corresponding to the grand thermodynamic potential have the following form

$$\nabla (\epsilon(\mathbf{r}) \psi(\mathbf{r})) = -4\pi \left(q_+ \bar{n}_+(\mathbf{r}) + q_- \bar{n}_-(\mathbf{r}) \right), \tag{6.119}$$

$$\phi_{\alpha}(\mathbf{r}) = \sum_{\gamma} A_{\alpha\gamma} \bar{n}_{\gamma}(\mathbf{r}). \tag{6.120}$$

In this scenario, to compute the potentials ψ and ϕ_{α} we need to self-consistently solve the Poisson differential equation (6.119) with appropriate boundary conditions and the system of transcendental equations (6.120).

It is easy to show using Eq. (4.53), that within this model the stress tensor coincides with Eq. (6.96) with ϵ determined by Eq. (6.109) and local osmotic pressure

$$P = P_R(\bar{\mu}_+, \bar{\mu}_-, \bar{\mu}_0) - \frac{1}{2} \sum_{\alpha\gamma} A_{\alpha\gamma} \bar{n}_{\alpha} \bar{n}_{\gamma}. \tag{6.121}$$

6.6 Self-consistent Field Theory of Ionic Liquids at a Charged Electrode: Effects of Ion Polarizability

Consider an ionic liquid comprising ions with charges $\pm q$ under thermodynamic equilibrium conditions. We assume that each ion type possesses, in general, a constant dipole moment, $p_{\pm}$, and static polarizability, $\gamma_{\pm}$, where the symbols $\pm$ denote cations and anions, respectively. Furthermore, we neglect the electrostatic correlations leading to nonlocal contributions to the electrostatic free energy functional. Let's place a flat charged electrode with a fixed potential into the ionic liquid medium, forming an electric double layer at its surface. Our focus lies solely on the influence of electronic and orientational polarizabilities of ions on differential capacitance of the electric double layer, without taking into account the effects associated with the short-range specific and structural interactions between ions or their adsorption onto the electrode, leading to the formation of Stern layer. By placing the origin of the z-axis at the electrode surface, we can express the grand thermodynamic potential of the ionic liquid per unit area of the electrode as follows:

$$\Omega[\psi]/\mathcal{A} = -\int_{0}^{\infty} dz \left(\frac{\varepsilon \mathcal{E}^2(z)}{8\pi} + P(\bar{\mu}_+(z), \bar{\mu}_-(z)) \right), \tag{6.122}$$

where $\mathcal{A}$ represents the total area of the electrode, $\psi(z)$ stands for the local electrostatic potential, $\mathcal{E}(z) = -\psi'(z)$ denotes the local electric field, and ε represents the background (reference) dielectric permittivity of the medium, which does not account for the effects of ion polarizability and polarity,

$$\bar{\mu}_{\pm} = \mu_{\pm} \mp q\psi + \frac{\gamma_{\pm}\mathcal{E}^2}{2} + k_B T \ln \left(\frac{\sinh \beta p_{\pm}\mathcal{E}}{\beta p_{\pm}\mathcal{E}} \right) \tag{6.123}$$

are the intrinsic chemical potentials of the ions, where $\beta = (k_B T)^{-1}$ and k_B represents the Boltzmann constant. $P(\bar{\mu}_+, \bar{\mu}_-)$ denotes the local pressure of the ions as a function of their intrinsic chemical potentials. It's worth noting that Eq. (6.123) define the thermodynamic equilibrium condition for a heterogeneous system.

To calculate the potential profile, $\psi(z)$, we need to solve the self-consistent field equation

$$\left(\epsilon(z)\psi'(z) \right)' = -4\pi q \left(\bar{n}_+(z) - \bar{n}_-(z) \right), \tag{6.124}$$

which can be initially obtained by varying the thermodynamic potential (6.122) with respect to the electrostatic potential [56, 89], where

$$\epsilon(z) = \varepsilon + 4\pi \sum_{\alpha=\pm} \left(\gamma_\alpha + \frac{p_i^2}{k_B T} \frac{L(\beta p_i \mathcal{E}(z))}{\beta p_i \mathcal{E}(z)} \right) \bar{n}_i(z) \tag{6.125}$$

is the effective local dielectric permittivity of the liquid, incorporating the electronic and orientational polarizabilities of ions, $L(x) = \coth x - 1/x$ is the standard Langevin function, and

$$\bar{n}_{\pm}(z) = \frac{\partial P}{\partial \bar{\mu}_{\pm}(z)} \tag{6.126}$$

are equilibrium average ion concentrations that depend on the coordinates through the intrinsic chemical potentials, $\bar{\mu}_{\pm}(z)$. The self-consistent field equation (6.124) represents the well-known Poisson equation for the self-consistent electrostatic field potential. In the case of the bulk solution phase, where $n_{\pm} = n$, we derive the following "mean-field" expression for the bulk dielectric permittivity

$$\epsilon_b = \varepsilon + 4\pi \sum_{\alpha=\pm} \left(\gamma_\alpha + \frac{p_\alpha^2}{3k_B T} \right) n. \tag{6.127}$$

It should be noted separately that the dipole moments, $p_{\pm}$, polarizabilities, $\gamma_{\pm}$, and the reference dielectric permittivity, ε, in Eq. (6.127) can serve as effective fitting parameters for matching the experimentally measured dielectric permittivity of a specific ionic liquid in its bulk phase.

We would also like to note that in the case of the ideal gas reference system, which has the following equation of state:

$$P(\mu_+, \mu_-) = k_B T \left(\Lambda_+^{-3} e^{\beta\mu_+} + \Lambda_-^{-3} e^{\beta\mu_-} \right), \tag{6.128}$$

with $\lambda_{\pm}$ representing the thermal de Broglie wavelengths of ions, the general self-consistent field equation (6.124) converges to the Poisson-Boltzmann equation with polarizability incorporation [136].

The boundary conditions for the self-consistent field equation have the form

$$\psi(0) = \psi_0, \quad \mathcal{E}(\infty) = 0, \tag{6.129}$$

where ψ_0 denotes the potential at the electrode. The first integral of the self-consistent field equation, describing the mechanical equilibrium of the fluid, has the following form

$$-\frac{\varepsilon \mathcal{E}^2}{8\pi} - \sum_{\alpha=\pm} \mathcal{E} \left(\gamma_\alpha \mathcal{E} + p_\alpha L(\beta p_\alpha \mathcal{E}) \right) \bar{n}_\alpha + P(\bar{\mu}_+, \bar{\mu}_-) = P_0, \tag{6.130}$$

where $P_0 = P(\mu_+, \mu_-)$ represents the ion pressure in the bulk phase. It's worth noting that Eq. (6.130) can also be derived from the contact theorem discussed in Chap. 4 using stress tensor (6.96). One of the primary characteristics of interest of the electric double layer is the differential capacitance, $C = d\sigma/d\psi_0$. To determine it, we calculate the charge density on the electrode surface using the known expression [87] $\sigma = \epsilon_s \mathcal{E}_s / 4\pi$, with the local dielectric constant taking the following form:

$$\epsilon(0) = \epsilon_s = \varepsilon + 4\pi \sum_{\alpha=\pm} \left(\gamma_\alpha + \frac{p_\alpha^2}{k_B T} \frac{L(\beta p_\alpha \mathcal{E}_s)}{\beta p_\alpha \mathcal{E}_s} \right) n_{s,\alpha}, \tag{6.131}$$

while the local electric field, $\mathcal{E}_s = \mathcal{E}(0)$ at the electrode, can be determined from the mechanical equilibrium condition (6.130) at $z = 0$, i.e.

$$-\frac{\varepsilon \mathcal{E}_s^2}{8\pi} - \sum_{\alpha=\pm} \mathcal{E}_s \left(\gamma_\alpha \mathcal{E}_s + p_\alpha L(\beta p_\alpha \mathcal{E}_s) \right) n_{s,\alpha} + P_s = P_0, \tag{6.132}$$

where $n_{s\pm} = \bar{n}_\pm(0) = n_{s\pm}(\mu_{s+}, \mu_{s-})$ and $P_s = P(\mu_{s+}, \mu_{s-})$ are local ion concentrations and ion pressure at the electrode, respectively, which are the functions of the local intrinsic chemical potentials of the ions

$$\mu_{s\pm} = \mu_\pm \mp q\psi_0 + \frac{\gamma_\pm \mathcal{E}_s^2}{2} + k_B T \ln \left(\frac{\sinh \beta p_\pm \mathcal{E}_s}{\beta p_\pm \mathcal{E}_s} \right). \tag{6.133}$$

Thus, calculating the derivative

$$d\sigma/d\psi_0 = (\partial\sigma/\partial\psi_0)_{\mathcal{E}_s} + (\partial\sigma/\partial\mathcal{E}_s)_{\psi_0} (d\mathcal{E}_s/d\psi_0) \tag{6.134}$$

using Eqs. (6.131) and (6.132), and considering that the local ion concentrations, $n_{s\pm}$, depend on the surface potential, ψ_0, and the local electric field, $\mathcal{E}_s$, through the local intrinsic chemical potentials (6.133), after some (rather cumbersome) calculations, we obtain the following simple analytical expression for the differential capacitance

$$C = \frac{q\,(n_{s-} - n_{s+})}{\mathcal{E}_s} = -\frac{\rho_s}{\mathcal{E}_s}, \tag{6.135}$$

where the local electric field at the electrode, $\mathcal{E}_s = \mathcal{E}_s(\psi_0)$, can be calculated from the numerical solution of Eq. (6.132), and $\rho_s(\psi_0) = q\,(n_{s+}(\psi_0) - n_{s-}(\psi_0))$ represents the local ion charge density at the electrode. Equation (6.135), the main result of this section, provides an estimate for the differential capacitance within the mean-field approximation for an arbitrary equation of state, $P = P(\mu_+, \mu_-)$, of the reference model of an ionic liquid, considering constant dipole moments and static polarizability of the ions. It's also noteworthy that when $p_\pm = \gamma_\pm = 0$, Eq. (6.135) transforms into the expression (4.80) previously obtained in Chap. 3.

Symmetric lattice gas as a reference system. Before turning to the results of numerical calculations, we introduce a reference model for an ionic liquid with a specific equation of state $P = P(\mu_+, \mu_-)$. Here, we exclude the influence of asymmetric steric interactions on the differential capacitance, as it was recently investigated in Ref. [82] within the Carnahan-Starling approximation for a mixture of hard spheres and a two-component asymmetric lattice gas model. Instead, we adopt a simpler model of a two-component symmetric lattice gas as our reference model, with the following pressure dependence on the chemical potential of ions [13, 82, 137]

$$P(\mu_+, \mu_-) = \frac{k_B T}{v} \ln\left(1 + e^{\beta\mu_+} + e^{\beta\mu_-}\right), \tag{6.136}$$

where v represents the unit cell volume of the lattice gas. By calculating the derivative in (6.126) and utilizing the expression for the bulk chemical potentials, $\mu_\pm = k_B T \ln(nv/(1-2nv))$, where $n = n_\pm^{(b)}$, we derive the following expression for the average ion concentration

$$\bar{n}_\pm = \frac{ne^{\beta\left(\frac{\gamma_\pm \mathcal{E}^2}{2} \mp q\psi\right)} \frac{\sinh(\beta p_\pm \mathcal{E})}{\beta p_\pm \mathcal{E}}}{1 + nv\left(e^{\beta\left(\frac{\gamma_+ \mathcal{E}^2}{2} - q\psi\right)} \frac{\sinh(\beta p_+ \mathcal{E})}{\beta p_+ \mathcal{E}} + e^{\beta\left(\frac{\gamma_- \mathcal{E}^2}{2} + q\psi\right)} \frac{\sinh(\beta p_- \mathcal{E})}{\beta p_- \mathcal{E}} - 2\right)}. \tag{6.137}$$

Thus, the expression for the ion concentration at the electrode takes the form

$$n_{s\pm} = \frac{ne^{\beta\left(\frac{\gamma_\pm \mathcal{E}_s^2}{2} \mp q\psi_0\right)} \frac{\sinh(\beta p_\pm \mathcal{E}_s)}{\beta p_\pm \mathcal{E}_s}}{1 + nv\left(e^{\beta\left(\frac{\gamma_+ \mathcal{E}_s^2}{2} - q\psi_0\right)} \frac{\sinh(\beta p_+ \mathcal{E}_s)}{\beta p_+ \mathcal{E}_s} + e^{\beta\left(\frac{\gamma_- \mathcal{E}_s^2}{2} + q\psi_0\right)} \frac{\sinh(\beta p_- \mathcal{E}_s)}{\beta p_- \mathcal{E}_s} - 2\right)}. \tag{6.138}$$

Equations (6.132), (6.135), and (6.138) form a closed system of coupled equations, enabling the calculation of the differential capacitance as a function of potential. It's important to note that for the case of nonpolar and nonpolarizable ions ($\gamma_\pm = p_\pm = 0$), the Eq. (6.135) along with Eq. (6.132) yields the Kornyshev's expression (4.84).

In the following discussion, we concentrate on physically significant scenarios involving ionic liquids with cations possessing non-zero static polarizability or a constant dipole moment. Such situations can occur in pure ionic liquids and their solutions with organic cations, characterized by large static polarizability and strongly polar groups. For simplicity, in this problem formulation, we can neglect the static polarizability of inorganic anions.

Figure 6.21a illustrates the dependence of the differential capacitance of the ionic liquid on the potential, ψ_0, at a temperature of $T = 298\,K$ and an ion volume fraction of $nv = 0.4$ with $v^{1/3} = 5\text{Å}$, for various values of the static cation polarizability, γ_+. At $\gamma_+ = 0$, the differential capacitance exhibits a symmetric 'bell-shaped' profile, consistent with Kornyshev's field-theoretic model [60]. However, an increase in the cation's electric polarizability results in higher values of the differential capacitance in the negative voltage region and shifts the position of its maximum to more negative voltage values.

In Fig. 6.21b, a similar behavior of the differential capacitance for negative voltage values is observed for polar cations, where the differential capacitance increases with the constant dipole moment of the cation, and simultaneously the position of the maximum also shifts to a more negative potential region. It's worth noting that the increase in differential capacitance values at moderate negative voltages is more pronounced with increasing dipole moment of the cation than with its static

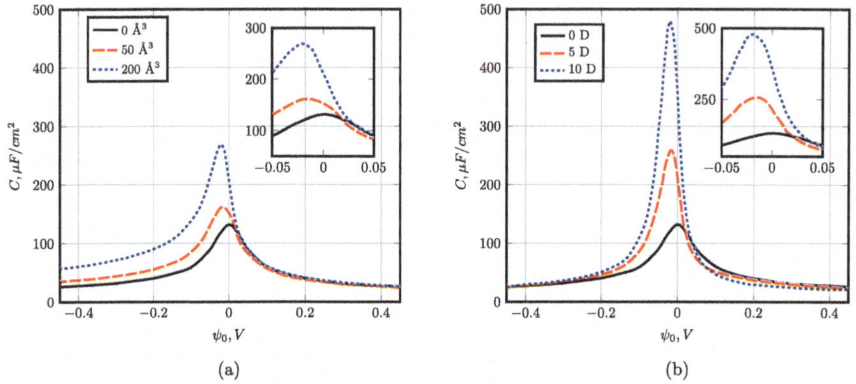

Fig. 6.21 Differential capacitance of the EDL in an ionic liquid as a potential drop function at different values of static polarizability γ_+ (**a**) and dipole moment p_+ (**b**). The data are shown for $\gamma_- = p_+ = p_- = 0$ (**a**), $\gamma_- = \gamma_+ = p_- = 0$ (**b**), $\varepsilon = 5$, $v^{1/3} = 5$ Å, $nv = 0.4$, $T = 298$ K

polarizability. Nevertheless, at sufficiently high negative potential values, the increase in dipole moment has little effect on the differential capacitance value (see also the discussion below).

The presence of static polarizability or a permanent dipole moment on cations enhances their attraction to the negatively charged electrode compared to the attraction of nonpolarized anions to the surface of a positively charged electrode. This stronger attraction of cations to the electrode stems from the additional electrophoretic attraction force acting on the induced or permanent dipole, which is located in an inhomogeneous electric field [138]. Figure 6.22 illustrates the dependencies of ion concentrations at the electrode as a function of surface potential under the same physical parameters used in Fig. 6.21 for the case of a polar cation ($p_+ \neq 0$, $\gamma_+ = 0$). As observed, when the dipole moment is sufficiently large ($p_+ = 10$ D), the competition between dielectrophoretic attraction and Coulomb repulsion acting on the polar cations results in a non-monotonic behavior of the local cation concentration profiles. This behavior can be explained as follows: at sufficiently small values of voltage, the Coulomb repulsion of cations from the electrode surface should outweigh their dielectrophoretic attraction to it, leading to a monotonic decrease in the local concentration of cations.

However, when the voltage exceeds a certain threshold value corresponding to the local minimum of the cation concentration on the electrode, the dielectrophoretic force becomes greater than the Coulomb one. Consequently, the local cation concentration increases with increasing voltage in absolute value. Nevertheless, at sufficiently high voltage values, after surpassing a certain threshold, the Coulomb repulsion of cations from the electrode surface once again becomes stronger than the attraction, causing the concentration of cations at the electrode to reach a local maximum before decreasing monotonically. It's important to note that the competition between the Coulomb and dielectrophoretic forces acting on the cations does not

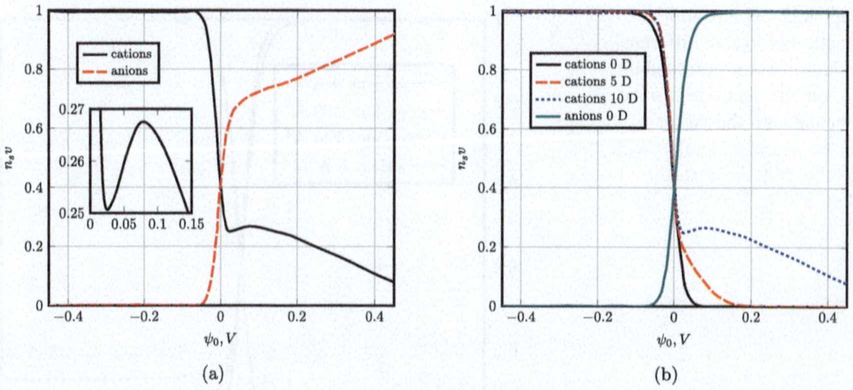

Fig. 6.22 The volume fraction of cations and anions at dipole moment of cation $p_+ = 10\ D$ (**a**) and the volume fraction of cations at different values of dipole moment p_+ (**b**) on the electrode. The data are shown for $\gamma_- = \gamma_+ = p_- = 0$, $\varepsilon = 5$, $v^{1/3} = 5$ Å, $nv = 0.4$, $T = 298$ K

disrupt the monotonic growth of the anion concentration at the electrode observed in the model with nonpolarizable and nonpolar ions and accompanied by the presence of an inflection point in its voltage dependence. However, the concentration of anions at the electrode increases much more slowly with increasing voltage than the local concentration of cations in the negative voltage region. The additional dielectrophoretic attraction of cations to the electrode makes it more difficult for anions to displace the cations from the near-surface layer at positive potentials, compared to the analogous case of replacing anions with cations at negative voltages. As negative voltage values increase, the concentration of cations at the electrode exhibits monotonic growth until reaching the point of densest packing (saturation) of cations in the near-surface layer. This saturation occurs at lower negative values when the dipole moment of the cation increases. This behavior is easily explained by the fact that an increase in the cation's dipole moment leads to a greater dielectrophoretic force, consequently increasing the attraction of cations to the negatively charged electrode. Additionally, it's noteworthy that the magnitude of the shift of the maximum of the differential capacitance in the region of more negative voltage values depends on the asymmetry in the local concentrations of cations in the region of negative voltages and anions in the region of positive voltages: the greater this asymmetry, the larger the magnitude of this shift.

In the case of nonpolar and polarizable cations ($p_+ = 0, \gamma_+ \neq 0$), one can observe qualitatively different behavior of ion concentrations at the electrode. As depicted in Fig. 6.23, for sufficiently small static polarizabilities of cations in the region of positive voltages, an increase in the voltage leads to a monotonic increase in the local concentration of anions up to complete saturation of the near-surface layer, along with a monotonic decrease in the concentration of cations until their complete displacement. However, when the static polarizability of cations exceeds a certain threshold value, the increase in voltage no longer results in complete displacement

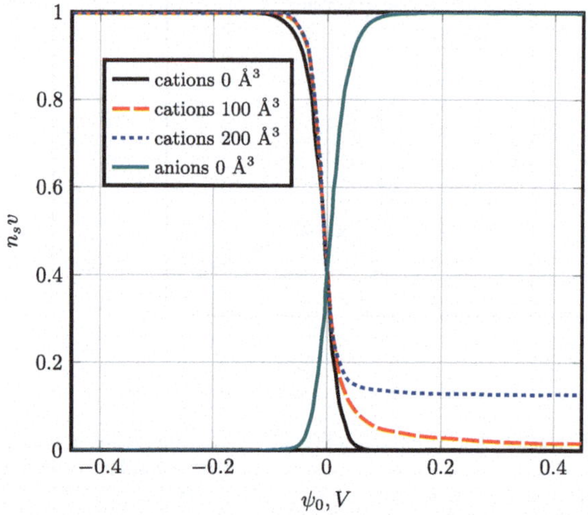

Fig. 6.23 The volume fraction of ions at different values of static polarizability γ_+ on the electrode surface. The data are shown for $\gamma_- = p_+ = p_- = 0$, $\varepsilon = 5$, $v^{1/3} = 5$ Å, $nv = 0.4$, $T = 298$ K

of cations from the electrode surface. Instead, a mixture of anions and cations in a fully saturated near-surface layer can be observed. This unexpected behavior cannot be observed in ionic liquids with polar and nonpolarizable cations because they carry a constant dipole moment, unlike polarized cations whose dipole moments grow linearly with increasing local electric field. Thus, a delicate balance between dielectrophoretic attraction of cations to the electrode, their Coulomb repulsion from it, and steric repulsion of anions and cations in the near-surface layer leads to a saturation regime of cation concentration near the positively charged electrode.

We should separately discuss the dependence of ion concentrations at the electrode on the dipole moment (Fig. 6.24a) and static polarizability (Fig. 6.24b) of cations at a fixed value of the applied voltage ($\psi_0 = 0.2\ V$). As illustrated in Fig. 6.24, for sufficiently small values of both variables and sufficiently large values of voltage, no cations accumulate near the electrode surface, while the local concentrations of anions reach their maximum value. However, when the dipole moment and static polarizability of cations exceed certain threshold values, the concentration of cations (anions) begins to increase (decrease), eventually reaching a plateau.

It is interesting to explore how the simultaneous consideration of the nonlinear polarization effect of both ions affects the differential capacitance within the framework of the proposed model. For this purpose, let's consider a scenario where cations and anions have equal electronic polarizabilities ($\gamma_\pm = \alpha$) and zero dipole moments. Figure 6.25a compares the profiles of the differential capacitance calculated using the proposed analytical theory, which accounts for the nonlinear polarization response of ions, with the theory that considers the dielectric effect using a constant permittivity defined by the expression: $\varepsilon_{IL} = \varepsilon + 2\gamma n/\varepsilon_0 \approx 9$ (obtained within the linear dielectric response approximation of the proposed theory) at $\gamma = 50\mathring{A}^3$ and $nv = 0.3$. As observed, in the first case, the present theory exhibits a symmetric "camel-shaped" profile similar to the results of [139], while in the second case, it shows

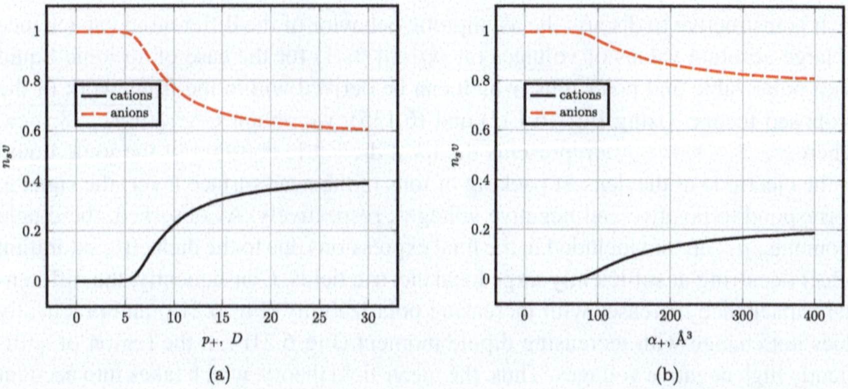

Fig. 6.24 The volume fraction of the cations and anions on the electrode surface as a function of the cation dipole moment p_+ (**a**) and the static polarizability γ_+ (**b**) at a fixed potential $\psi_0 = 0.2\ V$. The data are shown for $\gamma_- = \gamma_+ = p_- = 0$ (**a**), $\gamma_- = p_+ = p_- = 0$ (**b**), $\varepsilon = 5$, $v^{1/3} = 5$ Å, $nv = 0.4$, $T = 298$ K

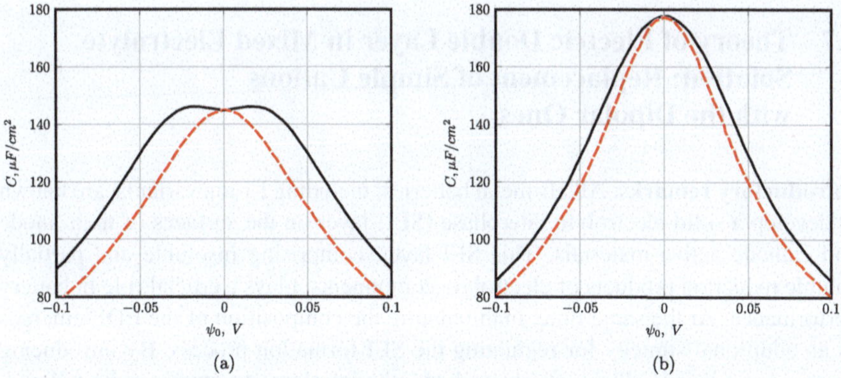

Fig. 6.25 Differential capacitance curves calculated within the analytical theory taking into account the nonlinear polarization response of ions (full black line) and the theory accounting for the dielectric effect via the constant dielectric permittivity (dashed red line) for two bulk ionic concentrations. The data are shown for $nv = 0.3$ (**a**) and $nv = 0.4$ (**b**), $p_+ = p_- = 0$, $\gamma_- = \gamma_+ = 50$ Å^3, $v^{1/3} = 5$ Å, $T = 298$ K

a "bell-shaped" profile in accordance with Kornyshev's mean-field theory presented in Chap. 3. However, at high bulk ion concentrations ($nv = 0.4$, see Fig. 6.25b), the nonlinear dielectric response of the ions does not disturb the "bell-shaped" shape of the differential capacitance, only increasing the width of the peak. We can conclude that the theory that explicitly considers the static polarizability of both ions predicts a transition from a "camel-shaped" profile to a "bell-shaped" profile at higher bulk ion concentrations compared to the theory with nonpolarizable ions. It's worth noting that a similar pattern is observed in the theory with polarized ions ($p_\pm = p, \gamma_\pm = 0$).

It is instructive to discuss the asymptotic behavior of the differential capacitance at large absolute values of voltages (at $\beta q|\psi_0| \gg 1$) for the case of an ionic liquid with polarizable and polar ions, which can be derived within the framework of the proposed theory. Using Eqs. (6.132) and (6.135), we obtain $C \simeq \sqrt{\pm q\varepsilon_{\mp}/8\pi v\psi_0}$, where $\varepsilon_{\mp} \simeq \varepsilon + 4\pi\gamma_{\mp}/v$ represents the local dielectric constant of the ionic liquid at the electrode at the densest packing of ions in the near-surface layer; the signs $\pm$ correspond to positive and negative voltages, respectively. As observed, the dipole moments, $p_{\pm}$, are not included in the final expressions due to the dielectric saturation effect occurring at sufficiently large local electric fields. Consequently, the differential capacitance increases with increasing polarizability (Fig. 6.21a) and practically does not change with increasing dipole moment (Fig. 6.21b) in the region of sufficiently high negative voltages. Thus, the mean-field theory, which takes into account static polarizabilities and dipole moments of ions, qualitatively predicts the same asymptotic voltage dependence of the differential capacitance at $\beta q|\psi_0| \gg 1$ as the theory without polarizability incorporation.

6.7 Theory of Electric Double Layer in Mixed Electrolyte Solution: Replacement of Simple Cations with the Dipolar Ones

Introductory remarks. Alkali-metal batteries, including Li-ion variants, are known to develop a solid-electrolyte interphase (SEI) layer on the surfaces of their anode and cathode active materials. This SEI layer, comprising insoluble and partially soluble reduction products of electrolyte components, plays a crucial role in battery performance. At the same time, manipulating the composition of the EDL emerges as an additional strategy for regulating the SEI formation process. By introducing organic cations into alkali-salt electrolytes, the interface structure can be tailored, owing to the unique characteristics of these molecules arising from their molecular complexity. These characteristics may include factors like steric interactions, the dipole moment of functional groups, and specific surface interactions, such as $\pi - \pi$ stacking. The diverse nature of organic cations suggests promising prospects for modulating the EDL structure through thoughtful compound design. Furthermore, the incorporation of both simple (alkali) and organic cations in mixed electrolyte compositions presents intriguing possibilities for various applications. Consequently, there arises a need for theoretical frameworks that account for the presence of both types of cations in describing the EDL.

From a physics standpoint, organic cations are expected to be more strongly attracted to the cathode than simple cations, such as Li^+. Moreover, this attraction can be further intensified by utilizing organic salts with highly polar cations, such as RTILs, whose organic cations typically possess a rather high dipole moment or electronic polarizability. This enhanced attraction is attributed to the well-known phenomenon when polar particles in an inhomogeneous electric field experience a dielectrophoretic force (see, for instance, [138]), directed towards the region of highest electric field value. In the system under consideration, the electric field inhomogeneity arises from the electrostatic screening of the electrode charge by mobile electrolyte ions, resulting in the highest electric field value on the electrode surface. Therefore, the presence of a dipole moment on the organic cations should lead to their stronger attraction to the electrode compared to simple cations. Consequently, this effect prompts the expulsion of simple cations from the electrode surface due to excluded volume interactions, ultimately replacing them with electrochemically inactive organic cations at sufficiently high electrode surface charge densities.

To ascertain the feasibility of replacing (at least partially) simple cations with dipolar organic ones at the cathode surface under physically reasonable system parameters, such as surface charge density, dielectric permittivity of the solvent, dipole moment of the cation, and bulk ionic concentrations, let's construct a simple analytical model of a two-component electrolyte solution with one type of anions and two types of cations interacting with a negatively charged metal electrode. In essence, this model utilizes SCF principles, employing the modified Poisson-Boltzmann equation to describe the electrostatic potential while accounting for steric interactions of ions and the polarity of one type of cations.

Self-consistent field theory. Let us consider an interface, occurring at a flat charged electrode with a surface charge density σ, immersed in an electrolyte solution containing one type of monovalent anions and two types of monovalent cations. We assume that the cations can be categorized into two types: simple cations and molecular cations. The simple cations carry an electric charge q, while the molecular cations, in addition to the same charge, possess a permanent dipole moment p. We start from the grand thermodynamic potential of the electrolyte solution

$$\Omega = \int\limits_0^\infty dz \left(-\frac{\varepsilon \mathcal{E}^2}{8\pi} + \rho\psi - c_+^{(2)}\Psi + f - \mu_+^{(1)} n_+^{(1)} - \mu_+^{(2)} n_+^{(2)} - \mu_- n_- \right), \quad (6.139)$$

where $f = f(n_+^{(1)}, n_+^{(2)}, n_-)$ denotes the Helmholtz free energy density of a reference system (a system without electrostatic interactions). We also introduce the following auxiliary function:

$$\Psi(z) = k_B T \ln \frac{\sinh \beta p \mathcal{E}(z)}{\beta p \mathcal{E}(z)} \quad (6.140)$$

with $\rho(z) = q\left(n_+^{(1)}(z) + n_+^{(2)}(z) - n_-(z)\right)$ being the local charge density of the ions. Applying the local Legendre transformation, we can rewrite the grand thermodynamic potential in the form

$$\Omega = -\int\limits_0^\infty dz \left(\frac{\varepsilon \mathcal{E}^2(z)}{8\pi} + P(\bar{\mu}_+^{(1)}, \bar{\mu}_+^{(2)}, \bar{\mu}_-) \right), \quad (6.141)$$

with the following local chemical potentials of the species

$$\bar{\mu}_+^{(1)} = \mu_+^{(1)} - q\psi, \quad \bar{\mu}_+^{(2)} = \mu_+^{(2)} - q\psi + \Psi, \quad \bar{\mu}_- = \mu_- + q\psi. \tag{6.142}$$

Varying the functional (6.141) with respect to the potential $\psi(z)$, we arrive at the the following self-consistent field equation [137]:

$$\left(\epsilon(z)\psi'(z)\right)' = -4\pi q\left(\bar{n}_+^{(1)}(z) + \bar{n}_+^{(2)}(z) - \bar{n}_-(z)\right), \tag{6.143}$$

where

$$\bar{n}_+^{(1,2)} = \frac{\partial P}{\partial \bar{\mu}_+^{(1,2)}}, \quad \bar{n}_- = \frac{\partial P}{\partial \bar{\mu}_-}, \tag{6.144}$$

represent the local equilibrium concentrations of the ions and

$$\epsilon(z) = \varepsilon + \frac{4\pi p^2}{k_B T}\frac{L(\beta p\mathcal{E})}{\beta p\mathcal{E}}\bar{n}_+^{(2)}(z) \tag{6.145}$$

is the local dielectric permittivity. The first integral of the self-consistent field equation, which governs the mechanical equilibrium condition of the solution, can be expressed as follows:

$$-\frac{\varepsilon\mathcal{E}^2}{8\pi} - p\mathcal{E}\bar{n}_+^{(2)}L(\beta p\mathcal{E}) + P(\bar{\mu}_+^{(1)}, \bar{\mu}_+^{(2)}, \bar{\mu}_-) = P(\mu_+^{(1)}, \mu_+^{(2)}, \mu_-). \tag{6.146}$$

Solving Eq. (6.146) for $\mathcal{E}$ at different values of ψ yields the implicit function $\mathcal{E} = \mathcal{E}(\psi)$. To compute the potential profile $\psi(z)$, we need to solve the first-order ordinary differential equation $\psi'(z) = -\mathcal{E}(\psi)$ with the initial condition $\psi(0) = \psi_0(\sigma)$. Here, $\psi_0(\sigma)$ represents the electrode potential corresponding to the fixed surface charge density σ. Obtaining the function $\psi_0(\sigma)$ requires utilizing the boundary condition $\epsilon_s\mathcal{E}_0 = \sigma$, where $\epsilon_s = \epsilon(0)$ denotes the local dielectric permittivity at the electrode, and $\mathcal{E}_0 = \mathcal{E}(0)$ represents the electric field at the electrode. The latter is implicitly determined by the surface potential ψ_0 through Eq. (6.146).

Let's examine the three-component symmetric lattice gas model as our reference system. Its equation of state is expressed as follows:

$$P(\mu_+^{(1)}, \mu_+^{(2)}, \mu_-) = \frac{k_B T}{v}\ln\left(1 + e^{\beta\mu_+^{(1)}} + e^{\beta\mu_+^{(2)}} + e^{\beta\mu_-}\right) \tag{6.147}$$

with $\beta = (k_B T)^{-1}$. The local concentrations of ions in the three-component lattice gas model can be obtained by the following relations:

$$\bar{n}_+^{(1)}(z)v = \frac{e^{\beta(\mu_+^{(1)} - q\psi(z))}}{1 + e^{\beta(\mu_+^{(1)} - q\psi(z))} + e^{\beta(\mu_+^{(2)} - q\psi(z) + \Psi(z))} + e^{\beta(\mu_- + q\psi(z))}}, \tag{6.148}$$

$$\bar{n}_{+}^{(2)}(z)v = \frac{e^{\beta(\mu_{+}^{(2)}-q\psi(z)+\Psi(z))}}{1+e^{\beta(\mu_{+}^{(1)}-q\psi(z))}+e^{\beta(\mu_{+}^{(2)}-q\psi(z)+\Psi(z))}+e^{\beta(\mu_{-}+q\psi(z))}}, \tag{6.149}$$

$$\bar{n}_{-}(z)v = \frac{e^{\beta(\mu_{-}+q\psi(z))}}{1+e^{\beta(\mu_{+}^{(1)}-q\psi(z))}+e^{\beta(\mu_{+}^{(2)}-q\psi(z)+\Psi(z))}+e^{\beta(\mu_{-}+q\psi(z))}}. \tag{6.150}$$

Adopting the electrical neutrality condition for the bulk solution, namely $n_{-,b} = n_{+,b}^{(1)} + n_{+,b}^{(2)}$, and introducing the bulk concentrations of electrolytes as $n_1 = n_{+,b}^{(1)}$ and $n_2 = n_{+,b}^{(2)}$, we derive the relations for the species' chemical potentials

$$\mu_{+}^{(1,2)} = k_B T \ln\left(\frac{n_{1,2}v}{1-2v(n_1+n_2)}\right), \quad \mu_{-} = k_B T \ln\left(\frac{(n_1+n_2)v}{1-2v(n_1+n_2)}\right). \tag{6.151}$$

Numerical results. Let's first investigate the predictions of the proposed model by examining the behavior of cation concentration profiles at the electrode surface for different values of the molecular cation's dipole moment (p) and the cathode surface charge density ($\sigma < 0$). We consider an equimolar electrolyte composition, where $n_1 = n_2 = 0.5$ M (total salt concentration of 1 M), falling within the typical range of 0.1 – 1.0 M used in practical electrochemical devices. The dielectric permittivity is set to $\varepsilon = 40$ to model common organic solvents such as dimethylsulfoxide or acetonitrile. Figure 6.26 depicts the theoretical concentration profiles of simple cations calculated for two values of the molecular cation's dipole moment, across varying cathode surface charge densities ($|\sigma|$). For sufficiently small $|\sigma|$, where there is enough free space at the electrode for both types of cations, the concentration of simple cations decreases monotonically with distance from the electrode, peaking at the electrode surface ($z = 0$). Higher absolute values of the surface charge density ($|\sigma|$) result in increased local concentrations of simple cations on the electrode. However, once $|\sigma|$ exceeds a certain threshold, molecular cations become more strongly attracted to the electrode than simple cations due to the additional dielectrophoretic force [85, 137, 138] acting on each dipolar cation. This leads to a monotonic decrease in the local concentration of simple cations, with concentration profiles exhibiting a pronounced maximum at a distance $z > 0$. Thus, at relatively large $|\sigma|$, the presence of molecular cations at the electrode becomes more thermodynamically favorable, leading to the expulsion of simple cations from the "near-surface" layer due to steric interactions. This explains the pronounced maxima observed on the concentration profiles of simple cations, which can be shifted by increasing the surface charge density (see Figs. 6.26). These effects are more pronounced at larger dipole moments of the molecular cations, resulting in sharper concentration profile maxima. Figure 6.26b also illustrates the concentration profile of molecular cations for one specific case. At $|\sigma| = 86\,\mu C/cm^2$, simple cations are depleted near the electrode surface, while the concentration of molecular cations reaches its maximum value there. This clearly demonstrates the replacement of simple cations by molecular cations, which is also in qualitative agreement with the results of MD simulation [140].

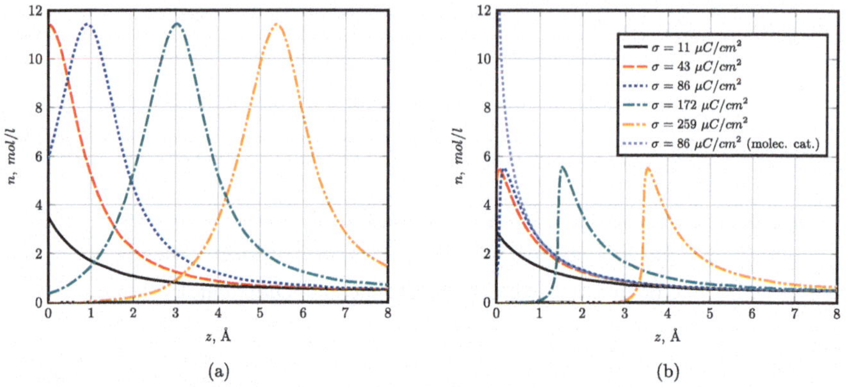

Fig. 6.26 The concentration profiles of the simple cations calculated within the mean field theory at different values of the cathode surface charge density. $|\sigma|$. The concentration profile of the molecular cations is also shown only for one case. The data are shown for the effective size $v^{1/3} = 3.7$ Å, temperature $T = 300$ K, bulk concentrations of the cations $n_1 = n_2 = 0.5$ M, and two dipole moments of the molecular cations $p = 2$ D (**a**) and $p = 4$ D (**b**)

Let's now analyze the local concentration of simple cations at the electrode surface as a function of the surface charge density for various electrolyte compositions. Figure 6.27 presents the results of the proposed mean-field theory. The grey dotted line represents the surface concentration of molecular cations. By comparing it with the concentration of simple cations in the same scenario (black line), we can discern the influence of electrode charge density on the surface ionic composition. For equimolar compositions ($n_1/n_2 = 0.5/0.5$, i.e., total concentration of 1 M, and $n_1/n_2 = 0.25/0.25$, 0.5 M total), it's noteworthy that the surface concentration of simple (and molecular) cations weakly depends on the total bulk electrolyte concentration. The higher dipole moment of molecular cations results in a lower surface concentration of simple cations, as the dielectrophoretic force in this case leads to stronger attraction of molecular cations to the cathode surface. Deviation from equimolar compositions, such as in the case of a higher bulk concentration of simple cations ($n_1/n_2 = 0.5/0.25$, 0.75 M total), leads to a higher surface concentration of simple cations, as there are fewer molecular cations to compete for space within the EDL. However, the first ionic layer still primarily consists of molecular cations. Additionally, it's important to note that the position of the maximum of the function $n_{surf} = n_{surf}(\sigma)$ depends on the dipole moment p, but remains independent of the bulk electrolyte composition.

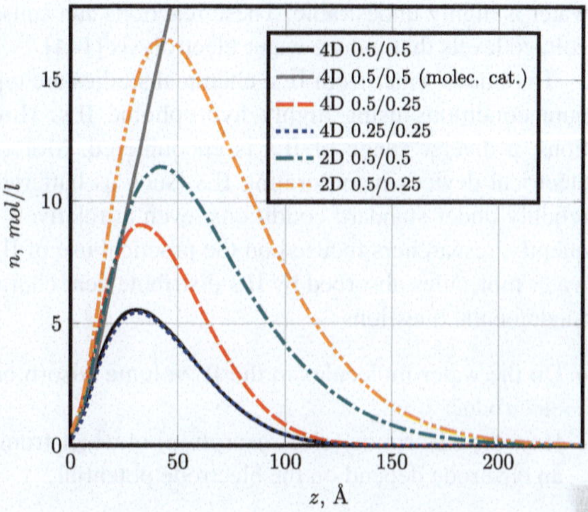

Fig. 6.27 The local concentration of the simple cations on the electrode as a function of the cathode surface charge density calculated within the mean field approximation. The local concentration of molecular cations is also shown for only one case. The data are shown for different dipole moments and concentrations of the simple and molecular cations in the bulk

6.8 Theory of Electric Double Layer in Ionic Liquids with Polar Solvent Admixture: Effects of Short-Range Specific Interactions

Introductory remarks. The case of liquid electrolytes/ionic liquids containing a small number of polar solvent molecules as an impurity is of significant interest in modern electrochemical applications. This problem's importance arises primarily from ILs' nature as liquid electrolytes without solvent in their pure form, making them universal absorbers of molecules of any compounds (see, for example, [141, 142]). Adding a small amount of polar organic solvent to ILs is a standard procedure that enhances ion mobility and increases electrical conductivity [141]. However, ILs themselves readily absorb foreign molecules from the atmosphere, primarily water molecules [141]. For chemical engineers, the absorption of water from the atmosphere is highly undesirable as it leads to uncontrolled changes in the physicochemical properties of ILs due to the fluctuations in air humidity.

This issue is particularly critical in electrochemistry. Indeed, one of the advantages of ILs is associated with a wide electrochemical window (the interval of electrode potential values at which the solution molecules are stable) [143]. When ILs are confined between two electrodes and used as inert electrolytes to drive electrocatalytic reactions of dissolved molecules, it's assumed that they do not participate in electrochemical reactions themselves. However, the presence of water molecules as impurities can trigger unintended electrolysis reactions, complicating the study of fundamental electrochemical processes at the electrodes. Such occurrences are particularly problematic in supercapacitors; for capacitors based on EDL rather than hybrid capacitors, the presence of unintended electrochemical reactions involving water is highly undesirable. These reactions can cause the capacitor to discharge at voltage levels that induce water electrolysis [144].

To remove water from ILs, chemical studies are typically conducted under ultra-pure conditions using highly hydrophobic ILs. However, in real-world applications, a diverse range of ILs is encountered. Moreover, it's desirable to develop electrical devices incorporating ILs (such as batteries and capacitors) that operate reliably under standard conditions, even at relatively high humidity levels. Consequently, researchers focused on the practical use of ILs are keen to understand how water molecules absorbed by ILs distribute near charged electrodes. This raises two fundamental questions

- Do the water molecules in the IL volume adsorb onto the surface of the charged electrode?
- How the adsorption (electrosorption) of water from a volume of ionic liquid onto an electrode depend on the electrode potential?

The first systematic investigation of these questions was conducted in Ref. [145]. The primary aim of this study was to explore the distribution of water molecules near a metal electrode across varying surface charge densities. Molecular dynamic simulations were employed using well-tested force fields for ions, water molecules, and atoms of the electrode. The authors observed that in the vicinity of zero charge potential (in its rather broad vicinity), the concentration of water molecules near the electrode was lower than in the bulk phase. However, as the potential increased in the positive region, there was a significantly sharper rise in water concentration at the electrode compared to the negative potential region. This disparity arises from the stronger attraction of water molecules to anions than to cations, attributed to the hydrophobicity of the latter.

Recently, both experimental studies [146, 147] and their molecular dynamic modeling [145] have been conducted to investigate these systems. However, a simple statistical theory able to phenomenologically describe how specific interactions affect the water electrosorption isotherms was not available until recently. Such a theory is presented in the present section.

We take into account short-range specific interactions between ions, impurity molecules, and the electrode. Our theory enables us to describe the behavior of EDL as water concentration increases in the ILs, as observed in experiments [146, 147].

Self-consistent field theory. Consider an IL with a small amount of polar solvent. Assume that the ions possess charges $\pm q$ (where q is the elementary charge), while the polar solvent molecules have a constant dipole moment p. To characterize the thermodynamic properties of this system, we employ the local density functional theory. Consequently, we express the grand thermodynamic potential of the system as:

$$\Omega = -\int d\mathbf{r} \left(\frac{\varepsilon (\nabla \psi)^2}{8\pi} + \frac{1}{2} \sum_{\alpha\gamma} A^{-1}_{\alpha\gamma} \phi_\alpha \phi_\gamma + P(\bar{\mu}_+, \bar{\mu}_-, \bar{\mu}_0) \right), \tag{6.152}$$

where $P(\mu_+, \mu_-, \mu_0)$ represents the pressure created by the particles of the mixture; the intrinsic chemical potentials of the particles have the form:

$$\bar{\mu}_\pm = \mu_\pm - \phi_\pm \mp q\psi - w_\pm, \quad \bar{\mu}_0 = \mu_0 - \phi_0 + \Psi - w_0, \tag{6.153}$$

where we introduce the following auxiliary function

$$\Psi(\mathbf{r}) = k_B T \ln\left(\frac{\sinh \beta p|\nabla\psi(\mathbf{r})|}{\beta p|\nabla\psi(\mathbf{r})|}\right). \tag{6.154}$$

where ε represents the dielectric constant of the pure ionic liquid. Consequently, by varying the grand thermodynamic potential (6.152) with respect to the functions ψ and ϕ_α, we derive a set of self-consistent equations:

$$\nabla(\epsilon(\mathbf{r})\nabla\psi(\mathbf{r})) = -4\pi q\left(\bar{n}_+(\mathbf{r}) - \bar{n}_-(\mathbf{r})\right), \tag{6.155}$$

and

$$\phi_\alpha(\mathbf{r}) = \sum_\gamma A_{\alpha\gamma}\bar{n}_\gamma(\mathbf{r}), \tag{6.156}$$

where

$$\epsilon(\mathbf{r}) = \varepsilon + \frac{4\pi p^2}{k_B T}\bar{n}_0(\mathbf{r})\frac{L(\beta p|\nabla\psi(\mathbf{r})|)}{\beta p|\nabla\psi(\mathbf{r})|} \tag{6.157}$$

is the local dielectric permittivity; $L(x) = \coth x - 1/x$ is the Langevin function.

If the explicit dependence $P = P(\mu_+, \mu_-, \mu_0)$ is known, the self-consistent field equations for the electrostatic potential $\psi(\mathbf{r})$ and the potentials of specific short-range interactions ϕ_α can be obtained explicitly by calculating the component concentrations as follows:

$$\bar{n}_\pm = \frac{\partial P}{\partial \mu_\pm}, \quad \bar{n}_0 = \frac{\partial P}{\partial \mu_0}. \tag{6.158}$$

The potentials of specific interactions in the bulk phase are determined using the following expression

$$\phi_\alpha^{(b)} = \sum_\gamma A_{\alpha\gamma} n_\gamma^{(b)}. \tag{6.159}$$

Ionic liquid with a small addition of polar solvent on a flat charged electrode. Let's examine the scenario of a flat infinite electrode. By placing the origin at the electrode, we express the grand thermodynamic potential per unit area of the electrode in the following manner:

$$\Omega/\mathcal{A} = -\int_0^\infty dz\left(\frac{\varepsilon\psi'^2}{8\pi} + \frac{1}{2}\sum_{\alpha\gamma} A^{-1}_{\alpha\gamma}\phi_\alpha\phi_\gamma + P(\bar{\mu}_+, \bar{\mu}_-, \bar{\mu}_0)\right), \tag{6.160}$$

Fig. 6.28 Schematic representation of an IL model with a small admixture of polar solvent

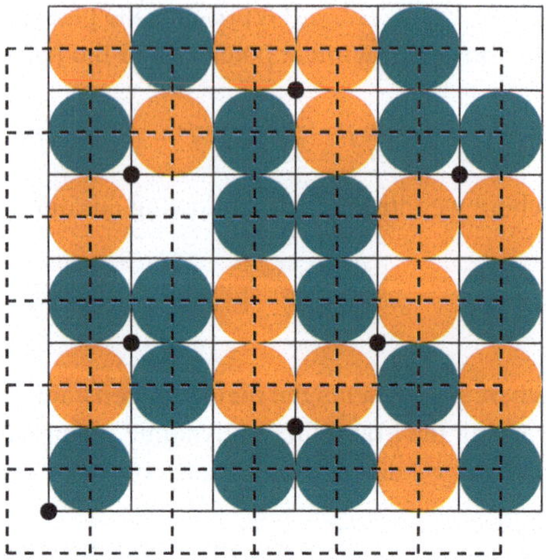

where $\psi' = d\psi/dz$.

We select two independent lattice gase models with equal effective cell volumes v as our reference system. This allows us to express the relationship between pressure and the chemical potentials of the components in the following manner:

$$P(\mu_+, \mu_-, \mu_0) = \frac{k_B T}{v}\left(\ln\left(1 + e^{\beta\mu_+} + e^{\beta\mu_-}\right) + \ln\left(1 + e^{\beta\mu_0}\right)\right). \quad (6.161)$$

The selection of such a reference system stems from the observation that, typically, the molecules of polar solvents added to the ILs possess effective sizes much smaller than that of the ions. This allows us to model the solvent molecules as lattice gas particles, with lattice nodes situated in the spaces between the IL ions. The existence of significant gaps between ions in ILs is indirectly supported by the data on the solubility of compounds like carbon dioxide within them. Indeed, the solubility of carbon dioxide in ILs is notably high, despite minimal changes in the partial molar volume of ILs upon the dissolution of CO_2 [148] (Fig. 6.28).

Further, let's introduce the parameters of the specific interactions: $A_{++} = vA_c$, $A_{--} = vA_a$, $A_{+-} = A_{-+} = vA_{ca}$, and $A_{00} = vA_s$, $A_{0+} = A_{+0} = vB_{cs}$ and $A_{0-} = A_{-0} = vB_{as}$. Thus, the set of Eq. (6.156) takes the form:

$$\phi_0 = A_s v\bar{n}_0 + B_{cs} v\bar{n}_+ + B_{as} v\bar{n}_-, \quad (6.162)$$

$$\phi_+ = B_{cs} v\bar{n}_0 + A_c v\bar{n}_+ + A_{ac} v\bar{n}_-, \quad (6.163)$$

$$\phi_- = B_{as} v\bar{n}_0 + A_{ac} v\bar{n}_+ + A_a v\bar{n}_-. \quad (6.164)$$

By solving this system of equations, we can derive the specific interaction potentials as functions of the local electrostatic potential $\psi(z)$ and electric field $\mathcal{E} = -\psi'(z)$, i.e., $\phi_\alpha = \phi_\alpha(\psi, \mathcal{E})$. With these relationships established, we can then calculate the components concentrations using the following relations:

$$\bar{n}_+ v = \frac{e^{\beta(\mu_+ - \phi_+ - q\psi - w_+)}}{1 + e^{\beta(\mu_+ - \phi_+ - q\psi - w_+)} + e^{\beta(\mu_- - \phi_- + q\psi - w_-)}}, \tag{6.165}$$

$$\bar{n}_- v = \frac{e^{\beta(\mu_- - \phi_- + q\psi - w_-)}}{1 + e^{\beta(\mu_+ - \phi_+ - q\psi - w_+)} + e^{\beta(\mu_- - \phi_- + q\psi - w_-)}}, \tag{6.166}$$

$$\bar{n}_0 v = \frac{\frac{\sinh(\beta p \mathcal{E})}{\beta p \mathcal{E}} e^{\beta(\mu_0 - \phi_0 - w_0)}}{1 + \frac{\sinh(\beta p \mathcal{E})}{\beta p \mathcal{E}} e^{\beta(\mu_0 - \phi_0 - w_0)}}. \tag{6.167}$$

The relations (6.165)–(6.167), in turn, allow us to define relations for chemical potentials. Indeed, assuming $w_\pm(\infty) = w_0(\infty) = 0$ and $n_i(\infty) = n_{i,b}$, we arrive at the following relations

$$\mu_\pm = k_B T \ln\left(\frac{nv}{1 - 2nv}\right) + \phi_\pm^{(b)}, \quad \mu_0 = k_B T \ln\left(\frac{n_0^{(b)} v}{1 - n_0^{(b)} v}\right) + \phi_0^{(b)}. \tag{6.168}$$

In order to calculate the electrostatic potential at each point, we need to solve the self-consistent field equation, which can be written as:

$$\left(\epsilon(z)\psi'(z)\right)' = -\frac{4\pi q n (e^{\beta(\phi_{+,b} - \phi_+ - q\psi - w_+)} - e^{\beta(\phi_{-,b} - \phi_- + q\psi - w_-)})}{1 - 2nv + nv[e^{\beta(\phi_{+,b} - \phi_+ - q\psi - w_+)} + e^{\beta(\phi_{-,b} - \phi_- + q\psi - w_-)}]}, \tag{6.169}$$

where we introduce the local dielectric permittivity

$$\epsilon(z) = \varepsilon + \frac{4\pi p^2 n_0^{(b)}}{k_B T} \frac{\mathcal{G}(\beta p \mathcal{E}) e^{\beta(\phi_{0,b} - \phi_0 - w_0)}}{1 - n_0^{(b)} v \left(1 - \frac{\sinh(\beta p \mathcal{E})}{\beta p \mathcal{E}} e^{\beta(\phi_{0,b} - \phi_0 - w_0)}\right)}, \tag{6.170}$$

with the auxiliary function $\mathcal{G}(x) = (\sinh(x) L(x))/x^2$. Equation (6.169) needs to be solved with the following boundary conditions

$$\psi(0) = \psi_0, \ \psi'(\infty) = 0, \tag{6.171}$$

where ψ_0 is the potential at the electrode.

The surface charge density of the electrode can be calculated using the standard relation

$$\sigma = \frac{\epsilon_{el} \mathcal{E}_s}{4\pi}, \tag{6.172}$$

where $\epsilon_{el} = \epsilon(0)$ and $\mathcal{E}_s = \mathcal{E}(0)$ are the values of dielectric permittivity and electric field at the electrode, respectively. Differential capacitance is calculated in the standard way:

$$C = \frac{d\sigma}{d\psi_0}. \tag{6.173}$$

Unfortunately, incorporating specific interactions with the electrode prevents the derivation of an analytical expression for DC within proposed theory. Thus, it needs to be computed numerically.

For numerical computations, we assume that $w_{\pm}(z) = w_{c/a}\theta(l - z)$ and $w_0(z) = w_0\theta(l - z)$, where $\theta(z)$ is the Heaviside step function and $w_{c/a} \leq 0$ is the energy of specific adsorption of ions onto the electrode; w_0 represents the solvophobicity of the electrode if $w_0 > 0$ or solvophilicity when, on the contrary, $w_0 < 0$. The parameter l denotes the characteristic thickness within which specific interactions of ions and solvent molecules with the electrode are significant. To simplify the model, we disregard specific interactions between molecules of the polar solvent, setting $A_s = 0$.[1]

Discussing the numerical results, let's assume that the thickness within which particles' specific interactions with the electrode are significant is approximately $l \simeq v^{1/3}$, i.e., on the order of molecular size. For numerical computations, we adopt the following physical parameters $\varepsilon = 3.5$, $p = 1.8\ D$, $T = 338\ K$, $v^{1/3} = 0.7\ nm$, $nv = 0.4$, $n_0^{(b)}v = 0.025$, $A_c = A_a = 50\ k_BT$, and $A_{ca} = 0$, which approximate the parameters of real ILs with water admixture. We set the values of the parameters for specific cation-cation and anion-anion interactions to align the theoretical DC values with experimental observations for pure ILs (refer to [90]).

Figure 6.29a illustrates the dependencies of the local water concentration at the electrode $n_{0,s}/n_{0,b}$, expressed in units of bulk water concentration, on the electrode potential (electrosorption curves) for three different cases: (1) IL without specific interactions ($B_{cs} = B_{as} = 0$ - solid line), (2) hydrophobic IL ($B_{cs} > 0$ $(12k_BT)$, $B_{as} = 0$ - dashed line) and (3) hydrophilic IL ($B_{cs} = 0$, $B_{as} < 0$ $(-12k_BT)$—dotted line). As can be seen, in the absence of specific ion-water interactions, the electrosorption curve is symmetric. Here, the electric field gradient, resulting from the screening of the electrode charge by oppositely charged ions, induces the attraction of polar molecules to the electrode (known as the dielectrophoretic force [138]). Thus, an increase in the electrode potential consistently leads to a monotonic increase in the water concentration. However, the effects of cation hydrophobicity or anion hydrophilicity lead to strongly asymmetric behavior of the DC. At positive electrode potentials, the water concentration at the electrode rises in both cases. However, at negative potentials, the water concentration decreases, reaches a minimum, and then gradually begins to slowly rise again.

In the case of a hydrophobic IL at positive potentials, there are more "neutral" anions (in terms of specific ion-water interactions) in the EDL compared to the number of hydrophobic cations. Consequently, even a low applied voltage results in

[1] This assumption is justified by the limited effect of specific interactions, such as hydrogen bonding in water or alcohols, on the DC at sufficiently low concentrations of the polar solvent, as typically encountered in practice.

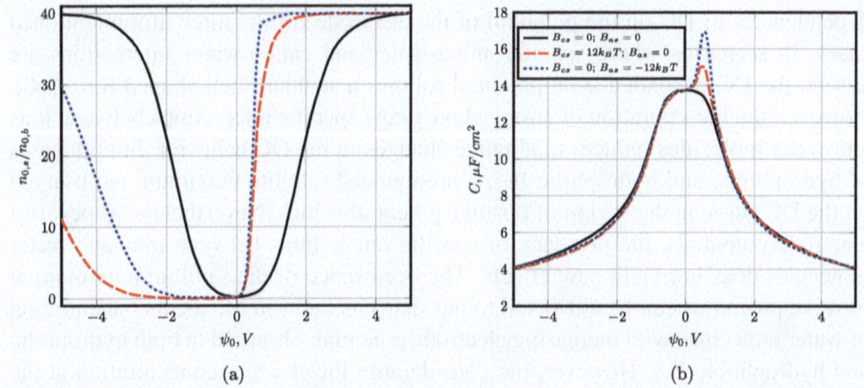

Fig. 6.29 Dependencies of the reduced water concentration $n_{0,s}/n_{0,b}$ at the electrode (**a**) (electrosorption curves) and the DC (**b**) on the electrode potential for three different cases: (1) ionic liquid without specific interactions (solid line), (2) hydrophobic ionic liquid (dashed line), and (3) hydrophilic ionic liquid (dashed line). The data are shown for $nv = 0.4$, $n_0^{(b)}v = 0.025$, $A_c = A_a = 50\,k_BT$, $A_{cs} = 0$, $A_s = 0$, $w_0 = 2\,k_BT$

a notable increase in water concentration at the electrode. Conversely, at negative voltages, the predominance of hydrophobic cations over the "neutral" anions causes water molecules to be expelled from the electrode. Hence, the interplay between the dielectrophoretic force and cation hydrophobicity leads to non-monotonic behavior of the local water concentration as a function of the electrode potential in the region of its negative values. Similarly, in the case of hydrophilic ionic liquids, we observe a rapid increase in water concentration at the electrode for positive potentials and a minimum on the electrosorption curve at negative potentials. This behavior of the local water concentration at positive potentials is due to the prevalence of hydrophilic anions over "neutral" cations in the EDL. Consequently, even a small electrode potential triggers a sharp rise in water concentration at the electrode. In this case, both the dielectrophoretic force and the hydrophilicity of the anion contribute to the effective attraction of water molecules to the electrode. However, at negative potentials, electrostatic repulsion between the negatively charged electrode and hydrophilic anions results in a decrease in the local concentration of the latter. Consequently, hydrophilic anions leave the EDL, carrying water molecules with them. As a result, the water concentration decreases at small negative potentials. However, as the potential's absolute value increases, the dielectrophoretic attraction of water molecules to the electrode becomes comparable in energy to the hydrophilicity of the anion. This leads to a minimum in local water concentration, followed by its further monotonic growth. It's worth noting that in all the cases considered, we accounted for the effect of electrode hydrophobicity (assuming $w_0 = 2\,k_BT$). Consequently, even at zero potential, the water concentration at the electrode is lower than in the bulk phase.

Now it is instructive to investigate the impact of the described behavior of local water concentration at the electrode on the DC profile. Figure 6.29b illustrates the dependencies of DC on the potential of the electrode for the three aforementioned cases. In scenarios where specific anion-water and cation-water interactions are absent, the DC dependence on potential follows a standard bell-shaped form [60]. However, the incorporation of strong short-range specific interactions between ions and water molecules induces qualitative changes in the DC behavior. In both cases of hydrophobic and hydrophilic ILs, a pronounced satellite maximum is observed on the DC curve in the region of positive potential values. Nevertheless, at negative electrode potentials, the presence of specific interactions between ions and water molecules does not yield new effects. The occurrence of the satellite maximum at positive potentials can be attributed to the sharp increase in the local concentration of water molecules with increasing electrode potential, observed in both hydrophilic and hydrophobic ILs. However, the considerably lower water concentration at the electrode compared to its bulk concentration at small negative potentials, and the slow growth of water concentration at sufficiently high magnitudes of potentials, results in the standard bell-shaped DC profile.

Figures 6.30a, b display the electrosorption curves alongside their corresponding DC profiles. It's evident that an increase in the bulk water concentration results in a monotonic rise in the water concentration at the electrode, accompanied by a more pronounced satellite maximum on the DC curves. This observation aligns qualitatively with the experimental findings regarding DC behavior in ILs with small water admixture, as reported in recent studies [146, 147].

Now, let's delve into the discussion regarding the influence of electrode hydrophobicity on electrosorption and DC behavior. Figure 6.31a illustrates the electrosorption curves at various values of the hydrophobicity parameter w_0 of the electrode, focusing on the hydrophobic IL case. It is evident that an increase in the electrode hydrophobicity parameter w_0 results in an anticipated behavior of the electrosorption curves: a monotonic decrease in water concentration at the electrode across all electrode potential values. Correspondingly, Fig. 6.31b showcases the dependencies of DC on

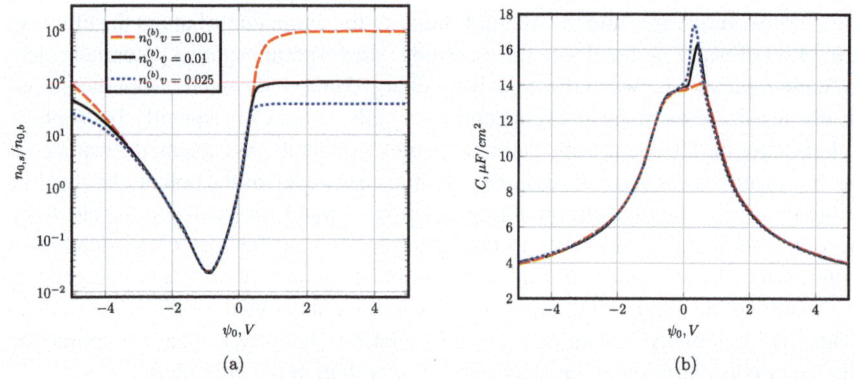

Fig. 6.30 Electrosorption curves (**a**) and differential capacitance profiles (**b**) at different bulk water concentrations. The data are shown for $B_{as} = 0$, $B_{cs} = 12\,k_BT$, $A_s = 0$, $w_0 = w_a = w_c = 0$

the electrode potential. Similar trends are observed for the case of hydrophilic ILs (refer to Fig. 6.32a, b).

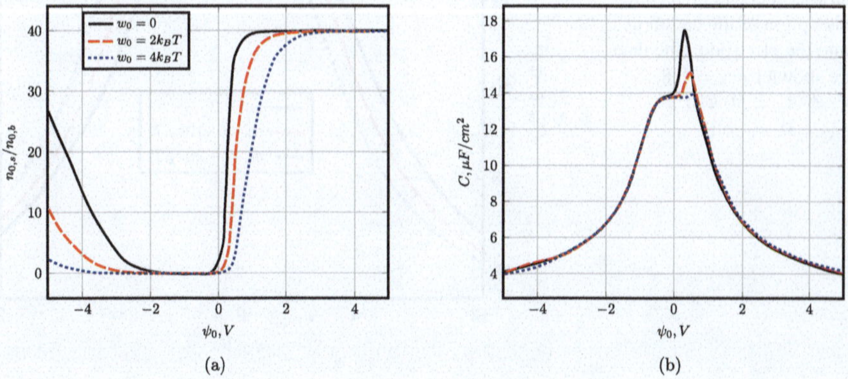

Fig. 6.31 Electrosorption curves (**a**) and differential capacitance profiles (**b**) for hydrophobic IL with the admixture of water at different values of hydrophobicity of the electrode w_0. The data are shown for $A_c = A_a = 50\,k_BT$, $A_s = 0$, $A_{ca} = 0$, $B_{cs} = 12\,k_BT$, $B_{as} = 0$, $nv = 0.4$, $n_0^{(b)}v = 0.025$

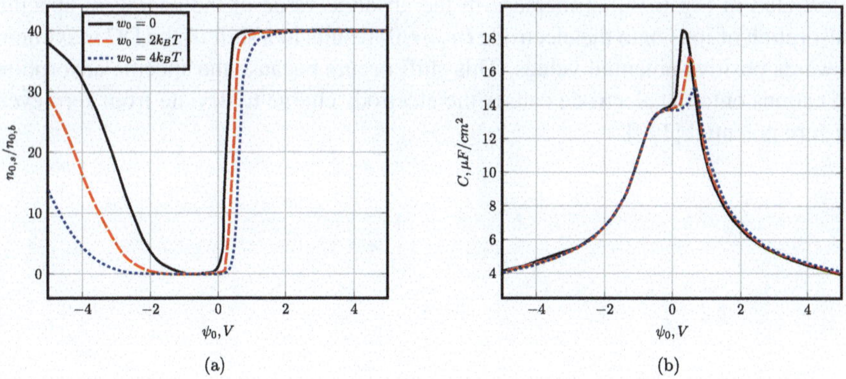

Fig. 6.32 Electrosorption curves (**a**) and differential capacitance profiles (**b**) for hydrophilic IL with the admixture of water at different values of hydrophobicity of the electrode w_0. The data are shown for $A_c = A_a = 50\,k_BT$, $A_{ca} = 0$, $B_{cs} = 0$, $B_{as} = -12\,k_BT$, $\gamma = 0.8$, $n_0^{(b)}v = 0.025$

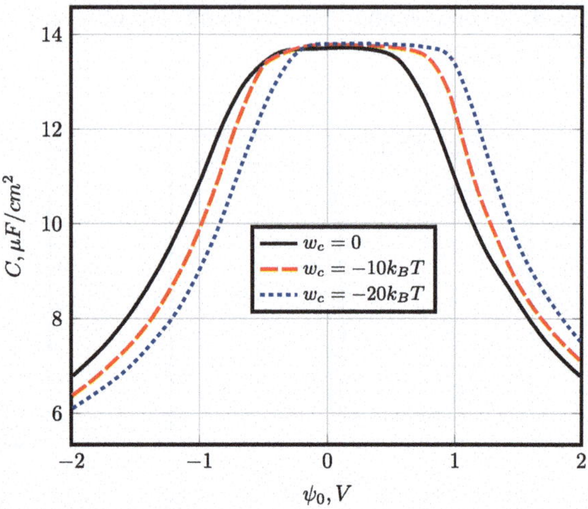

Fig. 6.33 Differential capacitance as a function of electrode potential plotted for a pure ionic liquid at various energies of specific adsorption of the cation w_c onto the electrode. The data are shown for $\gamma = 0.8$, $A_c = A_a = 50\ k_BT$, $A_{ca} = 0$, $w_a = 0$

Finally, we would like to discuss the impact of specific ion adsorption onto the electrode on DC behavior. For simplicity, let's focus on the case of an IL without impurities, considering the specific adsorption of cations onto the electrode. As illustrated in Fig. 6.33, an increase in the absolute value of the energy of specific adsorption of ions onto the electrode ($w_c < 0$) results in a shift of the DC maximum towards positive potential values. This shift occurs because the specific adsorption of cations onto the electrode causes the electrode charge to deviate from zero even at zero potential [149].

References

1. J. Zinn-Justin, *Path Integrals in Quantum Mechanics* (Oxford University Press, 2010)
2. R. Kubo, Generalized cumulant expansion method. J. Phys. Soc. Jpn. **17**(7), 1100–1120 (1962)
3. Y.A. Budkov, P.E. Brandyshev, Variational field theory of macroscopic forces in coulomb fluids. J. Chem. Phys. **159**(17) (2023)
4. R. Rivers, *Path Integral Methods in Quantum Field Theory* (Cambridge University Press, 1988)
5. C. Itzykson, J.-B. Zuber, *Quantum Field Theory* (Courier Corporation, 2006)
6. S. Buyukdagli, Schwinger-Dyson equations for composite electrolytes governed by mixed electrostatic couplings strengths. J. Chem. Phys. **152**(1) (2020)
7. M. Dineykhan, G. Efimov, G. Ganbold, S. Nedelko, *Oscillator Representation in Quantum Physics*, vol. 26 (Springer Science & Business Media, 2008)
8. G.V. Efimov, E.A. Nogovitsin, The partition functions of classical systems in the gaussian equivalent representation of functional integrals. Physica A: Stat. Mech. Appl. **234**(1–2), 506–522 (1996)
9. S.A. Baeurle, G.V. Efimov, E.A. Nogovitsin, Calculating field theories beyond the mean-field level. Europhys. Lett. IOP Publishing **75**(3), 378 (2006)
10. H. Kleinert, Variational perturbation theory for path integrals. *Functional Integration: Basics and Applications*, pp. 93–95 (1997)
11. R.P. Feynman, *Statistical Mechanics: A Set of Lectures* (CRC Press, 2018)
12. L.D. Landau, E.M. Lifshitz, *Statistical Physics*, vol. 5 (Elsevier, 2013)
13. T. Hill, *An Introduction to Statistical Thermodynamics* (Courier Corporation, 1986)
14. K. Binder, D.M. Ceperley, J.-P. Hansen et al, *Monte Carlo Methods in Statistical Physics*, vol. 7 (Springer Science & Business Media, 2012)
15. R. Dobrushin, Study of conditions for the asymptotic existence of the configuration integral of the gibbs distribution. Theory Probab. Appl. **9**(4), 566–581 (1964)
16. M.E. Fisher, D. Ruelle, The stability of many-particle systems. J. Math. Phys. **7**(2), 260–270 (1966)
17. Y. Levin, Electrostatic correlations: from plasma to biology. Rep. Prog. Phys. **65**(11), 1577 (2002)
18. C.G. Gray, K.E. Gubbins, C.G. Joslin, *Theory of Molecular Fluids: Volume 2: Applications*, vol. 10 (Oxford University Press, 2011)
19. J. Hansen, *I*, 3rd edn. (McDonald, Theory of Simple Liquids, 2006)
20. R. Evans, Density functional theory for inhomogeneous fluids i: simple fluids in equilibrium. *Lectures at 3rd Warsaw School of Statistical Physics, Kazimierz Dolny*, vol. 27 (2009)

Y. A. Budkov and N. N. Kalikin, *Statistical Field Theory of Ion-Molecular Fluids*,
https://doi.org/10.1007/978-3-031-68364-0

21. J.-L. Barrat, J.-P. Hansen, *Basic Concepts for Simple and Complex Liquids* (Cambridge University Press, 2003)
22. Y. Rosenfeld, Free-energy model for the inhomogeneous hard-sphere fluid mixture and density-functional theory of freezing. Phys. Rev. Lett. **63**(9), 980 (1989)
23. R. Roth, R. Evans, A. Lang, G. Kahl, Fundamental measure theory for hard-sphere mixtures revisited: the white bear version. J. Phys.: Condens. Matter **14**(46), 12063 (2002)
24. R. Roth, Fundamental measure theory for hard-sphere mixtures: a review. J. Phys.: Condens. Matter **22**(6), 063102 (2010)
25. J.S. Rowlinson, B. Widom, *Molecular Theory of Capillarity* (Courier Corporation, 2013)
26. A.Y. Grosberg, A.R. Khokhlov, *Statistical Physics of Macromolecules* (American Institute of Physics, 1994)
27. P.-G. De Gennes, P.-G. Gennes, *Scaling Concepts in Polymer Physics* (Cornell University Press, 1979)
28. I. Lifshitz, Some problems of the statistical theory of biopolymers. Sov. Phys. JETP **28**(6), 1280–1286 (1969)
29. N.N. Bogoliubov, N.N. Bogoliubov (Jr.), *Statistical Mechanics and the Theory of Dynamical Systems: Collection of Papers*, vol. 191 (American Mathematical Society, 1992)
30. J.N. Israelachvili, *Intermolecular and Surface Forces* (Academic, 2015)
31. C.N. Likos, Effective interactions in soft condensed matter physics. Phys. Rep. **348**(4–5), 267–439 (2001)
32. M. Dijkstra, R. van Roij, Vapour-liquid coexistence for purely repulsive point-Yukawa fluids. J. Phys.: Condens. Matter **10**(6), 1219 (1998)
33. D. Zubarev, Computation of configuration integrals for a system of particles with Coulomb interaction. Dokl. AN SSSR **95**, 757–760 (1954)
34. I. Iukhnovskii, M. Golovko, Statistical theory of classical equilibrium systems. (Kiev Izdatel Naukova Dumka, 1980)
35. R. Storer, Statistical mechanics of simple dense fluids using functional integration. Aust. J. Phys. **22**(6), 747–756 (1969)
36. F.H. Stillinger, T.A. Weber, Study of melting and freezing in the Gaussian core model by molecular dynamics simulation. J. Chem. Phys. **68**(8), 3837–3844 (1978)
37. A. Louis, P. Bolhuis, J. Hansen, Mean-field fluid behavior of the Gaussian core model. Phys. Rev. E **62**(6), 7961 (2000)
38. A.Y. Zakharov, I. Loktionov, Classical statistics of one-component systems with model potentials. Theor. Math. Phys. **119**(1), 532–539 (1999)
39. Z.-G. Wang, Fluctuation in electrolyte solutions: the self energy. Phys. Rev. E **81**(2), 021501 (2010)
40. Y.M. Ivanchenko, A.A. Lisyansky, *Physics of Critical Fluctuations* (Springer Science & Business Media, 2012)
41. N.V. Brilliantov, J.M. Rubí, Y.A. Budkov, Molecular fields and statistical field theory of fluids: application to interface phenomena. Phys. Rev. E **101**(4), 042135 (2020)
42. N. Brilliantov, Phase transitions in solutions of variably ionizable particles. Phys. Rev. E **48**(6), 4536 (1993)
43. N.V. Brilliantov, Effective magnetic Hamiltonian and Ginzburg criterion for fluids. Phys. Rev. E **58**(2), 2628 (1998)
44. N. Brilliantov, C. Bagnuls, C. Bervillier, Peculiarity of the coulombic criticality? Phys. Lett. A **245**(3–4), 274–278 (1998)
45. S.-K. Ma, *Modern Theory of Critical Phenomena* (Routledge, 2018)
46. A. Patashinskii, V. Pokrovskii, The renormalization-group method in the theory of phase transitions. Soviet Physics Uspekhi **20**(1), 31 (1977)
47. G.E. Uhlenbeck, G.W. Ford, et al., *Lectures in Statistical Mechanics* (1963)
48. S. Samuel, Grand partition function in field theory with applications to sine-Gordon field theory. Phys. Rev. D **18**(6), 1916 (1978)
49. S. Edwards, The statistical thermodynamics of a gas with long and short-range forces. Philosoph. Mag. **4**(46), 1171–1182 (1959)

50. A. Kholodenko, A. Beyerlein, Theory of symmetric electrolyte solutions: field-theoretic approach. Phys. Rev. A **34**(4), 3309 (1986)
51. R. Podgornik, Electrostatic correlation forces between surfaces with surface specific ionic interactions. J. Chem. Phys. **91**(9), 5840–5849 (1989)
52. A. Naji, M. Kanduč, J. Forsman, R. Podgornik, Perspective: Coulomb fluids-weak coupling, strong coupling, in between and beyond. J. Chem. Phys. **139**(15), 150901 (2013)
53. S. Buyukdagli, C. Achim, T. Ala-Nissila, Ion size effects upon ionic exclusion from dielectric interfaces and slit nanopores. J. Stat. Mech.: Theory Exper. **2011**(05), P05033 (2011)
54. S. Buyukdagli, T. Ala-Nissila, Excluded volume effects in macromolecular forces and ion-interface interactions. J. Chem. Phys. **136**(7), 074901 (2012)
55. L. Lue, A variational field theory for solutions of charged, rigid particles. Fluid Phase Equilibria **241**(1–2), 236–247 (2006)
56. A. Abrashkin, D. Andelman, H. Orland, Dipolar Poisson-Boltzmann equation: ions and dipoles close to charge interfaces. Phys. Rev. Lett. **99**(7), 077801 (2007)
57. R. Wang, Z.-G. Wang, On the theoretical description of weakly charged surfaces. J. Chem. Phys. **142**(10) (2015)
58. G.H. Fredrickson, *The Equilibrium Theory of Inhomogeneous Polymers* (Clarendon, 2006)
59. D.C. Grahame, Effects of dielectric saturation upon the diffuse double layer and the free energy of hydration of ions. J. Chem. Phys. **18**(7), 903–909 (1950)
60. A.A. Kornyshev, Double-layer in ionic liquids: paradigm change? J. Phys. Chem. B (2007)
61. B.V. Derjaguin, N.V. Churaev, V.M. Muller, V. Kisin, *Surface Forces* (Springer, 1987)
62. L.D. Landau, E.M. Lifshits, *Mechanics*, vol. 1 (CUP Archive, 1960)
63. B.W. Ninham, V. Yaminsky, Ion binding and ion specificity: the Hofmeister effect and onsager and lifshitz theories. Langmuir **13**(7), 2097–2108 (1997)
64. I.E. Dzyaloshinskii, E.M. Lifshitz, L.P. Pitaevskii, The general theory of van der Waals forces. Adv. Phys. **10**(38), 165–209 (1961)
65. R.R. Netz, H. Orland, One and two-component hard-core plasmas. Eur. Phys. J. E **1**, 67–73 (2000)
66. J. Ortner, Equation of states for classical Coulomb systems: use of the Hubbard-Schofield approach. Phys. Rev. E **59**(6), 6312 (1999)
67. N.V. Brilliantov, Accurate first-principle equation of state for the one-component plasma. Contrib. Plasma Phys. **38**(4), 489–499 (1998)
68. W. Slattery, G. Doolen, H. DeWitt, Improved equation of state for the classical one-component plasma. Phys. Rev. A **21**(6), 2087 (1980)
69. A. Broyles, H. Sahlin, D. Carley, Radial distribution functions for long-range forces. Phys. Rev. Lett. **10**(8), 319 (1963)
70. D.A. Young, E.M. Corey, H.E. DeWitt, Analytic fit to the one-component-plasma structure factor. Phys. Rev. A **44**(10), 6508 (1991)
71. W. Slattery, G. Doolen, H. DeWitt, N dependence in the classical one-component plasma Monte Carlo calculations. Phys. Rev. A **26**(4), 2255 (1982)
72. A. Kakliugin, Correlation energy of non-debye plasma. High Temper. **23**(2), 217–223 (1985)
73. A. Kaklyugin, Nonideal plasma thermodynamics: separate account of short-and long-range interaction. Contrib. Plasma Phys. **41**(4), 351–355 (2001)
74. N. Brilliantov, D. Kuznetsov, R. Klein, Chain collapse and counterion condensation in dilute polyelectrolyte solutions. Phys. Rev. Lett. **81**(7), 1433 (1998)
75. A. Ermoshkin, M. Olvera de La Cruz, A modified random phase approximation of polyelectrolyte solutions. Macromolecules **36**(20), 7824–7832 (2003)
76. Y.A. Budkov, A. Frolov, M. Kiselev, N. Brilliantov, Surface-induced liquid-gas transition in salt-free solutions of model charged colloids. J. Chem. Phys. **139**(19), 194901 (2013)
77. Y.A. Budkov, A. Kolesnikov, N. Georgi et al., A new equation of state of a flexible-chain polyelectrolyte solution: phase equilibria and osmotic pressure in the salt-free case. J. Chem. Phys. **142**(17), 174901 (2015)
78. A.M. Rumyantsev, A. Johner, Salt-added solutions of markov polyampholytes: Diagram of states, antipolyelectrolyte effect, and self-coacervate dynamics. Macromolecules (2023)

79. Y.D. Gordievskaya, Y.A. Budkov, E.Y. Kramarenko, An interplay of electrostatic and excluded volume interactions in the conformational behavior of a dipolar chain: theory and computer simulations. Soft Matter **14**(17), 3232–3235 (2018)
80. N.V. Brilliantov, V.V. Malinin, R.R. Netz, Systematic field-theory for the hard-core one-component plasma. Eur. Phys. J. D-Atomic Mol. Opt. Plasma Phys. **18**(3), 339–345 (2002)
81. Y.A. Budkov, A.L. Kolesnikov, Modified poisson-boltzmann equations and macroscopic forces in inhomogeneous ionic fluids. J. Stat. Mech.: Theory Exper. **2022**(5), 053205 (2022)
82. A. Maggs, R. Podgornik, General theory of asymmetric steric interactions in electrostatic double layers. Soft Matter **12**(4), 1219–1229 (2016)
83. R. Blossey, A. Maggs, R. Podgornik, Structural interactions in ionic liquids linked to higher-order Poisson-Boltzmann equations. Phys. Rev. E **95**(6), 060602 (2017)
84. Y.A. Budkov, A. Kolesnikov, N. Kalikin, M. Kiselev, A statistical theory of coil-to-globule-to-coil transition of a polymer chain in a mixture of good solvents. EPL (Europhys. Lett.) **114**(4), 46004 (2016)
85. Y.A. Budkov, A.L. Kolesnikov, Z.A. Goodwin et al., Theory of electrosorption of water from ionic liquids. Electrochimica Acta **284**, 346–354 (2018)
86. I. Borukhov, D. Andelman, H. Orland, Steric effects in electrolytes: a modified Poisson-Boltzmann equation. Phys. Rev. Lett. **79**(3), 435 (1997)
87. L.D. Landau, J.S. Bell, M. Kearsley, L. Pitaevskii, E. Lifshitz, J. Sykes, *Electrodynamics of Continuous Media*, vol. 8 (Elsevier, 2013)
88. J.P. Mallarino, G. Téllez, E. Trizac, The contact theorem for charged fluids: from planar to curved geometries. Molecular Phys. **113**(17–18), 2409–2427 (2015)
89. Y.A. Budkov, A. Kolesnikov, M. Kiselev, A modified Poisson-Boltzmann theory: Effects of co-solvent polarizability. EPL (Europhys. Lett.) **111**(2), 28002 (2015)
90. Z.A. Goodwin, G. Feng, A.A. Kornyshev, Mean-field theory of electrical double layer in ionic liquids with account of short-range correlations. Electrochimica Acta **225**, 190–197 (2017)
91. J. Wu, T. Jiang, D.-E. Jiang, Z. Jin, D. Henderson, A classical density functional theory for interfacial layering of ionic liquids. Soft Matter **7**(23), 11222–11231 (2011)
92. M.Z. Bazant, B.D. Storey, A.A. Kornyshev, Double layer in ionic liquids: overscreening versus crowding. Phys. Rev. Lett. **106**(4), 046102 (2011)
93. V.A. Vasileva, D.A. Mazur, Y.A. Budkov, Theory of electrolyte solutions in a slit charged pore: effects of structural interactions and specific adsorption of ions. J. Chem. Phys. **159**(2), 123 (2023)
94. L.D. Landau, L. Pitaevskii, A.M. Kosevich, E.M. Lifshitz, *Theory of Elasticity*, vol. 7 (Elsevier, 2012)
95. P.E. Brandyshev, Y.A. Budkov, Noether's second theorem and covariant field theory of mechanical stresses in inhomogeneous ionic liquids. J. Chem. Phys. **158**(17) (2023)
96. J.P. de Souza, M.Z. Bazant, Continuum theory of electrostatic correlations at charged surfaces. J. Phys. Chem. C **124**(21), 11414–11421 (2020)
97. N.R. Agrawal, R. Wang, Electrostatic correlation induced ion condensation and charge inversion in multivalent electrolytes. J. Chem. Theory Comput. **18**(10), 6271–6280 (2022)
98. H. Green, S. Ono, S. Kondo, F.P. Buff, S. Ono, S. Kondo, Molecular theory of surface tension in liquids. Structure of Liquids/Struktur der Flüssigkeiten 134–280 (1960)
99. Y.S. Barash, V.L. Ginzburg, Electromagnetic fluctuations in matter and molecular (van-der-waals) forces between them. Soviet Physics Uspekhi **18**(5), 305 (1975)
100. Y.A. Budkov, N.N. Kalikin, P.E. Brandyshev, Surface tension of aqueous electrolyte solutions. A thermomechanical approach. J. Chem. Phys. **160**(16) (2024)
101. L. Onsager, N.N. Samaras, The surface tension of debye-hückel electrolytes. J. Chem. Phys. **2**(8), 528–536 (1934)
102. T. Markovich, D. Andelman, R. Podgornik, Surface tension of electrolyte interfaces: Ionic specificity within a field-theory approach. J. Chem. Phys. AIP Publishing **142**(4) (2015)
103. E. Nightingale Jr., Phenomenological theory of ion solvation. Effective radii of hydrated ions. J. Phys. Chem. **63**(9), 1381–1387 (1959)

104. N. Matubayasi, *Surface Tension and Related Thermodynamic Quantities of Aqueous Electrolyte Solutions* (CRC Press, 2014)
105. R. Aveyard, S.M. Saleem, Interfacial tensions at alkane-aqueous electrolyte interfaces. J. Chem. Soc. Faraday Trans. 1: Phys. Chem. Condens. Phases **72**, 1609–1617 (1976)
106. Y.A. Budkov, Statistical field theory of ion-molecular solutions. Phys. Chem. Chem. Phys. **22**(26), 14756–14772 (2020)
107. Y.A. Budkov, Nonlocal statistical field theory of dipolar particles in electrolyte solutions. J. Phys.: Condens. Matter **30**(34), 344001 (2018)
108. W. Keesom, The second viral coefficient for rigid spherical molecules, whose mutual attraction is equivalent to that of a quadruplet placed at their centre. Proc. R. Acad. Sci **18**, 636–646 (1915)
109. Y.A. Budkov, A statistical field theory of salt solutions of 'hairy' dielectric particles. J. Phys.: Condens. **32**(5), 055101 (2019)
110. Y. Avni, D. Andelman, R. Podgornik, Charge regulation with fixed and mobile charged macromolecules. Curr. Opin. Electrochem. **13**, 70–77 (2019)
111. R.I. Slavchov, Quadrupole terms in the maxwell equations: Debye-Hückel theory in quadrupolarizable solvent and self-salting-out of electrolytes. J. Chem. Phys. **140**(16), 074503 (2014)
112. R.I. Slavchov, T.I. Ivanov, Quadrupole terms in the maxwell equations: born energy, partial molar volume, and entropy of ions. J. Chem. Phys. **140**(7) (2014)
113. J.G. Kirkwood, J.B. Shumaker, The influence of dipole moment fluctuations on the dielectric increment of proteins in solution. Proc. Natl. Acad. Sci. U.S.A. **38**(10), 855 (1952)
114. Y.A. Budkov, A.I. Victorov, Simple analytical theory for micelles with widespread radial distribution of charged heads. J. Mol. Liq. **341**, 117438 (2021)
115. V.Y. Borue, I.Y. Erukhimovich, A statistical theory of weakly charged polyelectrolytes: fluctuations, equation of state and microphase separation. Macromolecules **21**(11), 3240–3249 (1988)
116. R. Kjellander, D.J. Mitchell, An exact but linear and Poisson-Boltzmann-like theory for electrolytes and colloid dispersions in the primitive model. Chem. Phys. Lett. **200**(1–2), 76–82 (1992)
117. R. Kjellander, D.J. Mitchell, Dressed-ion theory for electrolyte solutions: a Debye-Hückel-like reformulation of the exact theory for the primitive model. J. Chem. Phys. **101**(1), 603–626 (1994)
118. L. Varela, M. Perez-Rodriguez, M. Garcia, F. Sarmiento, V. Mosquera, Static structure of electrolyte systems and the linear response function on the basis of a dressed-ion theory. J. Chem. Phys. **109**(5), 1930–1938 (1998)
119. Y.A. Budkov, Statistical theory of fluids with a complex electric structure: application to solutions of soft-core dipolar particles. Fluid Phase Equilibria **490**, 133–140 (2019)
120. M.E. Fisher, Y. Levin, Criticality in ionic fluids: Debye-Hückel theory, Bjerrum, and beyond. Phys. Rev. Lett. **71**(23), 3826 (1993)
121. Y. Levin, M.E. Fisher, Criticality in the hard-sphere ionic fluid. Phys. A: Stat. Mech. Appl. **225**(2), 164–220 (1996)
122. C.D. Daub, G. Patey, P.J. Camp, Liquid-vapor criticality in a fluid of charged hard dumbbells. J. Chem. Phys. **119**(15), 7952–7956 (2003)
123. A. Kolesnikov, Y. Budkov, G. Barbosa, J. Möllmer, F. Tavares, Water adsorption on planar interfaces: classical density functional study. Fluid Phase Equilibria **564**, 113567 (2023)
124. D. Gurina, E. Odintsova, A. Kolesnikov, M. Kiselev, Y. Budkov, Disjoining pressure of room temperature ionic liquid in charged slit carbon nanopore: molecular dynamics study. J. Mol. Liquids **366**, 120307 (2022)
125. D.L. Gurina, E.G. Odintsova, Y.A. Budkov, Disjoining pressure decay length in room-temperature ionic liquids: a molecular simulation study. J. Phys. Chem. B (2024)
126. J.P. de Souza, Z.A. Goodwin, M. McEldrew, A.A. Kornyshev, M.Z. Bazant, Interfacial layering in the electric double layer of ionic liquids. Phys. Rev. Lett. **125**(11), 116001 (2020)
127. A. Ciach, Simple theory for oscillatory charge profile in ionic liquids near a charged wall. J. Mol. Liquids **270**, 138–144 (2018)

128. R.R. Netz, D. Andelman, Neutral and charged polymers at interfaces. Phys. Rep. **380**(1–2), 1–95 (2003)
129. N.N. Kalikin, A.L. Kolesnikov, Y.A. Budkov, Polymerized ionic liquids on charged electrodes: new prospects for electrochemistry. Curr. Opin. Electrochem. **36**, 101134 (2022)
130. M. Muthukumar, Double screening in polyelectrolyte solutions: limiting laws and crossover formulas. J. Chem. Phys. **105**(12), 5183–5199 (1996)
131. D. Andelman, J.-F. Joanny, Polyelectrolyte adsorption. Comptes Rendus de l'Académie des Sciences-Series IV-Physics **1**(9), 1153–1162 (2000)
132. X. Châtellier, J.-F. Joanny, Adsorption of polyelectrolyte solutions on surfaces: Debye-Hückel theory. Journal de Physique II **6**(12), 1669–1686 (1996)
133. A.C. Maggs, R. Podgornik, General theory of asymmetric steric interactions in electrostatic double layers. Soft Matter **12**(4), 1219–1229 (2016)
134. Y.A. Budkov, N.N. Kalikin, Macroscopic forces in inhomogeneous polyelectrolyte solutions. Phys. Rev. E **107**(2), 024503 (2023)
135. R.D. Coalson, A. Duncan, Statistical mechanics of a multipolar gas: a lattice field theory approach. J. Phys. Chem. **100**(7), 2612–2620 (1996)
136. D. Frydel, Polarizable poisson-boltzmann equation: the study of polarizability effects on the structure of a double layer. J. Chem. Phys. **134**(23), 234704 (2011)
137. Y.A. Budkov, A. Kolesnikov, M. Kiselev, On the theory of electric double layer with explicit account of a polarizable co-solvent. J. Chem. Phys. **144**(18), 184703 (2016)
138. T.B. Jones, Dielectrophoretic force calculation. J. Electrostat. **6**(1), 69–82 (1979)
139. Y. Lauw, M. Horne, T. Rodopoulos, F. Leermakers, Room-temperature ionic liquids: excluded volume and ion polarizability effects in the electrical double-layer structure and capacitance. Phys. Rev. Lett. **103**(11), 117801 (2009)
140. Y.A. Budkov, A.V. Sergeev, S.V. Zavarzin, A.L. Kolesnikov, Two-component electrolyte solutions with dipolar cations on a charged electrode: theory and computer simulations. J. Phys. Chem. C **124**(30), 16308–16314 (2020)
141. J.P. Hallett, T. Welton, Room-temperature ionic liquids: solvents for synthesis and catalysis. 2. Chem. Rev. **111**(5), 3508–3576 (2011)
142. T. Welton, Room-temperature ionic liquids. Solvents for synthesis and catalysis. Chem. Rev. **99**(8), 2071–2084 (1999)
143. M.C. Buzzeo, C. Hardacre, R.G. Compton, Extended electrochemical windows made accessible by room temperature ionic liquid/organic solvent electrolyte systems. ChemPhysChem **7**(1), 176–180 (2006)
144. A.M. O'Mahony, D.S. Silvester, L. Aldous et al., Effect of water on the electrochemical window and potential limits of room-temperature ionic liquids. J. Chem. & Eng. Data **53**(12), 2884–2891 (2008)
145. G. Feng, X. Jiang, R. Qiao, A.A. Kornyshev, Water in ionic liquids at electrified interfaces: the anatomy of electrosorption. ACS Nano **8**(11), 11685–11694 (2014)
146. J. Friedl, I.I. Markovits, M. Herpich et al., Interface between an au (111) surface and an ionic liquid: the influence of water on the double-layer capacitance. ChemElectroChem **4**(1), 216–220 (2017)
147. K. Wippermann, J. Giffin, S. Kuhri et al., The influence of water content in a proton-conducting ionic liquid on the double layer properties of the pt/pil interface. Phys. Chem. Chem. Phys. **19**(36), 24706–24723 (2017)
148. H. Weingärtner, Understanding ionic liquids at the molecular level: facts, problems, and controversies. Angewandte Chemie International Edition **47**(4), 654–670 (2008)
149. Y. Uematsu, R.R. Netz, D.J. Bonthuis, The effects of ion adsorption on the potential of zero charge and the differential capacitance of charged aqueous interfaces. J. Phys.: Condens. Matter **30**(6), 064002 (2018)

GPSR Compliance

The European Union's (EU) General Product Safety Regulation (GPSR) is a set of rules that requires consumer products to be safe and our obligations to ensure this.

If you have any concerns about our products, you can contact us on ProductSafety@springernature.com

In case Publisher is established outside the EU, the EU authorized representative is:

Springer Nature Customer Service Center GmbH
Europaplatz 3
69115 Heidelberg, Germany

The manufacturer's authorised representative in the EU is Springer Nature Customer Service Centre GmbH, Europaplatz 3, 69115 Heidelberg, Germany. If you have any concerns regarding our products, please contact ProductSafety@springernature.com

Printed and bound by CPI Group (UK) Ltd, Croydon, CR0 4YY
01/10/2026
02225898-0018